BARRON'S

ISTEP+ GQE MATH

INDIANA HIGH SCHOOL
GRADUATION QUALIFYING EXAM

Donna Marie Osborn, M.S.
Formerly, Mathematics Department
Jefferson High School
Lafayette, Indiana

BARRON'S

All inquiries should be addressed to:
Barron's Educational Series, Inc.
250 Wireless Boulevard
Hauppauge, New York 11788
www.barronseduc.com

Library of Congress Control No. 2007009654

ISBN-13: 978-0-7641-3620-7
ISBN-10: 0-7641-3620-8

Library of Congress Cataloging-in-Publication Data
Osborn, Donna Marie.
    Barron's Indiana ISTEP & GQE : high school graduation qualifying exam in math /
Donna Marie Osborn.
        p. cm.
    Includes index.
    ISBN-13: 978-0-7641-3620-7
    ISBN-10: 0-7641-3620-8
    1. Indiana Statewide Testing for Educational Progress—Study guides. 2. Indiana
Graduation Qualifying Exam—Study guides. 3. Mathematics—Study and teaching
(Secondary)—United States. 4. Mathematics—Examinations—Study guides.
5. Mathematical ability—Testing. 6. Examinations—Study guides. I. Title.
II. Title: Indiana ISTEP & GQE.

    QA43.O4845 2007
    510.76—dc22   2007009654

PRINTED IN THE UNITED STATES OF AMERICA

9 8 7 6 5 4 3 2 1

Contents

Preface

INDIANA ACADEMIC STANDARDS IN MATHEMATICS

In the summer of 2000, the Indiana State Board of Education adopted a new set of mathematics standards to promote a more demanding set of courses. These standards provide both broad topics and specific benchmarks. They are written for every grade level in the middle and elementary levels, and for every course at the high school level. Copies of the Indiana Academic Standards can be obtained from the Indiana Department of Education's Web site at *www.doe.state.in.us* or directly from the Indiana Department of Education.

STANDARDS-BASED EXAMS

As students were learning the September 2000 standards in the classroom, new examinations were being developed. The new exams, ISTEP+, are standards-based, meaning that the test measures how well students understand the material included in the mathematics standards. Old statewide exams were norm-referenced, meaning that students were rated according to how they performed compared to all of the other students in the state.

The Graduation Qualifying Examination

In addition to changing all of the exams that are given from second grade to ninth grade, the test given to sophomores was changed. In order to be eligible to graduate from high school, all students must either pass the ISTEP+ Graduation Qualifying Examination (GQE) or qualify for a GQE waiver. The requirement to pass the GQE means that a student must score a minimum number of points on the standards-based exam.

Date of Testing

The GQE is administered to all sophomores in Indiana for a total of 6 hours on the same 3 days in September. No make-up tests are allowed for the sophomore exam or for any retake exams. If a student misses the GQE exam, the student automatically fails. A student retakes the mathematics part of the GQE only if the student failed the mathematics part. A student who fails can take the failed portion twice during the junior year and twice during the senior year. The dates for the retake exams are the same 3 days in September as the sophomore test and 3 days in March.

Special Needs Students

Students who have special needs must pass the GQE. Some special needs students are provided instructional accommodations in classroom testing situations. Those accommodations can be provided during the GQE.

Limited English Proficiency Students

The GQE is given in English only. Students with limited English proficiency can be given extra time. The school makes decisions concerning additional time. If a student has an Individualized Education Program or a Section 504 plan, additional accommodations can be made.

Standards for the GQE

The standards tested on the mathematics part of the GQE include all of the eighth grade mathematics standards and all of the Algebra 1 standards. Ninth grade standards do not exist because at that time students use high school course standards. Of course, it is assumed that a student has mastered all of the standards from grades 1 through 8, so any mathematics used in those standards can also be included on the GQE.

Each standard is a broad statement followed by several benchmarks. The eighth grade standards and the Algebra 1 standards are listed in the Appendix. The broad statement describes the mathematics topic. The benchmarks provide more specific concepts and skills that will be tested on the GQE. Each question appearing on the GQE must match one of the benchmarks listed in the eighth grade standards or the Algebra 1 standards.

Waivers

Waivers are available for the GQE requirement. An evidenced-based waiver involves completing certain Core 40 courses with a C or better. A work-readiness waiver requires earning a C average in certain courses and completing other criteria. A student should work with a school counselor to determine qualifications for the waiver process. Details on the qualifications for a waiver are available from the Indiana Department of Education at 888-544-7837 or through E-mail at istep@doe.state.in.us or at the Web site *www.doe.state.in.us/istep/welcome.html.*

Introduction

PASSING THE GQE

The minimum number of points to pass the GQE, the cut score, was set by the Indiana State Board of Education. The cut score on the mathematics part of the GQE is 586. If a student scores 585 or lower on the exam, then that student fails the mathematics portion of the GQE.

The cut score of 586 does not mean that a certain number of problems must be correct. Each problem on the exam is worth a number of points determined by a complicated formula using the pilot test results. Those results are not released to the public, the schools, or the teachers. The goal is to learn all of the mathematics covered by the standards tested.

TYPES OF QUESTIONS ON THE GQE

The mathematics portion of the GQE has several different types of questions: multiple-choice questions, gridded response questions, and open-ended questions. Gridded response questions and the multiple-choice questions appear in the same part of the GQE.

Multiple-Choice Questions

Multiple-choice questions provide options for answers. The multiple-choice items on the GQE are not released to the public, the schools, or the teachers.

Gridded Response Questions

Gridded response questions involve bubbling the correct answer into a grid, shown below.

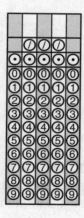

No answer choices are provided. When a gridded response is a mixed number, it must be bubbled in as an improper fraction. For example, if an answer is $2\frac{2}{3}$, then $\frac{8}{3}$ must be bubbled. There is a sample gridded response quiz online at *www.doe.state.in.us/istep/2002/02grd8_changes.html*. Gridded response questions are not released to the public, the schools, or the teachers.

Open-ended Questions

Open-ended questions are problems where the student's work and answer are given points. Partial credit is available on most open-ended problems. The correct answer with no work receives partial credit. Very seldom does a student receive full credit for only an answer. The open-ended questions are released to the public after the exam has been given and graded. Items released to the public will not be used again on the GQE.

REFERENCE SHEET

You are given a reference sheet to use on the entire GQE. Become familiar with the information that is provided. When you need the reference sheet for a problem on the GQE, there is an icon beside the problem. The icon is shown below.*

The icon is used on the sample GQE questions at the end of each chapter and the two practice tests at the end of the book.

CALCULATORS

On some sections of the GQE, a calculator can be used. You must bring the calculator. A calculator is not provided for you. A scientific calculator is best. Calculators in cell phones or calculators that make noise are not allowed. Contact your school or go to the Web site *http://www.doe.state.in.us/istep* for more information.

You will not be allowed to use a calculator on some sections of the GQE. When you cannot use a calculator, the teacher will tell you. The icon shown below* will appear beside the section.

The icon is used on the sample GQE questions in each chapter. The icon is also used on entire sections of the two practice tests.

QUESTIONS ON THE GQE

The exact questions chosen for each version of the GQE vary. The standards that are tested on the GQE are the same. No teacher or administrator knows what questions will be asked or what standards will be tested on a certain version. The author of this book does not know what questions will be on the GQE.

---

*Icons from CTB/McGraw-Hill LLC, *ISTEP + GQE Item Sampler*, © copyright 2001 by the State of Indiana Department of Education. Reproduced with permission of **The McGraw-Hill Companies, Inc.**

YOU AND THE GQE

Passing the GQE is an important goal in your high school career. Be serious about the exam, but do not panic. It is very important to prepare yourself to pass.

The best preparation for the GQE is hard work in the mathematics classrooms. Do your homework. Do not just finish the homework problems. Try to get the answers correct. If you do not understand a problem or a concept, ask a teacher or a friend to explain it to you. Take pride in your mathematics work and grade.

Review with your teachers in the weeks before the GQE. And again, if you do not understand a problem or a concept, ask a teacher or a friend to explain it to you.

Get enough sleep the night before the GQE. Eat breakfast before you go to the testing room. Bring a couple of extra #2 pencils. Bring a calculator that you know how to use. Stay calm.

Check your work on the GQE. Once you have completed a section of the mathematics exam, check your computations in that section. Redo the problems on a separate sheet of paper. Be sure that you bubbled the answer in the correct space on the answer sheet.

Believe in yourself and your ability to do mathematics. An optimistic attitude about your ability to pass the GQE is a must. Think positively about your mathematical skills.

THIS BOOK CAN AND CANNOT

This book can explain the mathematics that will be tested on the GQE. Each standard is divided into smaller parts and explained.

This book can give exercises and their solutions before moving to another concept. More practice problems with solutions are provided at the end of each chapter.

This book can provide practice problems similar to those on the GQE. Each chapter has questions comparable to those on the GQE. Additionally, there are two practice tests with problems similar to those on the GQE. Similar wording is used. Similar concepts are tested.

This book can provide answers. Most problems show work and an explanation. Step-by-step work processes are shown so that you can truly understand what was done.

This book can give you the confidence needed to pass the GQE. After completing the work in this book, you will have studied the mathematics that will be tested on the GQE. You will have the ability to do your best on the exam.

This book cannot tell you what will be on the GQE. This book cannot give you exact problems from the GQE. This book cannot promise that you will pass the GQE.

ORGANIZATION OF THIS BOOK

This book has chapters that match each of the eighth grade and Algebra 1 standards, and two practice GQE tests.

The chapters directly match the standards. The title of each chapter is actually the name of one of the standards tested on the GQE. The beginning of the chapter lists the benchmarks addressed in the chapter. Each chapter then breaks down the mathematics into smaller topics. These topics are all listed in the table of contents for quick reference. Exercises with solutions are given at the end of each topic. Then practice exercises are given with the solutions at the end of the chapter. Finally, there are sample GQE questions that test the content of the entire chapter.

The two practice GQE tests provide sample questions covering the standards. There is no way to recreate a real GQE. These two practice tests provide problems that are written similar to those on the GQE and cover the standards that are tested on the GQE. Each question on the two practice tests is mapped to a chapter in the book.

USING THIS BOOK

Start at the beginning of the book and work through each chapter at your own pace. Spend more time in the chapters that cause you difficulty. If you know one of the topics is very difficult for you, begin there.

Take the practice tests before taking the actual GQE. Work for about one hour. Check your answers. Try to understand your mistakes. Take a break. Continue working, checking, and learning from your mistakes. The night before the GQE, do not cram.

Chapter 1 | **Computation**

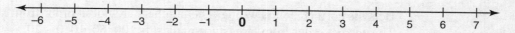

INDIANA'S ACADEMIC MATHEMATICS STANDARDS ADDRESSED:

8.2.1 Add, subtract, multiply, and divide rational numbers (integers, fractions, and terminating decimals).

8.2.2 Solve problems by computing simple interest.

8.2.3 Use estimation techniques to decide whether answers to computations on a calculator are reasonable.

INTRODUCTION TO INTEGERS

Definition: Integers are the numbers ..., –3, –2, –1, 0, 1, 2, 3, ...

As you can see by the definition, there are no fractions or decimals in the integers. There are positive integers, negative integers, and zero, which is neither positive nor negative.

Negative numbers describe situations like the loss of money or points, below the ground or below sea level, and temperatures below zero. Positive numbers describe conditions like a gain in yardage or money, an increase in salary, and a rise in altitude.

Look at the integers as they are placed on the number line shown below.

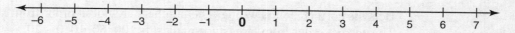

The larger numbers are on the right and the smaller numbers are on the left. This means that any positive number is bigger than any negative number. Also, any positive number is bigger than zero. To compare –6 and –2, remember that the larger numbers are on the right. This would mean that –2 is larger than –6 because –2 is farther to the right on the number line. To compare numbers that are too large to draw on a number line, try to imagine which one would be farther to the right.

Example
Which is larger, –45 or –100?

Think of a number line with zero on it, as shown below, with the negative numbers on the left and the positive numbers on the right.

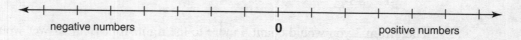

negative numbers   0   positive numbers

Start counting from zero in the negative direction (to the left) as shown below.

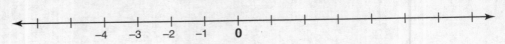

Which number, −45 or −100, would be the first one named?

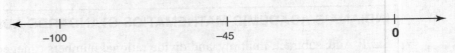

Look at the number line above to see that −45 would be the first number named. So, −45 is larger than −100.

Exercises

1. Which number is larger, −223 or −250?

2. Jon says that the numbers below are in order from largest to smallest.

$$100, -80, 60, -40, -20, 10, 0$$

Betsy is sure that Jon is wrong. She says that the numbers are not in order from largest to smallest. Who is correct? What is the smallest number?

Solutions to Exercises

1. Always think of the number line. The larger number is the number that is farther to the right. The number line below shows that −223 is farther to the right, so −223 is larger.

2. Betsy is correct. Jon is wrong. The numbers are not in order from largest to smallest. The correct order from largest to smallest is 100, 60, 10, 0, −20, −40, −80. The smallest number is −80.

COMPUTING WITH INTEGERS

Addition with Integers

It is easy to use the number line to understand adding integers. To add a positive number, you move to the right, the positive direction. So 3 + 4 would start at 0 and count 3 to the right, as shown below.

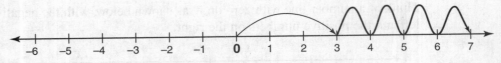

Then from 3 you would count 4 more to the right. Your final answer would be 3 + 4 = 7.

Of course that problem is extremely easy, so let's look to see if negative numbers are more difficult. To add a negative number, you move to the left, the negative direction. So –3 + –2 would mean to start at 0 and count 3 to the left, as shown below.

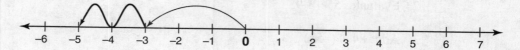

Then from –3 you would count –2 more, or two spaces to the left. Your final answer would be –3 + –2 = –5.

Now let's look at what happens when adding positive integers with negative integers. Consider the problem –3 + 5. Beginning at 0, count 3 to the left (–3), as shown below.

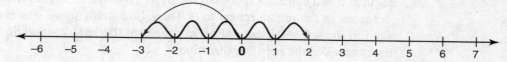

Then from –3 you would count 5 to the right. Your answer would be –3 + 5 = 2.

Try the problem 5 + –7. Beginning at 0, count 5 to the right, as shown below.

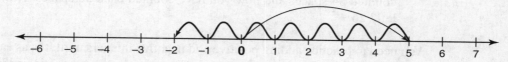

Then from 5 count 7 spaces to the left. Your answer would be 5 + –7 = –2.

Adding integers is really important in the study of Algebra, so you need to be really good at it. If you are just learning how to add integers, then practice adding on the number line until you feel confident.

Of course, you are not always going to draw a number line on paper. Try to understand the reasoning that you do on the number line, so that you can do the same reasoning in your head.

Example

With the problem –35 + 10, think of where you would be on a number line as you do the problem. Begin at 0 and count 35 places to the left. You are on the negative part of the number line. Now count to the right 10 places. You will still be on the negative side of the number line, so your answer is negative. You counted 10 spaces back from the first 35 spaces you counted. You would end up at –25. So –35 + 10 = –25.

Example

With the problem 52 + –80, think of a number line. Begin at 0 and count 52 spaces to the right. You are now on the positive side of the number line. You need to move –80, which is 80 spaces to the left. If you count 52 spaces to the left, you will be back at 0. You will need to count even more spaces to the left. How many more spaces will you need to move to the left? You will need to move 80 – 52 = 28 spaces to the left.

You will finish at –28.

Exercises

1. Simplify: $-41 + -38$

2. Evaluate: $55 + -93$

3. Add: $84 + (-58)$

Solutions to Exercises

1. $-41 + -38 = -79$
   Begin at 0 and count 41 spaces to the left. Now, count 38 more spaces to the left. What is the total number of spaces that you have counted? -79

2. $55 + -93 = -38$
   Begin at 0 and count 55 spaces to the right. To move 93 spaces to the left, first count 55 spaces to the left to return to 0. Then, how many more spaces will you need to count to the left to get a total of 93 spaces to the left? You would need to count 38 more spaces in the negative direction.

3. $84 + (-58) = 26$
   Begin at 0. Count 84 spaces to the right. Now, count -58, which is 58 spaces to the left. You will still be on the positive side of the number line. In the beginning you counted 84 spaces and now you have counted back 58 spaces. How many spaces are left? 26

You need to practice adding positive and negative integers until it is as easy as adding positive numbers. Before you move on to subtracting integers, practice addition.

Subtraction with Integers

Subtraction is called the inverse operation of addition. We know that $12 - 3 = 9$ because $9 + 3 = 12$. Addition and subtraction are related in another way. Every subtraction problem can be rewritten as an addition problem. For example, $15 - 8$ is the same as $15 + (-8)$; subtracting a number is the same as adding its opposite.

> When subtracting integers, rewrite the subtraction problem as an addition problem by keeping the first number the same, change the sign of the addition sign to subtraction, and change the sign of the second number.
> $a - b = a + (-b)$

Example
The problem $10 - 35$ can be rewritten as $10 + (-35)$, and then worked as an ordinary addition problem.

$$10 - 35 = 10 + (-35) = -25$$

Example
$-16 - (-9)$ can be rewritten as $-16 + (+9)$ by changing the subtraction sign to addition and changing the sign of the second number from negative nine to positive nine.

$$-16 - (-9) = -16 + (+9) = -7$$

Exercises

1. Subtract: $23 - (-15)$

2. Simplify: $-19 - 21$

3. Subtract: $20 - 39$

4. Subtract: $-16 - (-40)$

Solutions to Exercises

1. $23 - (-15) = 38$
   Rewrite $23 - (-15)$ as $23 + (+15)$.

   $$23 - (-15) = 23 + (+15) = 38$$

2. $-19 - 21 = -40$
   Rewrite $-19 - 21$ as $-19 + (-21)$.

   $$-19 - 21 = -19 + (-21) = -40$$

3. $20 - 39 = -19$
   Rewrite $20 - 39$ as $20 + (-39)$.

   $$20 - 39 = 20 + (-39) = -19$$

4. $-16 - (-40) = 24$
   Rewrite $-16 - (-40)$ as $-16 + (+40)$.

   $$-16 - (-40) = -16 + (+40) = 24$$

Multiplication with Integers

The rules for multiplying integers are simple.

> positive times positive = positive
> positive times negative = negative
> negative times positive = negative
> negative times negative = positive

Example

$-7 \cdot 5$ is a negative times a positive, so your answer will be negative.

$$-7 \cdot 5 = -35$$

Example

$-8 \cdot -3$ is a negative times a negative, so your answer will be positive.

$$-8 \cdot -3 = 24$$

Exercises

1. Multiply: $-2 \cdot 10$

2. Simplify: $-6 \cdot -2$

3. Simplify: $-8 \cdot -3 \cdot -2$

Solutions to Exercises

1. $-2 \cdot 10 = -20$

2. $-6 \cdot -2 = 12$

3. $-8 \cdot -3 \cdot -2 = -48$

    When you have three integers to multiply together, multiply two at a time. First multiply $-8 \cdot 3$. Then multiply that answer by -2.

    $$-8 \cdot -3 \cdot -2 = (-8 \cdot -3) \cdot -2 = 24 \cdot -2 = -48$$

Division with Integers

The rules for multiplication and division of integers are similar.

> positive divided by positive = positive   negative divided by positive = negative
> positive divided by negative = negative   negative divided by negative = positive

Exercises

1. Divide: $56 \div (-7)$

2. Simplify: $-40 \div (-8)$

3. Divide: $-63 \div 9$

Solutions to Exercises

1. $56 \div (-7) = -8$

2. $-40 \div (-8) = 5$

3. $-63 \div 9 = -7$

Practice Problems

1. Simplify: $-8 \cdot 5$

2. Compute: $-18 \div -3$

3. Add: $-18 + 24$

4. Subtract: $-23 - 14$

5. A submarine is traveling at 300 feet below the surface of the water. If the submarine dives down 120 feet, how far below the surface of the water will the submarine be?

(Solutions are found at the end of the chapter.)

COMPUTING WITH FRACTIONS

Addition of Fractions

To add fractions, you must have the same denominator, called a common denominator. If the fractions already have the same denominator, then add the numerators and keep the denominator the same.

Example

To add $\frac{7}{8} + \frac{5}{8}$, add the numerators, $7 + 5$, and keep the denominator, 8, the same. Then follow the rules for simplifying the fraction.

$$\frac{7}{8} + \frac{5}{8} = \frac{12}{8} = 1\frac{4}{8} = 1\frac{1}{2}$$

Example

To add $\frac{3}{4} + \frac{5}{6}$, change $\frac{3}{4}$ and $\frac{5}{6}$ so that they have the same denominator.

Try to think of what number both 4 and 6 go into. 4 goes into 4, 8, 12, 16, 20, 24, and so on. 6 goes into 6, 12, 18, 24, 30, 36, and so on. The smallest number in both lists is 12, so the common denominator for $\frac{3}{4}$ and $\frac{5}{6}$ is 12.

Now change $\frac{3}{4}$ into $\frac{\text{something}}{12}$ by multiplying the numerator and the denominator by the same number. $\frac{3}{4} = \frac{3 \cdot 3}{4 \cdot 3} = \frac{9}{12}$. Use the same process to change $\frac{5}{6}$ into $\frac{\text{something}}{12}$.

$\frac{5}{6} = \frac{5 \cdot 2}{6 \cdot 2} = \frac{10}{12}$.

The problem now becomes $\frac{3}{4} + \frac{5}{6} = \frac{9}{12} + \frac{10}{12}$, which can be worked by adding the numerators, keeping the denominator the same, and simplifying.

$$\frac{3}{4} + \frac{5}{6} = \frac{9}{12} + \frac{10}{12} = \frac{19}{12} = 1\frac{7}{12}$$

Example

$$5\frac{7}{8} + 9\frac{11}{12}$$

Find the common denominator for the fractions $\frac{7}{8}$ and $\frac{11}{12}$ by thinking of the smallest number that both 8 and 12 go into. Both 8 and 12 go into 24, so the common denominator is 24.

$$5\frac{7}{8} + 9\frac{11}{12} = 5\frac{21}{24} + 9\frac{22}{24} = 14\frac{43}{24}$$

Since $\frac{43}{24}$ is the same as $1\frac{19}{24}$, the problem now changes to

$$5\frac{7}{8} + 9\frac{11}{12} = 5\frac{21}{24} + 9\frac{22}{24} = 14\frac{43}{24} = 14 + 1\frac{19}{24} = 15\frac{19}{24}.$$

> To add fractions, change the fractions so that they have a common denominator (the same bottom number). Then, add the numerators and keep the denominators the same. Finally, simplify your answer.

Exercises

1. Add: $\frac{19}{24} + \frac{15}{24}$

2. Simplify: $\frac{5}{6} + \frac{7}{12}$

3. Compute: $5\frac{7}{15} + 9\frac{5}{6}$

Solutions to Exercises

1. $\frac{19}{24} + \frac{15}{24} = \frac{34}{24} = 1\frac{10}{24} = 1\frac{5}{12}$

2. $\frac{5}{6} + \frac{7}{12} = \frac{10}{12} + \frac{7}{12} = \frac{17}{12} = 1\frac{5}{12}$

3. $5\frac{7}{15} + 9\frac{5}{6} = 5\frac{14}{30} + 9\frac{25}{30} = 14\frac{39}{30} = 14 + 1\frac{9}{30} = 15\frac{9}{30} = 15\frac{3}{10}$

Subtraction of Fractions

To subtract fractions, you must have a common denominator. Once the fractions have a common denominator, then subtract the numerators and keep the denominators the same.

Example

To subtract $\frac{7}{8} - \frac{5}{12}$, change $\frac{7}{8}$ and $\frac{5}{12}$ into fractions with the same denominator. $\frac{7}{8} = \frac{21}{24}$

and $\frac{5}{12} = \frac{10}{24}$. So, $\frac{7}{8} - \frac{5}{12}$ changes into $\frac{7}{8} - \frac{5}{12} = \frac{21}{24} - \frac{10}{24} = \frac{11}{24}$.

Example

Subtract $14\frac{7}{8} - 3\frac{5}{8}$. Since the fractions already have a common denominator, subtract the

whole numbers $14 - 3$ and subtract the fractions $\frac{7}{8} - \frac{5}{8}$. Your answer is $11\frac{2}{8}$, which simpli-

fies to $11\frac{1}{4}$.

Example

$22\frac{3}{4} - 9\frac{2}{3}$ is one step more difficult because you must find a common denominator.

$22\frac{3}{4} - 9\frac{2}{3} = 22\frac{9}{12} - 9\frac{8}{12}$. Then subtract.

$$22\frac{3}{4} - 9\frac{2}{3} = 22\frac{9}{12} - 9\frac{8}{12} = 13\frac{1}{12}$$

The most difficult kind of subtraction problem involves rewriting whole numbers as fractions and borrowing. Remember $\frac{12}{12} = 1$. Using that fact, you could rewrite 7 as $6\frac{12}{12}$.

Example

In the problem $7 - 4\frac{5}{12}$, rewrite 7 as $6\frac{12}{12}$ before you subtract. The problem becomes

$7 - 4\frac{5}{12} = 6\frac{12}{12} - 4\frac{5}{12}$.

$$7 - 4\frac{5}{12} = 6\frac{12}{12} - 4\frac{5}{12} = 2\frac{7}{12}$$

Sometimes you must rewrite mixed numbers so that you can subtract the fractions. $11\frac{1}{8}$ means $11 + \frac{1}{8}$ and can be broken into $10\frac{8}{8} + \frac{1}{8}$. Now, $11\frac{1}{8}$ can be changed into $10\frac{9}{8}$. This process is very helpful when subtracting $11\frac{1}{8} - 4\frac{3}{8}$. You cannot subtract $\frac{1}{8} - \frac{3}{8}$, so change $11\frac{1}{8}$ into $10\frac{9}{8}$.

$$11\frac{1}{8} - 4\frac{3}{8} = 10\frac{9}{8} - 4\frac{3}{8} = 6\frac{6}{8} = 6\frac{3}{4}$$

The most difficult of all of the subtraction problems involves finding a common denominator and then borrowing.

Example

Subtract: $13\frac{7}{10} - 4\frac{5}{6}$

Change the fractions so that they have a common denominator.

$$13\frac{7}{10} - 4\frac{5}{6} = 13\frac{21}{30} - 4\frac{25}{30}$$

Decide if borrowing is necessary. In this case, you cannot subtract $\frac{21}{30} - \frac{25}{30}$, so rewrite $13\frac{21}{30}$ as $13 + \frac{21}{30} = 12\frac{30}{30} + \frac{21}{30} = 12\frac{51}{30}$.

$$13\frac{7}{10} - 4\frac{5}{6} = 13\frac{21}{30} - 4\frac{25}{30} = 12\frac{51}{30} - 4\frac{25}{30} = 8\frac{26}{30} = 8\frac{13}{15}$$

> To subtract fractions, change the fractions so that they have a common denominator. Then, subtract the numerators and keep the denominators the same. Sometimes before subtracting you may have to borrow by rewriting the mixed number as an improper fraction. Finally, simplify your answer.

Exercises

1. Simplify: $7\frac{5}{8} - 2\frac{1}{4}$

2. Evaluate: $18 - \frac{3}{5}$

3. Compute: $17\frac{2}{3} - 5\frac{4}{5}$

4. Subtract: $13\frac{7}{10} - 5\frac{9}{10}$

5. Calculate: $28 - 19\frac{21}{25}$

Solutions to Exercises

1. Remember to find a common denominator, and then decide if borrowing is necessary.

$$7\frac{5}{8} - 2\frac{1}{4} = 7\frac{5}{8} - 2\frac{2}{8} = 5\frac{3}{8}$$

2. Rewrite 18 as $17\frac{5}{5}$ and then subtract.

$$18 - \frac{3}{5} = 17\frac{5}{5} - \frac{3}{5} = 17\frac{2}{5}$$

3. Find a common denominator $17\frac{2}{3} - 5\frac{4}{5} = 17\frac{10}{15} - 5\frac{12}{15}$ and then decide if it is

necessary to borrow. In this case, you cannot subtract $\frac{10}{15} - \frac{12}{15}$, so rewrite $17\frac{10}{15} =$

$17 + \frac{10}{15} = 16\frac{15}{15} + \frac{10}{15} = 16\frac{25}{15}$. Now the problem is easy.

$$17\frac{2}{3} - 5\frac{4}{5} = 17\frac{10}{15} - 5\frac{12}{15} = 16\frac{25}{15} - 5\frac{12}{15} = 11\frac{13}{15}$$

4. Since the fractions already have a common denominator, decide if it is necessary to

borrow. Since you cannot subtract $\frac{7}{10} - \frac{9}{10}$, rewrite $13\frac{7}{10} = 13 + \frac{7}{10} = 12\frac{10}{10} + \frac{7}{10} =$

$12\frac{17}{10}$. The problem now becomes $13\frac{7}{10} - 5\frac{9}{10} = 12\frac{17}{10} - 5\frac{9}{10} = 7\frac{8}{10} = 7\frac{4}{5}$.

5. Rewrite $28 = 27\frac{25}{25}$ before you subtract.

$$28 - 19\frac{21}{25} = 27\frac{25}{25} - 19\frac{21}{25} = 8\frac{4}{25}$$

Multiplication of Fractions

Multiplication of fractions is very different from addition and subtraction because it does not involve finding a common denominator. To multiply fractions, you must have only fractions. Each whole number or mixed number must be changed into an improper fraction before beginning the problem. For example, in the problem $3\frac{5}{8} \cdot \frac{2}{3}$, the $3\frac{5}{8}$ must be changed into an improper fraction. Remember that $3 = \frac{24}{8}$, so $3\frac{5}{8} = 3 + \frac{5}{8} = \frac{24}{8} + \frac{5}{8} = \frac{29}{8}$.

Once all of the numbers are changed into fractions, then look to see if the numerator of one fraction can be reduced with the denominator of the other fraction. In the problem $\frac{29}{8} \cdot \frac{2}{3}$, the 8 in the denominator of the first fraction can be reduced with the 2 in the numerator of the second fraction by dividing both numbers by 2.

$$\frac{29}{8} \cdot \frac{2}{3} = \frac{29}{\underset{4}{\cancel{8}}} \cdot \frac{\cancel{2}^{1}}{3}$$

Multiply the numerators, multiply the denominators, and simplify.

$$3\frac{5}{8} \cdot \frac{2}{3} = \frac{29}{8} \cdot \frac{2}{3} = \frac{29}{\underset{4}{\cancel{8}}} \cdot \frac{\cancel{2}^{1}}{3} = \frac{29}{12} = 2\frac{5}{12}$$

Example

To multiply $4\frac{9}{10} \cdot 2\frac{2}{7}$, first change both mixed numbers into improper fractions.

$$4\frac{9}{10} \cdot 2\frac{2}{7} = \frac{49}{10} \cdot \frac{16}{7}$$

Then, check the numerator of one fraction to see if it can be reduced with the denominator of the other fraction. In this case, 7 goes into both 49 and 7, while 2 goes into both 10 and 16.

$$4\frac{9}{10} \cdot 2\frac{2}{7} = \frac{\overset{7}{\cancel{49}}}{\underset{5}{\cancel{10}}} \cdot \frac{\overset{8}{\cancel{16}}}{\underset{1}{\cancel{7}}} = \frac{56}{5} = 11\frac{1}{5}$$

> To multiply fractions, change all numbers into fractions or improper fractions. Check to see if you can reduce a numerator and a denominator. Then, multiply the numerators, multiply the denominators, and simplify your answer.

Exercises

1. Simplify: $\frac{5}{8} \cdot 2\frac{7}{10}$

2. Compute: $\frac{9}{10} \cdot \frac{14}{27}$

3. Multiply: $3\frac{3}{4} \cdot 4\frac{4}{5}$

Solutions to Exercises

1. $\frac{5}{8} \cdot 2\frac{7}{10} = \frac{\overset{1}{\cancel{5}}}{8} \cdot \frac{27}{\underset{2}{\cancel{10}}} = \frac{27}{16} = 1\frac{11}{16}$

2. $\frac{\overset{1}{\cancel{9}}}{\underset{5}{\cancel{10}}} \cdot \frac{\overset{7}{\cancel{14}}}{\underset{3}{\cancel{27}}} = \frac{7}{15}$

3. $3\frac{3}{4} \cdot 4\frac{4}{5} = \frac{\overset{3}{\cancel{15}}}{\underset{1}{\cancel{4}}} \cdot \frac{\overset{6}{\cancel{24}}}{\underset{1}{\cancel{5}}} = \frac{18}{1} = 18$

Division of Fractions

Division of fractions is similar to multiplication of fractions. Each whole number or mixed number must be changed into an improper fraction. Once all of the numbers are changed into fractions, change the division problem into a multiplication problem. Keep the first fraction the same, change the division sign to a multiplication sign, and write the reciprocal of the second fraction.

> **Definition:** The "reciprocal" of a number means the number that you must multiply by the original number to get 1. For example, the reciprocal of $\frac{4}{3}$ is $\frac{3}{4}$ because $\frac{4}{3} \cdot \frac{3}{4} = 1$, and the reciprocal of 5 is $\frac{1}{5}$ because $5 \cdot \frac{1}{5} = 1$.

Example

To divide $3\frac{3}{5} \div 2\frac{2}{3}$, change the mixed numbers into improper fractions.

$$3\frac{3}{5} \div 2\frac{2}{3} = \frac{18}{5} \div \frac{8}{3}$$

Keep the first fraction the same, change the division sign to a multiplication sign, and change the second fraction to its reciprocal.

$$3\frac{3}{5} \div 2\frac{2}{3} = \frac{18}{5} \div \frac{8}{3} = \frac{18}{5} \cdot \frac{3}{8}$$

Follow the rules for multiplication of fractions.

$$3\frac{3}{5} \div 2\frac{2}{3} = \frac{18}{5} \div \frac{8}{3} = \frac{\overset{9}{18}}{5} \cdot \frac{3}{\underset{4}{8}} = \frac{27}{20} = 1\frac{7}{20}$$

Exercises

1. Divide: $4\frac{2}{5} \div \frac{6}{7}$

2. Simplify: $\frac{14}{15} \div 1\frac{3}{4}$

3. Compute: $8\frac{2}{3} \div 2\frac{4}{5}$

4. Evaluate: $\frac{5}{8} \div 10$

Solutions to Exercises

1. $4\frac{2}{5} \div \frac{6}{7} = \frac{22}{5} \div \frac{6}{7} = \frac{\overset{11}{22}}{5} \cdot \frac{7}{\underset{3}{6}} = \frac{77}{15} = 5\frac{2}{15}$

2. $\frac{14}{15} \div 1\frac{3}{4} = \frac{14}{15} \div \frac{7}{4} = \frac{\overset{2}{14}}{15} \cdot \frac{4}{\underset{1}{7}} = \frac{8}{15}$

3. $8\frac{2}{3} \div 2\frac{4}{5} = \frac{26}{3} \div \frac{14}{5} = \frac{\overset{13}{26}}{3} \cdot \frac{5}{\underset{7}{14}} = \frac{65}{21} = 3\frac{2}{21}$

4. $\frac{5}{8} \div 10 = \frac{5}{8} \div \frac{10}{1} = \frac{\overset{1}{5}}{8} \cdot \frac{1}{\underset{2}{10}} = \frac{1}{16}$

Be Careful! $\frac{1}{16}$ is not the same as 16. $\frac{16}{1}$ is the same as 16.

Practice Problems

6. Simplify: $3\frac{3}{8} \cdot 2\frac{2}{3}$

7. Subtract: $24\frac{5}{8} - 8\frac{3}{4}$

8. Divide: $\frac{7}{8} \div 1\frac{3}{4}$

9. Add: $41\frac{5}{6} + 18\frac{3}{4}$

10. Funny Fanny's Flea Market needs to create $\frac{1}{3}$-acre booths out of $12\frac{1}{2}$ acres of land. How many booths will Funny Fanny be able to create?

(Solutions are found at the end of the chapter.)

COMPUTING WITH DECIMALS

Addition of Decimals

To add decimals, place the decimal points of the numbers directly under each other. For the problem $6.83 + 18.5$, line up the problem. Since there is no number under the 3, place a 0 there. After adding the numbers together, bring the decimal point straight down.

$$\begin{array}{r} 6.83 \\ + 18.50 \\ \hline 25.33 \end{array}$$

There is only one difficulty with addition problems. When one of the numbers does not have a decimal point, $82 + 17.943$, then place a decimal point at the end of the number.

> **Remember:** When using money, $27 can be written as $27.00. You cannot write $27 as $.27 because that means only 27 cents.

$82 + 17.943$ is written as $\begin{array}{r} 82.000 \\ + 17.943 \\ \hline \end{array}$ and then added to get 99.943.

> To add decimals, line up the decimal points. If a number does not have a decimal point shown, then the decimal point goes at the end of the number. Add the numbers and bring the decimal point straight down.

Exercise

Add: $4.9 + 18.36 + 7$

Solution to Exercise
Line up the decimal points when you rewrite $4.9 + 18.36 + 7$ vertically. Add.

$$\begin{array}{r} 4.90 \\ 18.36 \\ + 7.00 \\ \hline 30.26 \end{array}$$

Subtraction of Decimals

Subtracting decimals is similar to adding decimals. Line up the decimal points and subtract. If there are empty spaces when you line up the decimal points, fill in the spaces with zeros. For example, 17.1 − 8.462 would be rewritten vertically. The empty spaces would then be filled in with zeros.

$$\begin{array}{r} \overset{6\ 10\ 9_1}{\cancel{17}.\cancel{1}\cancel{0}0} \\ -8.462 \\ \hline 8.638 \end{array}$$

Be Careful! Many students are careless when they have to borrow. Be sure that you only borrow when necessary.

Example

To subtract 15 − 2.84, remember that 15 is the same as 15.00. Line up the decimal points and subtract.

$$\begin{array}{r} \overset{4\ 9_1}{1\cancel{5}.\cancel{0}\cancel{0}} \\ -2.84 \\ \hline 12.16 \end{array}$$

> To subtract decimals, line up the decimal points. If a number does not have a decimal point shown, then the decimal point goes at the right end of the number. If there are empty spaces in either of the numbers, fill the empty spaces with zeros. Subtract the numbers and bring the decimal point straight down.

Exercise

Compute: 5.86 − .019

Solution to Exercise

Rewrite 5.86 − .019 vertically by lining up the decimal points. Remember to place zeros in the empty places and then subtract. Bring the decimal point straight down.

$$\begin{array}{r} \overset{5_1}{5.8\cancel{6}0} \\ -.019 \\ \hline 5.841 \end{array}$$

Multiplication of Decimals

In multiplication, you line up the numbers, not the decimal points. To multiply 783 · .4, line up the numbers and multiply.

$$\begin{array}{r} 7.83 \\ \times\ .4 \\ \hline 3132 \end{array}$$

Since the decimal points are not lined up, you cannot bring the decimal point straight down. Instead, there must be the same number of numbers after the decimal point in the answer as there are numbers after the decimal point in the problem. With this problem, you count the number of numbers that are after both of the decimal points in the problem. In 7.83 there are two numbers after the decimal point, and in .4 there is one number after the decimal point. Together, there are three numbers after the decimal point. In the answer there needs to be three numbers after the decimal point.

$$7.83 \cdot .4 = 3.132$$

Example

To multiply $1.01 \cdot .015$, line up the numbers and multiply.

$$
\begin{array}{r}
1.01 \\
\times\,.015 \\
\hline
505 \\
1010 \\
\hline
1515
\end{array}
$$

Count the number of numbers after both of the decimal points in the original problem. Together, there are five numbers after the decimal points. In the answer, you must have five numbers after the decimal point. In this case, you must put a 0 in front of the first number to have five numbers after the decimal point.

$$1.01 \cdot .015 = .01515$$

To multiply decimals, line up the numbers. Multiply the numbers. Count the total number of numbers after both of the decimal points in the original problem. The answer must have the same total number of numbers after the decimal point.

Exercise

Evaluate: $(.065)(.0042)$

Solution to Exercise

Line up the numbers and multiply. Once you find the answer, find the total number of numbers after both of the decimal points in the original problem. Be sure there are that many numbers after the decimal point in your answer.

$$
\begin{array}{r}
.065 \\
\times\,.0042 \\
\hline
130 \\
2600 \\
\hline
.0002730
\end{array}
$$

Division of Decimals

Division of decimals is complicated only when deciding the placement of the decimal point. When dividing by a whole number, the decimal point in the answer is directly above the decimal point in the long division problem.

To divide $47.7 \div 18$, write the numbers as a long division problem. Move the decimal point straight up from its place in 47.7 and that is where it will be in the answer. Now divide to get your answer.

$$
\begin{array}{r}
2.65 \\
18\overline{)47.70} \\
-36 \\
\hline
117 \\
-108 \\
\hline
90 \\
-90 \\
\hline
\end{array}
$$

Sometimes the division is easy but the placement of the decimal point is tricky.

Example

To divide .36 ÷ 4, rewrite the problem as a long division problem. Place the decimal point directly above the decimal point in the problem. Since 4 does not go into 3, place a 0 above the 3 and continue dividing.

$$4\overline{)\,.36}\quad\text{(.0)}$$

Since 4 goes into 36, then place a 9 above the 6 in the problem.

$$4\overline{)\,.36}\quad\text{(.09)}$$

The problems are one step more difficult when you are not dividing by a whole number. To divide 3.84 ÷ .3, set up the problem as a long division problem. Before dividing, change the number you are dividing by into a whole number. In this case, change .3 into 3. To keep the problem the same, you must also change 3.84 into 38.4. However many places you move the decimal point to make the number you are dividing by into a whole number, you must also move the decimal point in the other number.

Now you are dividing by a whole number, so the problem is the same as those previously done.

```
     12.8
  3)38.4
    -3
     8
    -6
     24
    -24
```

Exercises

1. Divide: 135 ÷ .5

2. Divide: .135 ÷ 5

Solutions to Exercises

1. Change .5)135 to 5)1350.

When you change .5 into a whole number, change 135 into 1350. Move the decimal point straight up and divide.

```
     270.
  5)1350.
   -10
     35
    -35
```

2. $5\overline{).135}$ does not change because you are dividing by a whole number. Move the decimal point straight up and divide.

$$
\begin{array}{r}
.027 \\
5\overline{).135} \\
-10 \\
\hline
35 \\
-35 \\
\end{array}
$$

Practice Problems

11. Multiply: $4.36 \cdot .002$

12. Add: $14.8 + 231 + 1.965$

13. Divide: $.3498 \div .06$

14. Compute: $17.5 - 9.55$

(Solutions are found at the end of the chapter.)

SIMPLE AND COMPOUND INTEREST

Introduction to Interest

When you put money into a savings account at a bank, the bank gives you money for keeping your money in their bank. The amount of money that you put into the bank is called the principal. The amount of money the bank gives you is called the interest.

The amount of interest you receive from the bank is calculated according to how much money you have in your account. The bank gives you a certain percent of the amount of money that you keep in their bank. The percent that the bank gives you is called the rate of interest.

When you are putting money into the bank, you are getting interest and you are earning money. You want the bank to give you lots of money, so you want your interest rate to be high.

When you are borrowing money from the bank, the bank charges you money for using their money. The amount of money that you borrow is called the principal. The amount of money the bank charges you is called the interest.

The amount of interest that the bank charges you is calculated according to the amount of money you borrow. The percent that the bank charges you is called the rate of interest.

When you are borrowing money from the bank, you are paying interest and you are increasing the amount of money that you owe the bank. You do not want to have to pay the bank a lot of extra money, so you want the interest rate to be low.

Simple Interest

To calculate the extra money that you earn from the bank or owe to the bank, use the simple interest formula. The formula is given to you on the reference sheet.

Simple Interest Formula
$i = prt$ where i = interest, p = principal, r = rate, and t = time

i is the amount of interest in dollars that you earn from the bank or you owe the bank. p is the amount of money in dollars that you place or deposit in the bank or borrow from the bank. r is the interest rate expressed as a decimal. t is the time the money is kept in years.

Example

Big N. Debt borrowed $3,000 from Greedy Bank at a simple interest rate of 2.45%. What is the total amount of money Big N. Debt will have to repay Greedy Bank if he borrows the money for 5 years without making any payments or withdrawals?

Solution

The principal is $3,000, the interest rate is 2.45% changed to .0245, and the time is 5 years. Substitute all of those numbers into the formula.

$$i = 3,000 \cdot (.0245) \cdot 5 = 367.50$$

Big N. Debt will have to pay $367.50 in interest plus the $3,000 he originally borrowed for a total of $3,367.50.

Exercise

You deposit $550,000 in the bank at 2% simple interest. At the end of 3 years you withdraw all of the money and the interest. How much money will you withdraw?

Solution to Exercise

Your principal is $550,000. The simple interest rate is 2%. The length of time that the money is invested is 3 years.

$$i = 550,000 \cdot (.02) \cdot 3 = 33,000$$

You would earn $33,000 in interest. You could withdraw a total of $550,000 + $33,000 = $583,000.

Practice Problem

15. You put $350 in the bank at 4.6% simple interest. Your friend put $300 in another bank giving 5.5% simple interest. At the end of 2 years, you both take out all of the money, including the interest. Who earned more interest?

(Solution is found at the end of the chapter.)

ESTIMATING

Introduction to Estimation

Many times students work problems and get answers that are totally unreasonable. Suppose you worked the problem $5\frac{2}{3} \cdot 7\frac{1}{5}$ and you got the answer $10\frac{8}{15}$. Estimation will help you understand that your answer is not even close to the correct answer.

When you want to estimate what an answer should be, use a "nice" number to make the calculation. A "nice" number is not a mathematical term. It just means a number that you can use easily in a calculation. In the problem $5\frac{2}{3} \cdot 7\frac{1}{5}$, you could think of $5\frac{2}{3}$ as a number between 5 and 6 but closer to 6. Use 6. Reasoning the same way, think of $7\frac{1}{5}$ as a little more than 7. To estimate $5\frac{2}{3} \cdot 7\frac{1}{5}$, think of $6 \cdot 7 = 42$. Your answer will not be exactly 42 but should be closer to 42 than to $10\frac{8}{15}$. The exact calculation gives you $40\frac{4}{5}$.

$$5\frac{2}{3} \cdot 7\frac{1}{5} = \frac{17}{\underset{1}{\cancel{3}}} \cdot \frac{\overset{12}{\cancel{36}}}{5} = \frac{204}{5} = 40\frac{4}{5}$$

Estimation is extremely helpful in dealing with decimal computation. Many times students place the decimal point in the wrong place. Estimating to approximate an answer will tell you to check your work on the problem.

Example
Use estimation to choose the correct answer to $4.3561 \cdot 2.9$.

    A. 126.3269
    B. 12.63269
    C. 1.263269
    D. .1263269

Changing both 4.3561 and 2.9 into " nice" numbers would change the problem to $4 \cdot 3 = 12$. The best answer is choice B, 12.63269.

Exercises

1. Which answer is the best choice for $24.96 \div 1.2$?
    A. 2.8
    B. 2.08
    C. 28
    D. 20.8

2. Use estimation to choose the best answer for $3\frac{1}{3} \div \frac{1}{2}$.

 A. $6\frac{2}{3}$

 B. $1\frac{2}{3}$

 C. $\frac{3}{20}$

 D. $\frac{3}{5}$

Solutions to Exercises

1. Change $24.96 \div 1.2$ into "nice" numbers $25 \div 1 = 25$. You know that answers A and B cannot be correct because they are too small. Choices C and D are both close to 25, so we have to do better estimation. The answer to $24.96 \div 1.2$ will be somewhere between the answers to $24.96 \div 1 = 24.96$ and $24.96 \div 2$ which is around 12. The answer 28 is too big. The correct answer must be choice D, 20.8.

2. Fractions can be difficult to estimate when you are dealing with numbers smaller than 1. You estimate $3\frac{1}{3}$ as about 3, but there is not a " nice" number for $\frac{1}{2}$. The problem $3\frac{1}{3} \div \frac{1}{2}$ is now estimated to $3 \div \frac{1}{2}$. Now reason with fractions as pieces of pie. As pieces of pie, think $3 \div \frac{1}{2}$ means you have 3 pies and you want to divide them into half pies. How many half pies will you get? Your answer is 6, which is close to choice A, $6\frac{2}{3}$.

Practice Problem

16. You want to buy 46 squirt guns for a skit at a school pep session. Each squirt gun costs $2.89. Which choice is the best estimation for the total amount of money that you need to buy the squirt guns?

 A. $23
 B. $150
 C. $48
 D. $250

(Solution is found at the end of the chapter.)

SAMPLE GQE QUESTIONS FOR CHAPTER 1

1. Which problem gives you an answer that is different from all of the others?

   A. $-18 - 6$
   B. $-6 \cdot -4$
   C. $-6 + -18$
   D. $-6 \cdot 4$

2. A stock began the day valued at $42\frac{5}{8}$. The stock market prices went up $3\frac{5}{8}$ points then down $1\frac{1}{4}$ points. What was the stock worth at the end of the day?

   A. 44
   B. $44\frac{3}{8}$
   C. 45
   D. $46\frac{1}{2}$

3. For a skit in Social Studies class, you are going to make 7 Civil War uniforms. Each uniform uses $1\frac{7}{8}$ yards of material. How much material is needed to make all of the uniforms?

   A. $8\frac{7}{8}$ yards
   B. $14\frac{7}{8}$ yards
   C. $3\frac{1}{8}$ yards
   D. $13\frac{1}{8}$ yards

4. Which has the smallest answer, $-18 \div 2$ or $-2 + 18$ or $-20 + 0$?

   A. They all have the same answer.
   B. $-18 \div 2$ has the smallest answer.
   C. $-2 + 18$ has the smallest answer.
   D. $-20 + 0$ has the smallest answer.

5. Unlucky Gambler did not win the lottery, but he did win $2,500. Unlucky wants the bank to keep his money so he will not spend it. Tombstone Bank will give Unlucky 7.5% simple interest if he will leave the entire $2,500 in the bank for 8 years and 6 months. How much interest will Unlucky earn?

   A. $1,593.75
   B. $1,612.50
   C. $15,937.50
   D. $16,125.00

6. Ms. Seem Stress bought 15 yards of cloth to make placemats. Each placemat uses $\frac{3}{4}$ yard. How many placemats will Ms. Seem Stress be able to make?

   A. $15\frac{3}{4}$
   B. $11\frac{1}{4}$
   C. 20
   D. 12

7. Evaluate: $.162 \div 4.5$

   A. 3.6
   B. .36
   C. .036
   D. .0036

8. Ms. Keh Mist is combining 3 bottles into one large tub. The bottles hold .73 liter, 1.4 liters, and 5 liters. How many liters will be in the large tub?

   A. 7.13 liters
   B. 5.87 liters
   C. 10.13 liters
   D. 6.77 liters

SOLUTIONS TO SAMPLE GQE QUESTIONS FOR CHAPTER 1

Answer Key

1.	**B**	3.	**D**	5.	**A**	7.	**C**
2.	**C**	4.	**D**	6.	**C**	8.	**A**

Answers Explained

1. **B** $-6 \cdot -4$
   Choice A is $-18 - 6 = -18 + -6 = -24$.
   Choice B is $-6 \cdot -4 = 24$.
   Choice C is $-6 + -18 = -24$.
   Choice D is $-6 \cdot 4 = -24$

2. **C** 45
   The stock went up $3\frac{5}{8}$, so add the fractions.

   $$42\frac{5}{8} + 3\frac{5}{8} = 45\frac{10}{8} = 45 + 1\frac{2}{8} = 46\frac{2}{8} = 46\frac{1}{4}$$

   Then the stock went down $1\frac{1}{4}$ points, so subtract.

   $$46\frac{1}{4} - 1\frac{1}{4} = 45$$

3. **D** $13\frac{1}{8}$ yards

   Since each uniform uses $1\frac{7}{8}$ yards of material, multiply $7 \cdot 1\frac{7}{8}$ to find the total amount of material needed.

   $$7 \cdot 1\frac{7}{8} = \frac{7}{1} \cdot \frac{15}{8} = \frac{105}{8} = 13\frac{1}{8}$$

4. **D** $-20 + 0$ has the smallest answer.
   $-18 \div 2 = -9$
   $-2 + 18 = 16$
   $-20 + 0 = -20$

5. **A** $1,593.75
   The principal is $2,500, the simple interest rate is 7.5%, and the time is 8 years and 6 months.

   $$i = 2,500 \cdot .075 \cdot 8.5 = 1,593.75$$

> **Be Careful!**
> Time must always be given in years. Since a year is 12 months, the 6 months will be half of a year or .5.

6. **C** 20

$$15 \div \frac{3}{4} = \frac{\overset{5}{\cancel{15}}}{1} \cdot \frac{4}{\underset{1}{\cancel{3}}} = 20$$

7. **C** .036

$4.5\overline{)\,.162\,}$ must be changed to $45\overline{)\,1.62\,}$

$$\begin{array}{r} .036 \\ 45\overline{)1.620} \\ -\,135 \\ \hline 270 \\ -\,270 \\ \hline \end{array}$$

8. **A** 7.13 liters
    When you add decimals, line up the decimal points. Remember the decimal point is at the end of the 5.

$$\begin{array}{r} .73 \\ 1.40 \\ +\,5.00 \\ \hline 7.13 \end{array}$$

Solutions to Practice Problems

1. −40

2. 6

3. 6
    Starting at 0, count to the left 18 spaces. Then count to the right 24 spaces. You will be on the positive side of 0, at 6.

4. −37
    $$-23 - 14 = -23 + -14 = -37$$

5. −420
    The submarine is 300 feet below the surface of the water, −300, and then goes down another 120 feet.
    $$-300 - 120 = -300 + -120 = -420$$

6. 9
    $$3\frac{3}{8} \cdot 2\frac{2}{3} = \frac{\overset{9}{\cancel{27}}}{\cancel{8}} \cdot \frac{\overset{1}{\cancel{8}}}{\underset{1}{\cancel{3}}} = \frac{9}{1} = 9$$

7. $15\dfrac{7}{8}$

Change both fractions so that they have a common denominator.

$$24\dfrac{5}{8} - 8\dfrac{3}{4} = 24\dfrac{5}{8} - 8\dfrac{6}{8}$$

In this case you must borrow. Remember that $24\dfrac{5}{8} = 24 + \dfrac{5}{8} + 23\dfrac{8}{8} + \dfrac{5}{8} = 23\dfrac{13}{8}$.

$$24\dfrac{5}{8} - 8\dfrac{3}{4} = 24\dfrac{5}{8} - 8\dfrac{6}{8} = 23\dfrac{13}{8} - 8\dfrac{6}{8} = 15\dfrac{7}{8}$$

8. $\dfrac{1}{2}$

$$\dfrac{7}{8} \div 1\dfrac{3}{4} = \dfrac{7}{8} \div \dfrac{7}{4} = \dfrac{\overset{1}{\cancel{7}}}{\underset{2}{\cancel{8}}} \cdot \dfrac{\overset{1}{\cancel{4}}}{\underset{1}{\cancel{7}}} = \dfrac{1}{2}$$

9. $60\dfrac{7}{12}$

$$41\dfrac{5}{6} + 18\dfrac{3}{4} = 41\dfrac{10}{12} + 18\dfrac{9}{12} = 59\dfrac{19}{12} = 59 + 1\dfrac{7}{12} = 60\dfrac{7}{12}$$

10. Funny Fanny will be able to create 37 booths.

Funny Fanny needs to take $12\dfrac{1}{2}$ acres and cut it into $\dfrac{1}{3}$-acre pieces. She will divide $12\dfrac{1}{2}$ by $\dfrac{1}{3}$.

$$12\dfrac{1}{2} \div \dfrac{1}{3} = \dfrac{25}{2} \div \dfrac{1}{3} = \dfrac{25}{2} \cdot \dfrac{3}{1} = \dfrac{75}{2} = 37\dfrac{1}{2}$$

11. .00872

$$\begin{array}{r} 4.36 \\ \times\,.002 \\ \hline .00872 \end{array}$$

12. 247.765

$$\begin{array}{r} 14.800 \\ 231.000 \\ +\,1.965 \\ \hline 247.765 \end{array}$$

13. 5.83

$$\begin{array}{r} 5.83 \\ 6\overline{)34.98} \\ \underline{-30} \\ 4\ 9 \\ \underline{-4\ 8} \\ 18 \\ \underline{-18} \end{array}$$

14. 7.95
When subtracting decimals, line up the decimal points. If there are any empty spaces, fill them with zeros. Subtract normally. Bring the decimal point straight down.

$$\begin{array}{r} {}^{6\ 14} \\ 17.\cancel{8}\,0 \\ -\ 9.5\ 5 \\ \hline 7.9\ 5 \end{array}$$

15. Your friend earned more interest.
Your principal is $350, your interest rate is 4.6%, and your time is 2 years.

$$i = 350 \cdot .046 \cdot 2 = 32.2$$

You earned $32.20 in interest.
Your friend's principal is $300, your friend's interest rate is 5.5%, and your friend's time is 2 years.

$$i = 300 \cdot .055 \cdot 2 = 33$$

Your friend earned $33 in interest.

16. B . $150
To find the total cost of 46 squirt guns, multiply $46 \cdot 2.89$. Change each number into a "nice" number to get $50 \cdot 3 = 150$.

Chapter 2 | Number Sense

You CANNOT use your calculator on any of the problems in this chapter.

EXPONENTS: NEGATIVE AND INTEGER

Positive and Negative Exponents

Exponents (powers) are a shorthand way to write repeated multiplication. 3^4 means 3 multiplied by itself 4 times, $3 \cdot 3 \cdot 3 \cdot 3$. In 3^4, the 3 is the base and the 4 is the exponent or power.

Positive exponents describe large numbers. Negative exponents allow us to work with small numbers between 0 and 1.

Definition: Negative exponents are defined by $a^{-c} = \dfrac{1}{a^c}$ where $a \neq 0$.

Example: $5^{-3} = \dfrac{1}{5^3} = \dfrac{1}{5 \cdot 5 \cdot 5} = \dfrac{1}{125}$

Exercises

1. What is the difference between 3^5 and 5^3?

2. Rewrite 6^{-2} without exponents.

3. Write $\dfrac{1}{7 \cdot 7 \cdot 7 \cdot 7}$ as a power of 7.

Solutions to Exercises

1. 3^5 means 3 is multiplied by itself 5 times, $3 \cdot 3 \cdot 3 \cdot 3 \cdot 3 \cdot = 243$. 5^3 means that 5 is multiplied by itself 3 times, $5 \cdot 5 \cdot 5 \cdot = 125$.

2. $6^{-2} = \dfrac{1}{6^2} = \dfrac{1}{6 \cdot 6} = \dfrac{1}{36}$

3. $\dfrac{1}{7 \cdot 7 \cdot 7 \cdot 7} = \dfrac{1}{7^4} = 7^{-4}$

Multiplying and Dividing Numbers with Exponents

The rules for multiplying numbers with powers should make sense to you. Think of $5^3 \cdot 5^4 = (5 \cdot 5 \cdot 5) \cdot (5 \cdot 5 \cdot 5 \cdot 5) = 5^7$. Notice that you add the powers and the base does not change. This also works with variables having exponents.

$$x^3 \cdot x^5 = (x \cdot x \cdot x) \cdot (x \cdot x \cdot x \cdot x \cdot x) = x^8$$

$$a^b \cdot a^c = a^{b+c}$$

Example: $9^{10} \cdot 9^4 = 9^{10+4} = 9^{14}$

Many students make the mistake of multiplying the bases when multiplying numbers with powers. $2^5 \cdot 2^7 \neq 4^{12}$. Remember to keep the bases the same and add the exponents.

Students also make the mistake of multiplying numbers that do not have the same base. You cannot multiply numbers with powers unless the bases are the same.

$$2^5 \cdot 3^3 \neq 6^8$$

Dividing numbers with powers should also make sense to you. Think of $\dfrac{8^6}{8^2} =$

$$\dfrac{8^1 \cdot 8^1 \cdot 8 \cdot 8 \cdot 8 \cdot 8}{8^1 \cdot 8^1} = 8 \cdot 8 \cdot 8 \cdot 8 = 8^4.$$ Notice that you subtract the powers and the base does not change. This also works with variables having exponents.

$$\dfrac{w^6}{w^2} = \dfrac{w^1 \cdot w^1 \cdot w \cdot w \cdot w \cdot w}{w^1 \cdot w^1} = w^4$$

$$\dfrac{a^b}{a^c} = a^{b-c}$$

Example: $\dfrac{7^{10}}{7^2} = 7^{10-2} = 7^8$

Many students make the mistake of dividing the bases when dividing numbers with powers. Remember to keep the bases the same and subtract the exponents.

$$\dfrac{2^7}{2^3} \neq 1^4$$

Students also make the mistake of dividing numbers that do not have the same base. You cannot divide numbers with powers unless the bases are the same.

$$\dfrac{8^6}{2^3} \neq 4^3$$

Example
Multiply: $5^3 \cdot 5^{-3}$
One way to do this problem is by keeping the base the same and adding the exponents.

$$5^3 \cdot 5^{-3} = 5^{3+-3} = 5^0$$

Another way to do this problem is by evaluating each of the powers first. Remember that

$$5^3 = 5 \cdot 5 \cdot 5 = 125 \text{ and } 5^{-3} = \frac{1}{5^3} = \frac{1}{5 \cdot 5 \cdot 5} = \frac{1}{125}.$$

$a^0 = 1$ for every number $a \neq 0$

This means that $5^3 \cdot 5^{-3} = 125 \cdot \dfrac{1}{125} = \dfrac{125}{125} = 1.$

The two answers must be the same.

$$5^0 = 1$$

Exercises

1. Simplify: $13^3 \cdot 13^9$

2. Divide: $\dfrac{10^{20}}{10^5}$

3. Multiply: $10^{-5} \cdot 10^{-3}$

Solutions to Exercises

1. $13^3 \cdot 13^9 = 13^{3+9} = 13^{12}$

2. $\dfrac{10^{20}}{10^5} = 10^{20-5} = 10^{15}$

3. $10^{-5} \cdot 10^{-3} = 10^{-5 + -3} = 10^{-8}$

Powers of Powers

The rule for finding the power of a power should make sense to you. Think that $(4^3)^5$ means to multiply (4^3) by itself 5 times.

$$(4^3)^5 = (4^3) \cdot (4^3) \cdot (4^3) \cdot (4^3) \cdot (4^3) = 4^{15}$$

Notice that you keep the base the same and you multiply the powers. This also works with variables having exponents.

$$(y^{10})^4 = (y^{10}) \cdot (y^{10}) \cdot (y^{10}) \cdot (y^{10}) = y^{40}$$

$(a^b)^c = a^{bc}$

Example: $(3^{11})^3 = 3^{11 \cdot 3} = 3^{33}$

Exercises

1. Simplify: $(25^5)^2$

2. Simplify: $(5^3)^4 \cdot 5^7$

Solutions to Exercises

1. $(25^5)^2 = 25^{5 \cdot 2} = 25^{10}$

2. $(5^3)^4 \cdot 5^7 = 5^{12} \cdot 5^7 = 5^{12+7} = 5^{19}$

Power of a Product and Power of a Quotient

Power of a product means you are going to take the power of two things multiplied together. Think that $(5x)^4$ means to multiply $5x$ by itself 4 times.

$$(5x)^4 = (5x) \cdot (5x) \cdot (5x) \cdot (5x) = 5^4 x^4$$

Each number in the parentheses must be taken to the 4th power. It is not necessary to find $5^4 = 1,024$.

Power of a quotient means to take the power of two things divided. Think that $\left(\dfrac{4}{7}\right)^5$ means to multiply $\left(\dfrac{4}{7}\right)$ times itself 5 times.

$$(ab)^c = a^c b^c$$

Example: $(4wc)^5 = 4^5 w^5 c^5$

$$\left(\frac{4}{7}\right)^5 = \left(\frac{4}{7}\right) \cdot \left(\frac{4}{7}\right) \cdot \left(\frac{4}{7}\right) \cdot \left(\frac{4}{7}\right) \cdot \left(\frac{4}{7}\right) = \frac{4^5}{7^5}$$

Each number in the parentheses must be taken to the 5th power. It is not necessary to find $\dfrac{4^5}{7^5} = \dfrac{1,024}{16,807}$.

Exercises

1. Simplify: $\left(\dfrac{3}{4}\right)^3$

2. Evaluate: $(2x^2)^3$

3. Simplify: $(5x^{-4})^{-2}$

Solutions to Exercises

1. $\left(\dfrac{3}{4}\right)^3 = \dfrac{3^3}{4^3} = \dfrac{3 \cdot 3 \cdot 3}{4 \cdot 4 \cdot 4} = \dfrac{27}{64}$

2. $(2x^2)^3 = (2)^3 \cdot (x^2)^3 = 8x^6$

3. $(5x^{-4})^{-2} = (5)^{-2} \cdot (x^{-4})^{-2} = \dfrac{1}{25} \cdot x^8 = \dfrac{x^8}{25}$

Practice Problems

1. Simplify: $(8^3)^{10}$

2. Evaluate: 9^{-2}

3. Evaluate: $\dfrac{5^4 \cdot 5^{18}}{5^{10} \cdot 5^{14}}$

4. Simplify: $(7x^{10}c^{20})^2$

(Solutions are found at the end of the chapter.)

Scientific Notation

Scientific notation is a different way to write numbers, especially very large numbers or very small numbers. Before we learn to write complicated numbers in scientific notation, we will work with numbers that make sense. The number 500 is the same as 5×100 which is the same as 5×10^2. We say that 500 is written in scientific notation as 5×10^2.

> **Remember:** Scientific notation must have a number between 1 and 10 (not including 10) multiplied by a power of 10.

To change 700,000 into scientific notation, break 700,000 into $7 \times 100,000$. Then, think of 100,000 as 10^5. So, $700,000 = 7 \times 10^5$.

Not all numbers that you want to change into scientific notation are so simple. If you want to change 5,300 into scientific notation, you cannot make it 53×100. The first number in scientific notation must be between 1 and 10 (not including 10). You must change 5,300 into 5.3×1000 which is 5.3×10^3.

There is a way to jump directly from 5,300 to 5.3×10^3 by looking at the decimal point in 5,300. The decimal point in 5,300 is at the end of the number. For a number to be in scientific notation, there must be only one number between 1 and 10 (not including 10) in front of the decimal point. Move the decimal point in 5,300 to between the 5 and the 3. But you cannot just change 5,300 into 5.3. If you move the decimal point 3 places, then you need to multiply 5.3 by something to keep the number equal to 5,300. In this case, $5,300 = 5.3 \times 10^3$.

Sometimes problems can be even more difficult to put into scientific notation because they are already in a similar form. Consider placing 436×10^5 into scientific notation. First change 436 into scientific notation and then combine it with the 10^5.

$$436 \times 10^5 = (4.36 \times 10^2) \times 10^5 = 4.36 \times (10^2 \times 10^5) = 4.36 \times 10^7$$

Scientific notation with small numbers uses negative exponents.

$$10^2 = 100$$
$$10^1 = 10$$
$$10^0 = 1$$
$$10^{-1} = \frac{1}{10} = .1$$
$$10^{-2} = \frac{1}{10^2} = \frac{1}{100} = .01$$

As you can see, 10 to a negative power is a small number. To change a small number into scientific notation, use negative powers of 10.

Example

To change .0000563 into scientific notation, move the decimal point between the 5 and the 6. You will need to move the decimal point 5 places from the beginning of the number.

$$.0000563 = 5.63 \times 10^{-5}$$

Remember very large numbers involve positive powers of 10, while very small numbers involve negative powers of 10.

Exercises

1. Rewrite 1,430,000 in scientific notation.

2. What is .000011 in scientific notation?

3. Rewrite 6.145×10^{-2} as a regular number.

4. Rewrite $.043 \times 10^{-3}$ in scientific notation.

> **Be Careful!**
> Many students try to memorize which way to move the decimal point. It is better to think that a large number has positive powers of 10 and a small number has negative powers of 10.

Solution to Exercises

1. 1.43×10^6
   The number is very large, so positive powers of 10 will be used. Move the decimal point 6 places so it is between the 1 and the 4.

2. 1.1×10^{-5}
   The number is very small, so negative powers of 10 will be used. Move the decimal point 5 places to between the 1 and the 1.

3. .06145
   The number has a negative power of 10, so the number must be small. To make the number small, move the decimal point 2 places to the left.

4. 4.3×10^{-5}
   First, rewrite .043 in scientific notation.
   $$.043 = 4.3 \times 10^{-2}$$
   Now, put that back into the original problem.
   $$.043 \times 10^{-3} = (4.3 \times 10^{-2}) \times 10^{-3} = 4.3 \times (10^{-2} \times 10^{-3}) = 4.3 \times 10^{-5}$$

Practice Problems

5. What is .00000106 in scientific notation?

6. Change $5,700 \times 10^{-8}$ into scientific notation.

7. Rewrite 7.6×10^{-5} in regular notation.

(Solutions are found at the end of the chapter.)

Comparing Numbers in Scientific Notation

Usually comparing numbers in scientific notation is quite simple. To decide which is larger, begin by looking at the powers of 10. When comparing 4.6×10^8 and 9.99×10^{-3}, remember that the positive powers of 10 represent large numbers and the negative powers of 10 represent small numbers. The numbers 4.6 and 9.99 do not matter in this problem because 10^8 is a much larger number than 10^{-3}.

$$4.6 \times 10^8 > 9.99 \times 10^{-3}$$

Sometimes comparing numbers can be a bit more complicated. If you want to compare 4.59×10^{-6} and 5.2×10^{-6}, first look at the powers of 10. Since they are the same, you must look at the first numbers. Compare 4.59 and 5.2 and the comparison will be the same for 4.59×10^{-6} and 5.2×10^{-6}. Since $4.59 < 5.2$, then $4.59 \times 10^{-6} < 5.2 \times 10^{-6}$.

Exercise

Which is larger, 8.34×10^{-3} or 9.867×10^{-4}?

Solution to Exercise
To compare numbers in scientific notation, first compare the powers of 10. If the powers of 10 are the same, then compare the first numbers. In this case, the powers of 10 are not the same. Since $-3 > -4$, then $8.34 \times 10^{-3} > 9.867 \times 10^{-4}$.

Practice Problems

8. Place the numbers 7.6×10^{-4}, 5.8×10^{-3}, and 3.4×10^{-4} in order from largest to smallest.

(Solution is found at the end of the chapter.)

Adding and Subtracting Numbers in Scientific Notation

Adding or subtracting numbers in scientific notation can be done in two different ways. One way is to change the numbers out of scientific notation and work the problem normally. Change $5.3 \times 10^3 - 4.62 \times 10^2$ into $5,300 - 462$. Then subtract normally to get 4,838. If the problem asks for the answer in scientific notation, then change 4,838 into 4.838×10^3. This way of working the problem is easiest when the numbers are not very large or not very small.

The second way to add and subtract numbers in scientific notation is best when working with very large or very small numbers. To add $4.75 \times 10^{11} + 3.8 \times 10^{10}$, change the larger number to the same power as the smaller number or vice versa.

Change 3.8×10^{10} into $.38 \times 10^{11}$. The problem becomes $4.75 \times 10^{11} + .38 \times 10^{11}$. Add the first numbers of each number in scientific notation and keep the power of 10 the same.

$4.75 \times 10^{11} + .38 \times 10^{11} = 5.13 \times 10^{11}$, which is already in scientific notation.

> **Be Careful!**
> Many students make a mistake when changing 3.8×10^{10} into a number multiplied by 10^{11}.
>
> $$.38 \times 10^{11} = (3.8 \times 10^{-1}) \times 10^{11} = 3.8 \times (10^{-1} \times 10^{11}) = 3.8 \times 10^{10}$$

Exercises

1. Add: $4.3 \times 10^{-3} + 9.17 \times 10^{-4}$
   Write your answer in scientific notation.

2. Compute: $7.68 \times 10^{25} - 8 \times 10^{23}$
   Write your answer in scientific notation.

Solutions to Exercises

1. The easiest way to add $4.3 \times 10^{-3} + 9.17 \times 10^{-4}$ is to change each number out of scientific notation and add.

$$4.3 \times 10^{-3} = .0043$$

$$9.17 \times 10^{-4} = .000917$$

The problem now changes into $.0043 + .000917 = .005217 = 5.217 \times 10^{-3}$

2. You certainly do not want to change the numbers out of scientific notation. Instead, change one of the numbers so that it has the same power of 10 as the other number. The easiest way is to change 8×10^{23} into $.08 \times 10^{25}$. Then work the problem.

$$7.68 \times 10^{25} - 8 \times 10^{23} = 7.68 \times 10^{25} - .08 \times 10^{25} = 7.6 \times 10^{25}$$

Multiplying and Dividing Numbers in Scientific Notation

Multiplying and dividing numbers in scientific notation uses the laws of exponents explained in the beginning of this chapter.

To multiply $(4.6 \times 10^4) \cdot (2 \times 10^5)$, use the associative and commutative properties to regroup the numbers to multiply.

Associative property says that you can put different numbers in the parentheses to multiply first.

$$(a \cdot b) \cdot c = a \cdot (b \cdot c)$$

Example: $(2 \cdot 3) \cdot 4 = 2 \cdot (3 \cdot 4)$
Commutative property says that you can switch the order in which you multiply numbers.

$$a \cdot b = b \cdot a$$

Example: $5 \cdot 3 = 3 \cdot 5$

Change $(4.6 \times 10^4) \cdot (2 \times 10^5)$ into $(4.6 \cdot 2) \times (10^4 \cdot 10^5)$. Then, multiply the numbers in the parentheses.

$$(4.6 \cdot 2) \times (10^4 \cdot 10^5) = 9.2 \times 10^9$$

Your answer is already in scientific notation.

Example
To multiply $(6 \times 10^{-3}) \cdot (3.4 \times 10^5)$, regroup the numbers.

$$(6 \times 10^{-3}) \cdot (3.4 \times 10^5) = (6 \cdot 3.4) \times (10^{-3} \cdot 10^5)$$

Then multiply the numbers in the parentheses.

$$(6 \cdot 3.4) \times (10^{-3} \cdot 10^5) = 20.4 \times 10^2$$

Now change 20.4×10^2 into scientific notation.

$$20.4 \times 10^2 = (2.04 \times 10^1) \times 10^2 = 2.04 \times (10^1 \times 10^2) = 2.04 \times 10^3$$

To divide $\dfrac{7.2 \times 10^8}{3 \times 10^2}$, break the problem into two separate parts.

$$\frac{7.2 \times 10^8}{3 \times 10^2} = \frac{7.2}{3} \times \frac{10^8}{10^2} = 2.4 \times 10^6$$

Exercises

1. Multiply: $(6 \times 10^{-8}) \cdot (2 \times 10^{-4})$
   Write your answer in scientific notation.

2. Divide: $(1.98 \times 10^{-3}) \div (6 \times 10^5)$
   Write your answer in scientific notation.

Solutions to Exercises

1. $(6 \times 10^{-8}) \cdot (2 \times 10^{-4}) = (6 \cdot 2) \times (10^{-8} \cdot 10^{-4}) = 12 \times 10^{-2}$
   Change the answer 12×10^{-2} into scientific notation.

   $$12 \times 10^{-2} = (1.2 \times 10^1) \times 10^{-12} = 1.2 \times (10^1 \times 10^{-12}) = 1.2 \times 10^{-11}$$

2. $(1.98 \times 10^{-3}) \div (6 \times 10^5) = \left(\dfrac{1.98}{6}\right) \times \left(\dfrac{10^{-3}}{10^5}\right) = .33 \times 10^{-8}$

Be Careful!

When dividing $\dfrac{10^{-3}}{10^5}$, subtract the exponents. $-3 - 5 = -8$

Now change .33 into scientific notation and simplify.

$$\left(\frac{1.98}{6}\right) \times \left(\frac{10^{-3}}{10^5}\right) = .33 \times 10^{-8} = (3.3 \times 10^{-1}) \times 10^{-8} = 3.3 \times (10^{-1} \times 10^{-8}) = 3.3 \times 10^{-9}$$

Practice Problems

9. Add: $1.45 \times 10^8 + 2.3 \times 10^7$
   Write your answer in scientific notation.

10. Subtract: $1.63 \times 10^{-3} - 2.8 \times 10^{-4}$
    Write your answer in scientific notation.

11. Find the product of $(7.8 \times 10^{-5}) \cdot (2 \times 10^8)$.
    Write your answer in scientific notation.

12. Find the quotient of $(1.846 \times 10^{100}) \div (2 \times 10^4)$.
    Write your answer in scientific notation.

(Solutions are found at the end of the chapter.)

RATIONAL NUMBERS VS. IRRATIONAL NUMBERS

At this point in your mathematical career, there are only two kinds of numbers, rational and irrational. A number is either rational or it is irrational. It cannot be both rational and irrational. As a matter of fact, the prefix ir- means not, so ir-rational means not rational.

Definition: A rational number is any number that can be written as a fraction.

Lots of numbers that you know can be written as fractions.

$$8 = \frac{8}{1} \qquad\qquad .7 = \frac{7}{10}$$

$$14.3 = 14\frac{3}{10} = \frac{143}{10} \qquad\qquad .\overline{3} = \frac{1}{3}$$

$$.\overline{12} = \frac{4}{33} \qquad\qquad 0 = \frac{0}{10}$$

Can all numbers be written as fractions? No. Irrational numbers cannot be written as fractions.

Definition: An irrational number is a nonrepeating, nonterminating decimal.

An irrational number never repeats and never stops. A famous irrational number is π. We frequently use 3.14 or $\frac{22}{7}$ as approximations for π, but they are only approximations. π goes on for millions and millions of digits and never repeats the same order time after time.

There are lots of irrational numbers. $\sqrt{3}$ is an irrational number because it can never be written as a fraction. Sometimes we use 1.732 for $\sqrt{3}$. When you square 1.732, you get 2.999824, which is close to 3 but not exactly 3.

Remember: $\sqrt{25}$ means what number multiplied by itself gives you 25. $\sqrt{25} = 5$ because $5 \cdot 5 = 25$. $\sqrt{3}$ means what number multiplied by itself gives you 3.

Is $\sqrt{4}$ a rational or an irrational number? $\sqrt{4} = 2$ because $2 \cdot 2 = 4$. $\sqrt{4}$ is a rational number.

Is $\sqrt{5}$ rational or irrational? $\sqrt{5}$ is irrational because it cannot be written as a fraction. An approximation for $\sqrt{5}$ is 2.2360679. But when you multiply 2.2360679 by itself, you get 4.999999653, an approximation for 5.

Irrational numbers include $\sqrt{2}$, $\sqrt{3}$, $\sqrt{5}$, $\sqrt{6}$, $\sqrt{7}$, $\sqrt{8}$, $\sqrt{10}$, $\sqrt{11}$, $\sqrt{12}$, $\sqrt{13}$, $\sqrt{14}$, $\sqrt{15}$, and $\sqrt{17}$.

Rational numbers include $\sqrt{1}$, $\sqrt{4}$, $\sqrt{9}$, $\sqrt{16}$, $\sqrt{25}$, $\sqrt{36}$, $\sqrt{49}$, and $\sqrt{64}$.
All of the square roots listed under the rational numbers are perfect squares.

Definition: A perfect square is an integer that is a square of another integer. 289 is a perfect square because $17^2 = 289$.

Another type of irrational number is a decimal that never stops and never repeats. .12112111211112… is an irrational number. It never stops as the three dots at the end of the number indicate. Although it does have a pattern, it never repeats the numbers in exactly the same order time after time.

Exercise

Choose all of the irrational numbers from the list.

$$\sqrt{18}, 0, -\pi, .\overline{7}, -5\tfrac{3}{4}, 7.41441, -847.9$$

Solution to Exercise

1. $\sqrt{18}$, $-\pi$ are irrational.

   0 can be written as $\dfrac{0}{7}$.

   $.\overline{7}$ can be written as $\dfrac{7}{9}$. All repeating decimals are rational.

   $-5\dfrac{3}{4}$ can be written as $-\dfrac{23}{4}$.

   7.41441 can be written as $7\dfrac{41,441}{100,000} = \dfrac{741,441}{100,000}$. All decimals that stop are rational.

   -847.9 can be written as $-847\dfrac{9}{10} = -\dfrac{8,479}{10}$.

APPROXIMATING SQUARE ROOTS

It is easy to find the square root of a perfect square. We all know that $\sqrt{9} = 3$ and $\sqrt{100} = 10$. What is $\sqrt{7}$? You will not be able to use a calculator on these problems on the GQE. To approximate $\sqrt{7}$ you need to find out which two whole numbers $\sqrt{7}$ is between and which one of the whole numbers it is closer to.

A list of square roots between $\sqrt{1} = 1$ and $\sqrt{4} = 2$ includes $\sqrt{1} = 1$, $\sqrt{2}$, $\sqrt{3}$, and $\sqrt{4} = 2$. You know that $\sqrt{3}$ must be somewhere between 1 and 2 because $\sqrt{3}$ is between $\sqrt{1}$ and $\sqrt{4}$. Actually, $\sqrt{3}$ must be closer to 2 than it is to 1 because $\sqrt{3}$ is closer to $\sqrt{4}$ than it is to $\sqrt{1}$. A calculator will show that $\sqrt{3} \approx 1.732$, which is closer to 2 than to 1.

A list of square roots between $\sqrt{4}$ and $\sqrt{9}$ includes $\sqrt{4} = 2$, $\sqrt{5}$, $\sqrt{6}$, $\sqrt{7}$, $\sqrt{8}$, and $\sqrt{9} = 3$. You know that $\sqrt{5}$ is close to 2 and $\sqrt{8}$ is close to 3. You also know that $\sqrt{8}$ is greater than $\sqrt{5}$.

To approximate $\sqrt{51}$, writing out all of the square roots does not make sense. Think of what perfect square is close to 51 but smaller than 51. The next smaller perfect square is 49. So $\sqrt{49} = 7$ is a little smaller than $\sqrt{51}$. Of course, the next larger perfect square is 64. $\sqrt{64} = 8$

Exercises

1. Place the numbers $\sqrt{50}$, 7.5, 6.5, $\sqrt{47}$, and $\sqrt{49}$ in order from largest to smallest.

2. Jon and Betsy are having an argument. Jon says that $\sqrt{14}$ is a little more than 7 because $2 \cdot 7 = 14$. Betsy says that $\sqrt{14}$ is a little less than 4. Who is correct and why?

Solutions to Exercises

1. 7.5, $\sqrt{50}$, $\sqrt{49}$, $\sqrt{47}$, 6.5

   $\sqrt{50}$ is a little more than $\sqrt{49} = 7$, which is a little more than $\sqrt{47}$.

2. Betsy is correct.

   $\sqrt{14}$ is between $\sqrt{9} = 3$ and $\sqrt{16} = 4$. $\sqrt{14}$ is a little less than $\sqrt{16} = 4$.

Practice Problems

13. Graph the points π, –2.1, $-\sqrt{17}$, 4.1, and .2 on the number line shown below.

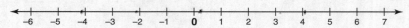

Be sure to label the points.

14. Which is the best approximation for $\sqrt{101}$?

   A. 50.1
   B. 49.1
   C. 10.1
   D. 9.1

(Solutions are found at the end of the chapter.)

SAMPLE GQE QUESTIONS FOR CHAPTER 2

 1. In an average 150-pound person, there are approximately .000021 pounds of gold. Which of the following equals .000021 in scientific notation?

    A. 2.1×10^5
    B. 0.21×10^{-5}
    C. 2.1×10^{-5}
    D. 2.1×10^{-6}

 2. Which of the following is the same as 2^{-4}?

    A. $\dfrac{1}{8}$

    B. $\dfrac{1}{16}$

    C. -8
    D. -16

 3. Simplify: $\dfrac{x^5 \cdot x^{10}}{x^3}$

    A. x^5

    B. $x^{\frac{50}{3}}$

    C. x^{47}
    D. x^{12}

 4. The human body is composed of many different elements. A 150-pound individual will have 3.75×10^{-1} pounds of oxygen and 7.5×10^{-2} pounds of zirconium. What is the combined weight of the oxygen and zirconium?

    A. 11.25×10^{-3}
    B. 1.125×10^{-1}
    C. 4.5×10^{-1}
    D. 4.5×10^{-2}

  5. To change from square inches into square miles, multiply the number of square inches by 2.491×10^{-10}. Convert 500,000,000 square inches into square miles.

    A. 1.2455×10^1
    B. 1.2455
    C. 1.2455×10^{-1}
    D. 1.2455×10^{-2}

 6. Simplify: $\dfrac{3^8}{3^2} \cdot (3^5)^4$

    A. 3^{20}
    B. 3^9
    C. 3^{26}
    D. 3^{24}

7. Which of the following is true?

    A. $\sqrt{49} + \pi$ is rational.

    B. $\sqrt{78+3}$ is rational.

    C. $7.\overline{4545} - 3$ is irrational.

    D. $\sqrt{103-54}$ is irrational.

8. Which of the following is false?

    A. $\sqrt{50} > 7$

    B. $4.5 < \sqrt{23}$

    C. $\sqrt{7} < 3$

    D. $9 > \sqrt{99}$

SOLUTIONS TO SAMPLE GQE QUESTIONS FOR CHAPTER 2

Answer Key

1. **C**	3. **D**	5. **C**	7. **B**
2. **B**	4. **C**	6. **C**	8. **D**

Answers Explained

1. **C** 2.1×10^{-5}

   In scientific notation, the first number must be between 1 and 10 (not including 10) and the second number must be the appropriate power of 10. Choice A can be eliminated because 2.1×10^5 is a large number and .000021 is a small number. Choice B can be eliminated because it does not have the first number between 1 and 10 (not including 10). Choice D can be eliminated because the power of 10 is –6 instead of –5. The number of places you moved the decimal point to get it between the 2 and the 1 is 5, so the power of 10 should be –5, choice C.

2. **B** $\dfrac{1}{16}$

$$2^{-4} = \frac{1}{2^4} = \frac{1}{2 \cdot 2 \cdot 2 \cdot 2} = \frac{1}{16}$$

3. **D** x^{12}

$$\frac{x^5 \cdot x^{10}}{x^3} = \frac{x^{15}}{x^3} = x^{15-3} = x^{12}$$

4. **C** 4.5×10^{-1}

   Change 3.75×10^{-1} into .375 and change 7.5×10^{-2} into .075. Add .375 + .075 = .450. Change back into scientific notation to get 4.5×10^{-1}.

5. **C** 1.2455×10^{-1}

   Change 500,000,000 to scientific notation, 5×10^8. Multiply.

$$(2.491 \times 10^{-10}) \times (5 \times 10^8) = (2.491 \times 5) \times (10^{-10} \times 10^8)$$
$$= 12.455 \times 10^{-2}$$
$$= (1.2455 \times 10^1) \times 10^{-2}$$
$$= 1.2455 \times (10^1 \times 10^{-2})$$

6. **C** 3^{26}

$$\frac{3^8}{3^2} \cdot (3^5)^4 = 3^{8-2} \cdot 3^{5 \cdot 4} = 3^6 \cdot 3^{20} = 3^{26}$$

7. **B** $\sqrt{78+3}$ is rational.

In choice A, $\sqrt{49} = 7$ is rational and π is irrational, so the sum is irrational.

In choice B, $\sqrt{78+3} = \sqrt{81} = 9$, which is a rational number.

In choice C, both $7.\overline{4545}$ and 3 are rational, so their sum is rational.

In choice D, $\sqrt{103-54} = \sqrt{49}$, which is rational.

8. **D** $9.4 > \sqrt{99}$

In choice A, $\sqrt{50} > \sqrt{49} = 7$.

In choice B, $4.5 = \sqrt{20.25} < \sqrt{23}$.

In choice C, $\sqrt{7} < \sqrt{9} = 3$.

In choice D, $9 = \sqrt{81} \not> \sqrt{99}$.

Solutions to Practice Problems

1. $(8^3)^{10} = 8^{3 \cdot 10} = 8^{30}$

2. $\dfrac{1}{81}$

   Rewrite 9^{-2} as $\dfrac{1}{9^2}$ and then simplify 9^2.

3. $\dfrac{1}{25}$

$$\frac{5^4 \cdot 5^{18}}{5^{10} \cdot 5^{14}} = \frac{5^{22}}{5^{24}} = 5^{-2} = \frac{1}{5^2} = \frac{1}{25}$$

4. $49x^{20}c^{40}$

$$(7x^{10}c^{20})^2 = (7)^2 \cdot (x^{10})^2 \cdot (c^{20})^2 = 49x^{20}c^{40}$$

5. 1.06×10^{-6}

   The decimal point in front of the number must be moved 6 spaces to get it between the 1 and the 0. This is a very small number so the power of 10 is negative.

6. 5.7×10^{-5}

   Rewrite 5,700 in scientific notation.

$$5,700 = 5.7 \times 10^3$$

   Then substitute your answer into the original problem.

$$5,700 \times 10^{-8} = (5.7 \times 10^3) \times 10^{-8} = 5.7 \times (10^3 \times 10^{-8}) = 5.7 \times 10^{-5}$$

7. .000076

   The power of 10 is negative, so 7.6×10^{-5} is a small number. Move the decimal point 5 places to make it a small number, which would be 5 places to the left.

8. 5.8×10^{-3}, 7.6×10^{-4}, 3.4×10^{-4}

To compare numbers in scientific notation, first compare the powers of 10. If those are the same, compare the numbers that are multiplied by the powers of 10. 5.8×10^{-3} is the largest number because it has the largest power of 10.

> **Be careful!**
> Remember that $-3 > -4$, so $10^{-3} > 10^{-4}$

7.6×10^{-4} and 3.4×10^{-4} have the same power of 10, so you compare the numbers multiplied by the power of 10. $7.6 > 3.4$, so $7.6 \times 10^{-4} > 3.4 \times 10^{-4}$.

9. 1.68×10^8

To add numbers in scientific notation, they must have the same power of 10. So change 2.3×10^7 into $.23 \times 10^8$.

> **Be careful!**
> $.23 \times 10^8 = (2.3 \times 10^{-1}) \times 10^8 = 2.3 \times (10^{-1} \times 10^8) = 2.3 \times 10^7$

$$1.45 \times 10^8 + 2.3 \times 10^7 = 1.45 \times 10^8 + .23 \times 10^8 = 1.68 \times 10^8$$

10. 1.35×10^{-3}

Change each number out of scientific notation, subtract, and then change your answer back into scientific notation.

$$1.63 \times 10^{-3} = .00163$$
$$2.8 \times 10^{-4} = .00028$$
$$.00163 - .00028 = .00135 = 1.35 \times 10^{-3}$$

11. 1.56×10^4

$$(7.8 \times 10^{-5}) \cdot (2 \times 10^8) = (7.8 \cdot 2) \times (10^{-5} \times 10^8) = 15.6 \times 10^3$$

Change your answer into scientific notation.

$$15.6 \times 10^3 = (1.56 \times 10^1) \times 10^3 = 1.56 \times 10^4$$

12. 9.23×10^{95}

$$(1.846 \times 10^{100}) \div (2 \times 10^4) = (1.846 \div 2) \times (10^{100} \div 10^4)$$

Now divide and change back into scientific notation.

$$(1.846 \div 2) \times (10^{100} \div 10^4) = .923 \times 10^{96} = (9.23 \times 10^{-1}) \times 10^{96} = 9.23 \times 10^{95}$$

13.

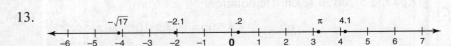

14. **C** 10.1

$\sqrt{101}$ is between $\sqrt{100} = 10$ and $\sqrt{121} = 11$. $\sqrt{101}$ is very close to $\sqrt{100} = 10$.

Chapter 3 | Algebra 1: Linear Equations and Inequalities

> **INDIANA'S ACADEMIC MATHEMATICS STANDARDS ADDRESSED:**
>
> A1.2.1 Solve linear equations.
> A1.2.2 Solve equations and formulas for a specified variable.
> A1.2.3 Find solution sets of linear inequalities when possible numbers are given for the variable.
> A1.2.4 Solve linear inequalities using properties of order.
> A1.2.5 Solve combined linear inequalities.
> A1.2.6 Solve word problems that involve linear equations, formulas, and inequalities.

SOLVING LINEAR EQUATIONS

Solving Single-Step Equations

"Solving an equation" means you want to find what number should be put in place of the variable so that the equation is true. For example, to solve the equation $x + 5 = 12$, you need to find what number plus 5 equals 12. It is easy to see that the solution is $x = 7$ because $7 + 5 = 12$.

The GQE has many problems that are not so easy to solve. We need a process that can be used to solve every kind of problem. We will start with equations that can be solved in one step and then progress to equations that require many steps. Remember: When solving one-step equations you may be able to "see" the answer, but you need to learn the processes that will give you the skills necessary to solve more difficult equations.

The most important rule when solving an equation is whatever operation you do to one side of the equal sign, you must do exactly the same thing to the other side of the equal sign.

> **Definition:** Addition Property of Equality: If $a = c$, then $a + b = c + b$
>
> **Example:** If $x = 6$, then $x + 4 = 6 + 4$

> **Definition:** Multiplication Property of Equality: If $a = c$, then $ba = bc$
>
> **Example:** If $x = 8$, then $\frac{1}{2} \cdot x = \frac{1}{2} \cdot 8$

Example

Solve: $x + 12 = 7$

Isolate x on the left side of the equal sign; isolate the variable. That means that you want to eliminate the 12 that is added to x. So, you add the opposite of 12 to both sides of the equation.

$$\begin{array}{r} x + 12 = 7 \\ \underline{-12 = -12} \\ x = -5 \end{array}$$

The solution is $x = -5$. Check: $-5 + 12 = 7$

Example

Solve: $x - 15 = -2$

First, change the subtraction problem into an addition problem. Now solve $x + -15 = -2$, which is similar to the original addition problem.

Get the x by itself on the left side of the equal sign; isolate the variable. That means you want to eliminate the -15 added to x. So, add the opposite of -15 to both sides of the equation.

$$\begin{array}{r} x + -15 = -2 \\ \underline{+15 = +15} \\ x = 13 \end{array}$$

The solution is $x = 13$. Check: $13 - 15 = 13 + -15 = -2$

Example

Solve: $\frac{3}{5}x = -12$

Isolate x on the left side of the equal sign; isolate the variable. That means that you want to eliminate the $\frac{3}{5}$ that is multiplied by x. So, multiply by the reciprocal of $\frac{3}{5}$ on both sides of the equation.

Remember: The reciprocal of $\frac{4}{3}$ is $\frac{3}{4}$ because $\frac{4}{3} \cdot \frac{3}{4} = 1$, and the reciprocal of 5 is $\frac{1}{5}$ because $5 \cdot \frac{1}{5} = 1$.

$$\frac{3}{5}x = -12$$

$$\frac{5}{3} \cdot \frac{3}{5}x = \frac{5}{3} \cdot -12$$

$$x = -20$$

The solution is $x = -20$. Check: $\frac{3}{5} \cdot -20 = \frac{3}{\cancel{5}} \cdot \frac{\overset{4}{\cancel{-20}}}{1} = -12$

Example

Solve: $\frac{x}{7} = -3$

$\frac{x}{7}$ means the same thing as $\frac{1}{7} \cdot x$ or $\frac{1}{7}x$. Solve $\frac{1}{7}x = -3$, which is equivalent to a multiplication type of problem. Isolate the x on the left side of the equal sign. That means you want to eliminate the $\frac{1}{7}$ that is multiplied by x. So, multiply by the reciprocal of $\frac{1}{7}$, which is $\frac{7}{1}$ or 7 on both sides of the equal sign.

$$\frac{x}{7} = -3$$

$$\frac{1}{7}x = -3$$

$$7 \cdot \frac{1}{7}x = 7 \cdot -3$$

$$x = -21$$

The solution is $x = -21$. Check: $\frac{-21}{7} = -3$

Exercises

1. Solve: $-5 + x = 20$

2. Solve: $-5x = 20$

3. Solve: $\frac{x}{-5} = 20$

4. Solve: $x - 5 = 20$

Solutions to Exercises

1. Isolate the variable by adding 5 to both sides of the equal sign.

$$-5 + x = 20$$
$$\underline{+5 \qquad = +5}$$
$$x = 25$$

2. Isolate the variable by multiplying both sides by $-\frac{1}{5}$, the reciprocal of -5.

$$-5x = 20$$

$$-\frac{1}{5} \cdot -5x = -\frac{1}{5} \cdot 20$$

$$x = -4$$

3. Isolate the variable by multiplying by -5 on both sides of the equal sign.

$$\frac{x}{-5} = 20$$

$$-\frac{1}{5}x = 20$$

$$-5 \cdot -\frac{1}{5}x = -5 \cdot 20$$

$$x = -100$$

4. Isolate the variable by adding 5 to both sides of the equal sign.

$$x - 5 = 20$$
$$x + -5 = 20$$
$$\underline{+5 = +5}$$
$$x = 25$$

Practice Problems

1. Solve: $28 = 2G$

2. Solve: $14 = 2 + R$

3. Solve: $-14 = E - 8$

4. Solve: $\dfrac{A}{-3} = 9$

(Solutions are found at the end of the chapter.)

Solving Multi-Step Equations

The GQE questions are not as simple as the one-step equations; the GQE problems involve many steps. However, each problem can be solved using the two processes we just learned.

Example
Solve: $4x + 3 = 31$
First, isolate the $4x$ on the left side of the equal sign. You need to eliminate the 3 added to $4x$. So add -3 to both sides of the equal sign.

$$4x + 3 = 31$$
$$\underline{-3 = -3}$$
$$4x = 28$$

Now the problem looks similar to our one-step problems. Isolate the variable by multiplying by $\dfrac{1}{4}$ on both sides of the equal sign.

$$4x = 28$$
$$\frac{1}{4} \cdot 4x = \frac{1}{4} \cdot 28$$
$$x = 7$$

So the solution is $x = 7$. Check: $4 \cdot 7 + 3 = 31$

> When solving a problem of the type $ax + b = c$, first isolate the term with the variable by adding the opposite of b to both sides and then isolate the variable by multiplying by the reciprocal of a.

Example
Solve: $5x + 9 = 3x - 5$
First, get all of the variables on the same side of the equal sign. You need to eliminate the $3x$ on the right side of the equal sign, so add the opposite of $3x$, $-3x$, to both sides.

$$5x + 9 = 3x - 5$$
$$\underline{-3x \qquad = -3x}$$
$$2x + 9 = 15$$

Now the problem is similar to the last example. Isolate the term with the variable by adding -9 to both sides of the equal sign.

$$2x + 9 = -5$$
$$\underline{-9 = -9}$$
$$2x = -14$$

Isolate the variable by multiplying by $\frac{1}{2}$ on both sides of the equal sign.

$$\frac{1}{2} \cdot 2x = \frac{1}{2} \cdot -14$$
$$x = -7$$

So the solution is $x = -7$. Check: $5 \cdot -7 + 9 = 3 \cdot -7 - 5$

When you finish solving an equation on the GQE, check your work by substituting your answer for the variable in the original equation. After you simplify both sides, you should have a true statement.

$$5x + 9 \stackrel{?}{=} 3x - 5$$
$$5 \cdot (-7) + 9 \stackrel{?}{=} 3 \cdot (-7) - 5$$
$$-35 + 9 \stackrel{?}{=} -21 - 5$$
$$-26 = -26$$

When solving a problem of the type $ax + b = cx + d$, first get all of the variables on the same side of the equal sign. Isolate the term with the variable by adding the opposite of b. Then isolate the variable by multiplying by the reciprocal.

Exercise

Solve: $5x + 6 = -2x - 15$

Solution to Exercise

Collect the variables on the same side of the equal sign by adding $2x$ to both sides of the equal sign.

$$5x + 6 = -2x - 15$$
$$\underline{+2x = +2x}$$

Isolate the term with the variable by adding -6 to both sides of the equal sign.

$$7x + 6 = -15$$
$$\underline{-6 = -6}$$

Isolate the variable by multiplying by $\frac{1}{7}$ on both sides of the equal sign.

$$7x = -21$$

$$\frac{1}{7} \cdot 7x = \frac{1}{7} \cdot -21$$

$$x = -3$$

Practice Problems

5. Solve: $8s - 16 = 24 - 2s$

6. Solve: $15 + p - 12 = 4p + 15$

7. Solve: $17 - e = 12 - 2e$

(Solutions are found at the end of the chapter.)

Distributive Property

Distributive Property

$$a(b + c) = ab + ac$$

Example: $4(3 + x) = 4 \cdot 3 + 4 \cdot x = 12 + 4x$

Example: $-(3x + 4) = -1(3x + 4) = -1 \cdot 3x + -1 \cdot 4 = -3x - 4$

The Distributive Property can be used in an equation that is to be solved. The problem does not become more difficult; it just requires one more step of work. As you begin to solve the problem, use the distributive property to simplify.

Example

Solve: $5(x + 3) = 2x - 9$

Use the distributive property to simplify the left side of the equal sign.

$$5x + 15 = 2x - 9$$
$$\underline{-2x} \qquad = \underline{-2x}$$

Get the variables on the same side of the equal sign by adding $-2x$ to both sides of the equal sign.

$$3x + 15 = -9$$
$$\underline{-15} = \underline{-15}$$

Isolate the term with the variable by adding -15 to both sides of the equal sign.

$$3x = -24$$

Isolate the variable by multiplying by $\frac{1}{3}$ on both sides of the equal sign.

$$\frac{1}{3} \cdot 3x = \frac{1}{3} \cdot -24$$

$$x = -8$$

Check:

$$5(-8 + 3) \stackrel{?}{=} 2 \cdot -8 - 9$$
$$5(-5) \stackrel{?}{=} -16 - 9$$
$$-25 = -25$$

Exercise

Solve: $4x - 13 = -2(x + 5)$

Solution to Exercise

Simplify, using the Distributive Property.

$$4x - 13 = -2(x + 5)$$
$$4x - 13 = -2x - 10$$
$$\underline{+2x \quad\quad = +2x}$$

Get the variables on the same side of the equal sign by adding $2x$ to both sides of the equal sign.

$$6x - 13 = -10$$
$$\underline{+13 = +13}$$

Isolate the term with the variable by adding 13 to both sides of the equal sign.

$$6x = 3$$

Isolate the variable by multiplying by $\frac{1}{6}$ on both sides of the equal sign.

$$\frac{1}{6} \cdot 6x = \frac{1}{6} \cdot 3$$

$$x = \frac{3}{6} = \frac{1}{2}$$

Practice Problems

8. Solve: $18 = 2 + 2(3I + 5)$

9. Solve: $11 - 3E = -4(6 - E)$

(Solutions are found at the end of the chapter.)

SOLVING EQUATIONS AND FORMULAS FOR A VARIABLE

Sometimes you may be asked to solve an equation that does not have numbers, only variables. To solve this type of problem, use the same process that you would use to solve an equation with numbers. Below are two problems that are solved the same way—one has a variable and numbers and one only has variables.

Solve for x.

$$2x + 3 = 19$$
$$\underline{-3 = -3}$$
$$2x = 16$$

Add the opposite to both sides

Multiply by the reciprocal on both sides

$$\frac{1}{2} \cdot 2x = \frac{1}{2} \cdot 16$$

$$x = 8$$

Solve for x.

$$ax + b = c$$
$$\underline{-b = -b}$$
$$ax = c - b$$

$$\frac{1}{a} \cdot ax = \frac{1}{a}(c - b)$$

$$x = \frac{1}{a}(c - b)$$

Example

Solve for c: $c + a = r$

When a problem asks you to solve for c, you want to get c by itself. Add the opposite of a to both sides of the equal sign.

$$c + a = r$$
$$\underline{-a = -a}$$
$$c = r - a$$

Example

Solve for p: $p - e = n$

$$p - e = n$$

$$p + (-e) = n$$
$$\underline{+e = +e}$$

The problem wants you to get p by itself. You must eliminate $-e$. Add the opposite of $-e$ to both sides of the equal sign.

$$p = n + e$$

Example

Solve for v: $va = n$

You are asked to isolate the variable v. In this problem you need to eliminate the a. Multiply both sides of the equal sign by the reciprocal of a.

$$va = n$$

$$\frac{1}{a} \cdot va = \frac{1}{a} \cdot n$$

$$v = \frac{1}{a} \cdot n$$

$$v = \frac{n}{a}$$

Be Careful! $\frac{1}{a} \cdot n = \frac{1}{a} \cdot \frac{n}{1} = \frac{n}{a}$

Example $\frac{1}{4} \cdot 7 = \frac{1}{4} \cdot \frac{7}{1} = \frac{7}{4}$

Example

Solve for r: $\frac{r}{u} = n$

To solve for r, you need to get the r alone. Remember $\frac{r}{u}$ means $\frac{1}{u} \cdot r$.

$$\frac{r}{u} = n$$

$$\frac{1}{u} \cdot r = n$$

Eliminate $\frac{1}{u}$ by multiplying both sides by the reciprocal of $\frac{1}{u}$.

$$u \cdot \frac{1}{u} \cdot r = u \cdot n$$

$$r = un$$

Exercises

1. Solve for i: $h(i + o) = y$ 2. Solve for e: $15 = ce - nt$

Solutions to Exercises

1. Use the Distributive Property on the left side of the equal sign.

$$h(i + o) = y$$
$$hi + ho = y$$
$$\underline{-ho = -ho}$$

$$hi = y - ho$$

$$\frac{1}{h} \cdot hi = \frac{1}{h} \cdot (y - ho)$$

$$i = \frac{1}{h} \cdot (y - ho)$$

$$i = \frac{y - ho}{h}$$

Be careful! A typical mistake is to try to cancel the h in the numerator and denominator. $\frac{y - ho}{h}$ is <u>not</u> the same as $\frac{y - o}{1}$. When you are not sure if two answers are the same, try a simpler problem without variables. Is $\frac{10 - 2 \cdot 3}{2}$ the same as $\frac{10 - \overset{1}{\cancel{2}} \cdot 3}{\underset{1}{\cancel{2}}} = \frac{10 - 3}{1}$? No, $\frac{10 - 2 \cdot 3}{2} = \frac{10 - 6}{2} = \frac{4}{2} = 2$ and $\frac{10 - \overset{1}{\cancel{2}} \cdot 3}{\underset{1}{\cancel{2}}} = \frac{10 - 3}{1} = \frac{7}{1} = 7$.

2. Switch the left and the right sides so that e is on the left side.

$$ce - nt = 15$$
$$\underline{+nt = +nt}$$

$$ce = 15 + nt$$

$$\frac{1}{c} \cdot ce = \frac{1}{c} \cdot (15 + nt)$$

$$e = \frac{1}{c} \cdot (15 + nt)$$

$$e = \frac{15 + nt}{c}$$

Be Careful! A typical mistake is to try to make $15 + nt$ into $15nt$. $15 + nt$ is not the same as $15nt$. Remember that $15nt$ means 15 times nt, not 15 plus nt. $15 + nt$ cannot be simplified.

Practice Problems

10. Bob Gibson's earned run average (ERA) was 1.13. ERA is calculated using the formula $ERA = \dfrac{9 \cdot e}{I}$ where e is the number of earned runs and I is the number of innings pitched. Solve the formula $ERA = \dfrac{9 \cdot e}{I}$ for e.

(Solution is found at the end of the chapter.)

SOLUTION SETS OF INEQUALITIES

An equation has to have an equal sign. An inequality is the same as an equation except it can have a less than sign ($<$), a greater than sign ($>$), a less than or equal to sign ($\leq$), or a greater than or equal to sign ($\geq$).

The solution set of an inequality is the set of numbers that make the inequality true.

Definition: A set is a collection of objects or numbers.

Sometimes you are given a group of numbers to try to find out which ones will work in the inequality; that means you want to find out which numbers make the inequality true.

You are given the set of numbers $\{1, 2, 3, 4, 5\}$ and the inequality $x + 3 \leq 6$. The solution set of $x + 3 \leq 6$ means to choose only the numbers from the set $\{1, 2, 3, 4, 5\}$ that will make the inequality $x + 3 \leq 6$ true.

To find these numbers, substitute each number for x in the inequality $x + 3 \leq 6$.

Test for 1
$1 + 3 \leq 6$
$4 \leq 6$
is true so 1 is in the solution set.

Test for 4
$4 + 3 \leq 6$
$7 \leq 6$
is false so 4 is not in the solution set.

Test for 2
$2 + 3 \leq 6$
$5 \leq 6$
is true so 2 is in the solution set.

Test for 5
$5 + 3 \leq 6$
$8 \leq 6$
is false so 5 is not in the solution set.

Test for 3
$3 + 3 \leq 6$
$6 \leq 6$
is true so 3 is in the solution set.

The only numbers from the set $\{1, 2, 3, 4, 5\}$ that make the inequality $x + 3 \leq 6$ true are $\{1, 2, 3\}$. There are many other numbers that work in the inequality $x + 3 \leq 6$, but the problem only wanted to know which numbers in the given set of numbers would make the inequality true.

Practice Problems

11. Using the inequality $4x - 3 \geq 5x - 1$, which numbers in the set $\{-1, -2, -3\}$ are in the solution set?

12. Nev R. Right says that an example of a number that is a solution to the inequality $-8 < x$ is -10. Is he right or wrong? List three numbers in the solution set of the linear inequality $-8 < x$.

(Solutions are found at the end of the chapter.)

SOLVING LINEAR INEQUALITIES

There is a difference between solving an inequality and solving an equation. Let's try to understand why there is difficulty solving certain kinds of inequalities.

What happens if we add -10 to both sides of the inequality sign?

$$15 > 2$$
$$15 + -10 > 2 + -10$$
$$5 > -8$$

You can add the same negative number to both sides of the inequality and the inequality is still true.

What happens when you multiply by -7 on both sides of the inequality sign?

$$8 > -5$$
$$-7 \cdot 8 > -7 \cdot -5$$
$$-56 > 35$$

This inequality is not true.

To make the inequality true when multiplying by a negative number, reverse the inequality sign.

$$8 > -5$$
$$-7 \cdot 8 < -7 \cdot -5$$
$$-56 < 35$$

> When you multiply or divide an inequality by a negative number, then you must reverse the inequality sign.

The major difference between solving an inequality and solving an equation is when you multiply or divide by a negative number, you must reverse the inequality sign.

Solve $-7x + 12 \leq 47$.

Go through the same process used in solving the equation $-7x + 12 = 47$. First, isolate the term with the variable. Add -12 to both sides of the inequality sign. The inequality sign does not switch.

$$-7x + 12 \leq 47$$
$$\underline{-12 = -12}$$
$$-7x \leq 35$$

Now, isolate the variable by multiplying both sides by $-\frac{1}{7}$. You multiplied by a negative number on both sides of the inequality sign so the inequality sign must reverse.

$$-7x \leq 35$$

$$-\frac{1}{7} \cdot -7x \geq -\frac{1}{7} \cdot 35$$

$$x \geq -5$$

So the solution is $x \geq -5$, which means that any number that is greater than or equal to -5 will make the inequality $-7x + 12 \leq 47$ true.

Exercises

1. Solve the following inequality: $-\frac{3}{4}x - 6 < -15$

2. Solve: $-14 > x - 3$

3. Solve: $-8(x - 2) \geq 40$

Solutions to Exercises

1. Isolate the term with the variable by adding 6 to both sides of the inequality.

$$-\frac{3}{4}x - 6 < -15$$

$$\underline{+6 = +6}$$

Isolate the variable by multiplying both sides of the inequality by $-\frac{4}{3}$. Since we multiplied by a negative number, the inequality sign must be reversed.

$$-\frac{3}{4}x < -9$$

$$-\frac{4}{3} \cdot -\frac{3}{4}x > -\frac{4}{3} \cdot -9$$

$$x > 12$$

The solution is $x > 12$. All numbers that are greater than 12 will satisfy the inequality $-\frac{3}{4}x - 6 < -15$.

2. This is a very tricky problem because the variable is on the right side of the inequality sign instead of the left side. The easiest way to fix this difficulty is move the $x - 3$ to the left side and the -14 to the right side. However, there is a problem when you do this.

Example

$12 > -5$ means twelve is greater than negative five. But if you put the -5 on the left and the 12 on the right, then the problem must be written $-5 < 12$, negative five is less than twelve.

If a is less than b, then that means that b is greater than a. In symbols, if $a < b$, then $b > a$ or if $c > d$, then $d < c$.

Think: If -14 is bigger than $x - 3$, then $x - 3$ must be smaller than -14.

Now back to solving the problem $-14 > x - 3$.

Move the $x - 3$ to the left side of the inequality sign and the -14 to the right side. Remember to reverse the inequality sign.

$$-14 > x - 3$$
$$x - 3 < -14$$
$$\underline{+3 = +3}$$

Isolate the variable, so add 3 to both sides of the equal sign.

$$x < -11$$

The solution is $x < -11$. All numbers that are smaller than -11 will work in the inequality $-14 > x - 3$.

3. The Distributive Property must be used first when solving this inequality.

$$-8(x - 2) \geq 40$$
$$-8x + 16 \geq 40$$

> **Be Careful!** A typical mistake is $-8(x - 2) = -8x - 16$ or $-8(x - 2) = -8x - 2$. Remember the Distributive Property says
>
> $$-8(x - 2) = -8(x + -2) = -8 \cdot x + -8 \cdot -2 = -8x + 16.$$

Isolate the term with the variable by adding -16 to both sides of the inequality.

$$-8x + 16 \geq 40$$
$$\underline{-16 = -16}$$
$$-8x \geq 24$$

Isolate the variable by multiplying by $-\dfrac{1}{8}$ on both sides of the inequality sign. Remember when multiplying both sides by a negative number, reverse the inequality sign.

$$-\frac{1}{8} \cdot -8x \leq -\frac{1}{8} \cdot 24$$
$$x \leq -3$$

The solution is $x \leq -3$, which means all numbers less than or equal to -3 will make the equation $-8(x - 2) \geq 40$ true.

Practice Problems

13. Solve: $-14 - 3x < 2x + 26$

14. Solve: $\dfrac{p}{-3} + 5 \leq 1$

15. Solve: $5(r + 3) \geq 7r - 19$

16. Solve: $18 < -\dfrac{3}{4}t + 6$

(Solutions are found at the end of the chapter.)

SOLVING COMBINED INEQUALITIES

One kind of combined inequality, $2 < x < 5$, is read x is greater than 2 <u>and</u> x is less than 5. It is a combination of the two inequalities $x > 2$ and $x < 5$. The numbers that satisfy $2 < x < 5$ are all of the numbers that are between 2 and 5, not including 2 and not including 5.

Change the combined inequality $-5 > x \geq -7$ into two separate inequalities. Use the middle of the inequality twice, as shown in the figures below.

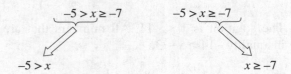

You know that $-5 > x$ is the same as $x < -5$. Now you have broken apart $-5 > x \geq -7$ into $x < -5$ and $x \geq -7$.

The combined inequality $-4 \leq x \leq 2$ breaks into $x \geq -4$ and $x \leq 2$.

Example

Solve: $3 \leq 2x + 5 < 17$

One way to solve this kind of combined inequality is to break it apart, solve each separate inequality, and then put the two parts back together. Separate $3 \leq 2x + 5 < 17$ into $2x + 5 \geq 3$ and $2x + 5 < 17$. Now solve each part.

$$
\begin{array}{ll}
2x + 5 \geq 3 & \qquad\qquad 2x + 5 < 17 \\
\underline{-5 = -5} & \qquad\qquad \underline{-5 = -5} \\
2x \geq -2 & \qquad\qquad 2x < 12 \\
\\
\dfrac{1}{2} \cdot 2x \geq \dfrac{1}{2} \cdot -2 & \qquad\qquad \dfrac{1}{2} \cdot 2x < \dfrac{1}{2} \cdot 12 \\
\\
x \geq -1 & \qquad\qquad x < 6
\end{array}
$$

Combine $x \geq -1$ and $x < 6$ into a solved combined inequality $-1 \leq x < 6$.

Another way to solve the inequality $3 \leq 2x + 5 < 17$ is to keep the three parts together and solve it all at once. Look at the middle part of the inequality and try to isolate the term with the variable.

$$
\begin{array}{ccc}
3 \leq & 2x + 5 & < 17 \\
\underline{-5 =} & \underline{-5} & = \underline{-5} \\
& -2 \leq 2x < 12 &
\end{array}
$$

Now isolate the variable by multiplying all three parts by $\dfrac{1}{2}$.

$$
\frac{1}{2} \cdot -2 \leq \frac{1}{2} \cdot 2x < \frac{1}{2} \cdot 12
$$

$$
-1 \leq x < 6
$$

Either way of solving a combined inequality is acceptable. Choose the method that makes sense to you.

Another kind of combined inequality is $x > 2$ or $x < 0$. Numbers that satisfy $x > 2$ or $x < 0$ are all numbers that are either greater than 2 or less than 0. This type of combined inequality cannot be joined together into one statement; the parts must be kept separate.

Example

Solve: $19 - 4x > 31$ or $5x + 17 > 52$

To solve this type of combined inequality, solve each part separately.

$$
\begin{aligned}
19 - 4x &> 31 \\
\underline{-19} &= \underline{-19} \\
-4x &> 12
\end{aligned}
\qquad\qquad
\begin{aligned}
5x + 17 &> 52 \\
\underline{-17} &= \underline{-17} \\
5x &> 35
\end{aligned}
$$

$$
\begin{aligned}
-\frac{1}{4} \cdot -4x &< -\frac{1}{4} \cdot -12 \\
x &< -3
\end{aligned}
\qquad\qquad
\begin{aligned}
\frac{1}{5} \cdot 5x &> \frac{1}{5} \cdot 35 \\
x &> 7
\end{aligned}
$$

The answer is $x < -3$ or $x > 7$.

Exercises

1. Solve: $13 > 5x - 7 > -22$

2. Solve: $-6x - 7 \geq 29$ or $9 - 8x < -23$

Solutions to Exercises

1. Solving by separating

$13 > 5x - 7 > -22$ breaks
into $5x - 7 < 13$ and $5x - 7 > -22$

Solve each part separately.

$$
\begin{aligned}
5x - 7 &< 13 \\
\underline{+7} &= \underline{+7} \\
5x &< 20
\end{aligned}
\qquad\qquad
\begin{aligned}
5x - 7 &> -22 \\
\underline{+7} &= \underline{+7} \\
5x &> -15
\end{aligned}
$$

$$
\begin{aligned}
\frac{1}{5} \cdot 5x &< \frac{1}{5} \cdot 20 \\
x &< 4
\end{aligned}
\qquad\qquad
\begin{aligned}
\frac{1}{5} \cdot 5x &> \frac{1}{5} \cdot -15 \\
x &> -3
\end{aligned}
$$

Solving together

$$
\begin{aligned}
13 > 5x - 7 &> -22 \\
\underline{+7} = \underline{+7} &= \underline{+7} \\
20 > 5x &> -15
\end{aligned}
$$

$$
\frac{1}{5} \cdot 20 > \frac{1}{5} \cdot 5x > \frac{1}{5} \cdot -15
$$

$$
4 > x > -3
$$

Then combine $x < 4$ and $x > -3$ into $-3 < x < 4$ or $4 > x > -3$

2. To solve $-6x - 7 \geq 29$ or $9 - 8x < -23$, you solve each part separately.

$$
\begin{aligned}
-6x - 7 &\geq 29 \\
\underline{+7} &= \underline{+7} \\
-6x &\geq 36
\end{aligned}
\qquad\qquad
\begin{aligned}
9 - 8x &< -23 \\
\underline{-9} &= \underline{-9} \\
-8x &< -32
\end{aligned}
$$

$$
\begin{aligned}
-\frac{1}{6} \cdot -6x &\leq -\frac{1}{6} \cdot 36 \\
x &\leq -6
\end{aligned}
\qquad\qquad
\begin{aligned}
-\frac{1}{8} \cdot -8x &> -\frac{1}{8} \cdot -32 \\
x &> 4
\end{aligned}
$$

So your answer is $x \leq -6$ or $x > 4$. These cannot be joined into one combined inequality.

Practice Problems

17. Solve: $26 \leq 5 - 3x < 50$

18. Solve: $5 > 2x - 3$ or $-26 > 4 - 5x$

(Solutions are found at the end of the chapter.)

STORY PROBLEMS

Number Problems

The GQE has word problems that require you to both set up an equation and then solve the equation. Points will be given for writing the equation and points will be given for solving the problem.

The hardest part of solving word problems is writing the equation. We will start with word problems where you are looking for a number. With the easy problems, you may be able to "see" the answer, but we will go through a process that will give you the skills necessary to solve more difficult word problems.

Some key words give you a hint as to what operation should be used in the equation. The key words are shown in the table below.

Key Words and Their Operations			
Addition	*Subtraction*	*Multiplication*	*Division*
Sum of	Difference of	Product of	Quotient of
Increased by	Decreased by	Multiplied by	Divided by
More than	** Less than	Twice	
Plus	Minus	Times	
Added to	** Subtracted from		
	** Taken away from		
** means you must reverse the order of the terms (For example, 7 subtracted from 10 means 10 − 7, not 7 − 10.)			

Example

Write an equation that could be used to solve the problem.

The sum of a number and 6 is −27. What is the number?

Choose a variable to represent the number that you want to find. Let n stand for the number we want to find.

The *sum of* means you are going to add together two numbers.

A number we will represent with a variable, n.

and 6 means plus 6.

is −27 means equals −27.

So you put all the parts together to get

The sum of a number and 6 is −27 translates into $n + 6 = -27$.

Now solve the problem using the process learned in solving linear equations.

$$n + 6 = -27$$
$$\underline{-6 \quad\quad -6}$$
$$n = -33$$

Your answer is −33. That means that the sum of −33 and 6 is −27.

Exercises

1. Seven more than twice a number is 8 less than the number. Write an equation that could be used to find the number.

2. Write an equation that could be used to find a number such that the quotient of the number and –2 is 8.

Solutions to Exercises

1. This problem is a bit more complicated but can still be broken down into parts.
   Seven more than means 7 plus.
   twice means 2 times.
   a number we will represent with a variable, *n*.
   is means equals.
   8 less than the number means we are going to subtract 8 from the number, *n*.
   Seven more than twice a number is 8 less than the number translates to $7 + 2n = n - 8$.

2. Although this problem is stated in a different way, you still break it into parts.
   The quotient of means that you are going to divide two numbers.
   A number and –2 are the two numbers that you are going to divide.
   Is 8 means equals 8.
   The quotient of the number and –2 is 8 translates to $\frac{n}{-2}$ or $n \div -2 = 8$.

Practice Problems

19. Write the equation that could be used to solve the problem.
    A number, *F*, subtracted from –3 is 18. What is the number?
    Do not solve your equation.

20. Write the equation that could be used to find a number, *N*, such that the number increased by twice that number would give 27. Do not solve your equation.

(Solutions are found at the end of the chapter.)

Other Story Problems

Not all problems on the GQE are the type where you are looking for a number. Other word problems require you to determine what you are trying to find and to decide on the operations involved.

Example

Mr. I. M. Rich's checking account charges $3 each month and $.15 for every check. Mr. Rich was charged $7.50 for the month of July. Write an equation that could be used to determine the number of checks Mr. Rich wrote in July. Do not solve your equation.

First, read the problem several times so that you understand what the problem is talking about. In this case, Mr. I. M. Rich has a checking account that has two charges each month—one is a monthly fee and one is a charge for each check he writes.

Second, determine what you are trying to find. The easiest way is to find the part of the problem that asks the question. In this problem, the question is hidden in the phrase that says "determine the number of checks." Once you know what you are trying to find, choose a variable to represent what you want to find.

Choose a variable that helps you remember what you are trying to find. In this problem, let *C* represent the number of checks Mr. Rich wrote. If the problem on the GQE gives you a letter to use, be sure to use the letter you are given.

$$C = \text{number of checks}$$

Finally, translate your word problem into an equation by thinking what parts go together, as shown in the diagram below.

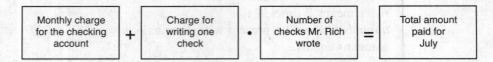

| Monthly charge for the checking account | + | Charge for writing one check | • | Number of checks Mr. Rich wrote | = | Total amount paid for July |

The diagram translates into the equation $3 + .15 \cdot C = 7.5$.

Exercise

Suzy Q. Tips is paid $7 per hour for washing cars at Scrub-a-Heap Carwash. Suzy Q. earned $4 in tips. Her total earnings one Saturday came to $60. Write an equation that could be used to find how many hours, *H*, Suzy worked that day. Do not solve your equation.

Solution to Exercise

We want to find the number of hours that Suzy worked. The problem tells us to let *H* represent the number of hours Suzy worked that day. Suzy earned $7 each hour that she worked. She also earned $4 in tips. Totally, Suzy earned $60. The figure below shows this problem in a diagram.

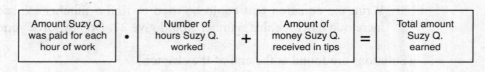

| Amount Suzy Q. was paid for each hour of work | • | Number of hours Suzy Q. worked | + | Amount of money Suzy Q. received in tips | = | Total amount Suzy Q. earned |

$$7 \cdot H + 4 = 60$$

Practice Problem

21. Kreigh Z. Driver is delivering pizzas for MaMa Joan's Pizzas. Kreigh earns $8.26 per hour plus tips. On Saturday, he worked $6\frac{1}{2}$ hours and earned a total of $73.19 in tips and wages. Write an equation that could be used to find how much Kreigh earned in tips, *T*. Do not solve your equation.

(Solution is found at the end of the chapter.)

STORY PROBLEMS AND FORMULAS

Remember that your mathematics reference sheet gives you a variety of formulas that can be used in word problems. Every time you should use the reference sheet you will see the symbol shown on the left.

Once you have found the correct formula, then you must put the numbers in for the appropriate variables and solve for the missing letter.

Example
A tank in the shape of a cylinder holds 3,165.12 cubic meters of liquid, as shown in the figure below.

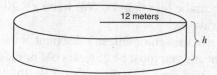

12 meters

h

If the radius of the tank is 12 meters, what is the height of the tank?

First, find the correct formula on the mathematics reference sheet. The problem tells you that the shape of the tank is a cylinder, so decide whether you need the volume formula or the surface area formula. The amount of liquid that a shape holds is the volume, so use the formula $V = \pi r^2 h$.

The radius is 12 meters, the volume is 3,165.12 cubic meters, and $\pi \approx 3.14$. The problem asks for the height of the tank. Put the numbers in the appropriate places in the formula $V = \pi r^2 h$.

$$3{,}165.12 = 3.14 \cdot 12^2 \cdot h$$

To solve the problem, move the $3.14 \cdot 12^2 \cdot h$ to the left side and the 3,165.12 to the right side

$$3.14 \cdot 12^2 \cdot h = 3{,}165.12$$

Simplify the left side by squaring 12.

$$3.14 \cdot 144 \cdot h = 3{,}165.12$$

Simplify the left side by multiplying $3.14 \cdot 144$.

$$452.16h = 3{,}165.12$$

Isolate the variable, h, by multiplying both sides by $\dfrac{1}{452.16}$.

$$\frac{1}{452.16} \cdot 452.16h = \frac{1}{452.16} \cdot 3{,}165.12$$
$$h = 7$$

So, the height of the tank is 7 meters.

Practice Problem

22. The diameter of a major league baseball is 7.4 centimeters. How much leather is needed to cover the baseball? Use 3.14 for π.

(Solution is found at the end of the chapter.)

STORY PROBLEMS AND INEQUALITIES

In the same way as equations can be used to solve word problems, inequalities can be written to solve word problems. On the GQE, you may be asked to write an inequality that can be used to solve a problem or you may be asked to write the inequality and then solve it.

Some students have difficulty with some of the words associated with inequalities. Carefully read the explanations so that you understand the meaning of "at most" and "at least."

To receive the children's discount at the movie theater, you must be <u>at most</u> 12 years old. That means you must be 12 years old or younger than 12 years old. If you have at most $20, then you have $20 or you have less than $20. In algebra, if x is at most 20, then x can be 20 or it can be less than 20. We write $x \leq 20$. The words "no more than" would be used the same way.

To receive the senior citizen's discount at the movie theater, you must be at least 65 years old. That means you must be 65 years old or older than 65. If you have at least $30, then you have $30 or you may have more than $30. In algebra, if c is at least 17, then c is 17 or c is more than 17. We write $c \geq 17$. The words "no less than" would be used the same way.

Example

Supa Shopper went to the mall to shop for clothes, shoes, and accessories for the school dance. She decided to spend at most $250 for everything. She bought a dress for $129 and shoes for $89. Write an inequality to represent how much Supa has left for accessories (A).

Remember to read the problem several times. Then rephrase the problem in your own words. Supa is buying a dress, shoes, and accessories. She can only spend $250. You want to find the amount of money Supa will have left to spend on accessories. Look at the summary in the figure below.

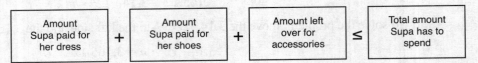

The diagram can be summarized by the inequality $129 + 89 + A \leq 250$.

Now solve your inequality to find how much Supa can spend for accessories.

$$129 + 89 + A \leq 250$$
$$A + 218 \leq 250$$
$$\underline{-218 \quad = -218}$$
$$A \leq 32$$

Supa could spend $32 or less for accessories.

Practice Problems

23. You want to buy some candy bars and gummy bears from Calories Abound Store. Candy bars costs $.52 each and gummy bears cost $.06 each. You buy three candy bars. Write an inequality that could be used to find the number of gummy bears (G) you could also purchase to spend no more than $2.90 total. Solve your inequality to find the number of gummy bears.

24. A school wants to buy sound equipment from Blastm Out Company. The cost of the equipment begins at $15,400 and increases according to the number of extras ordered. At this time, the school has raised $7,300 through bake sales. Each future bake sale will earn about $1,250. Write an inequality to determine the number of bake sales (B) necessary to earn the minimum amount for the sound system. Do not solve your inequality.

(Solutions are found at the end of the chapter.)

SAMPLE GQE QUESTIONS FOR CHAPTER 3

 1. Solve: $4(2x - 3) = 9x - 18$

Show All Work

 2. Solve: $\frac{x}{7} + 8 = 2$

Show All Work

 3. What is the value of x in the equation $-8(3x + 5) = -88$?

Show All Work

4. Wood N. Teeth earns $7.25 per hour at Dentures R Us. He also earns a bonus of $45 for Employee of the Mouth in July. His total earnings for July were $1,205 before deductions. Which of the following is an equation that could be used to find the number of hours (H) that Wood N. Teeth worked in July?

A. $7.25 + 45 + H = 1,205$
B. $7.25H + 45H = 1,205$
C. $7.25 + 45H = 1,205$
D. $7.25H + 45 = 1,205$

5. The formula for the surface area (SA) of a cylinder is $SA = 2\pi r^2 + 2\pi rh$ where r is the radius and h is the height. Solve the formula $SA = 2\pi r^2 + 2\pi rh$ for h.

6. Solve for c: $ac - w = 3a$

A. $c = \dfrac{3a + w}{a}$

B. $c = \dfrac{3 + w}{1}$

C. $c = \dfrac{3aw}{a}$

D. $c = \dfrac{3w}{1}$

  7. Jay is going to build a rectangular sandbox for his daughter, so he bought 9 cubic feet of sand. The sandbox must be 3 feet wide. He wants to place 9 inches or $\frac{3}{4}$ foot of sand in the sandbox. What should be the length of the sandbox?

A. 1 foot
B. 3 feet
C. 4 feet
D. 9 feet

8. In the inequality $-2x + 3 > 15$, which numbers from the set $\{-8, -6, -4\}$ are in the solution set?

 9. Solve for x: $-2x + 16 \geq 28$

A. $x \leq -14$
B. $x \leq -6$
C. $x \geq -6$
D. $x \geq -14$

10. Which inequality is the same as $12 > 10 - x$?

A. $x < 22$
B. $x > 22$
C. $x > -2$
D. $x < -2$

11. Sue's scores on the first three of four 100-point tests were 90, 96, and 86. Write an inequality that could be used to find the number of points (P) she must receive on the last test to have an average of at least 92 for all the tests.

12. Look at the statement below.

28 subtracted from a number is less than 18.

Which of the following is an inequality that represents this situation?

A. $28 - x < 18$
B. $28 - x > 18$
C. $x - 28 < 18$
D. $x - 28 > 18$

13. Solve: $5x - 13 > 9x + 27$ or $5(2x + 6) \geq 80$

Show All Work

14. Solve: $-12 < 7 - x < 15$

    A. $-5 < x < 22$
    B. $5 > x > -22$
    C. $19 > x > -8$
    D. $-19 < x < 8$

Answers Explained

1. $x = 6$

$$4(2x - 3) = 9x - 18$$
$$8x - 12 = 9x - 18$$
$$\underline{-9x \qquad = -9x}$$

$$-1x - 12 = -18$$
$$\underline{+12 = +12}$$

$$-1x = -6$$

$$-1 \cdot -1x = -1 \cdot -6$$
$$x = 6$$

2. $x = -42$

$$\frac{x}{7} + 8 = 2$$
$$\underline{-8 = -8}$$
$$\frac{x}{7} = -6$$

> **Be Careful!** $\frac{x}{7}$ is the same as $\frac{1}{7} \cdot x$ or $\frac{1}{7}x$.

$$\frac{1}{7}x = -6$$

$$7 \cdot \frac{1}{7}x = 7 \cdot -6$$

$$x = -42$$

3. $x = 2$

$$-8(3x + 5) = -88$$

Use Distributive Property to simplify the left side of the equal sign.

$$-24x - 40 = -88$$
$$\underline{+40 = +40}$$

Isolate the term with the variable by adding 40 to both sides of the equal sign.

$$-24x = -48$$

Isolate the variable by multiplying both sides by $-\frac{1}{24}$.

$$-\frac{1}{24} \cdot -24x = -\frac{1}{24} \cdot -48$$

$$x = 2$$

4. D.

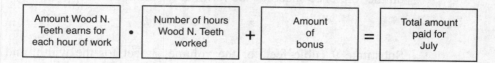

The diagram above translates into the equation $7.25 \cdot H + 45 = 1{,}205$.

5. $h = \frac{1}{2\pi r} \cdot (SA - 2\pi r^2)$ or $h = \frac{SA - 2\pi r^2}{2\pi r}$

The GQE is not asking you to use the formula to calculate the surface area of a cylinder. You need to solve the formula for h.

$$SA = 2\pi r^2 + 2\pi rh$$

Move the $2\pi r^2 + 2\pi rh$ to the left side and SA to the right side.

$$2\pi r^2 + 2\pi rh = SA$$
$$\underline{-2\pi r^2 = -2\pi r^2}$$

$$2\pi rh = SA - 2\pi r^2$$

$$\frac{1}{2\pi r} \cdot 2\pi rh = \frac{1}{2\pi r} \cdot (SA - 2\pi r^2)$$

$$h = \frac{1}{2\pi r} \cdot (SA - 2\pi r^2) = \frac{SA - 2\pi r^2}{2\pi r}$$

Be Careful! A typical mistake is to try to cancel the $2\pi r$ in the denominator with the $2\pi r^2$ in the numerator. You cannot cancel in this manner.

6. A.

$$ac - w = 3a$$
$$\underline{+w = +w}$$

$$ac = 3a + w$$

$$\frac{1}{a} \cdot ac = \frac{1}{a} \cdot (3a + w)$$

$$c = \frac{3a + w}{a}$$

Remember $\frac{1}{a} \cdot (3a + w) = \frac{1}{a} \cdot \frac{(3a + w)}{1} = \frac{3a + w}{a}$. You cannot cancel the a's.

7. **C** The sandbox is in the shape of a rectangular prism, as shown below.

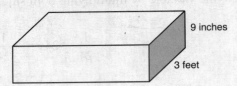

When you fill a rectangular prism, you are finding the volume of the prism. So look on the reference sheet for the formula for the volume of a rectangular prism.

$$V = l\,w\,h$$

Substitute 9 cubic feet for the volume, 3 feet for the width, and $\frac{3}{4}$ foot for the height to get $9 = l \cdot 3 \cdot \frac{3}{4}$. Solve for l.

$$9 = l \cdot 3 \cdot \frac{3}{4}$$

$$3 \cdot \frac{3}{4} \cdot l = 9$$

$$\frac{9}{4} l = 9$$

$$\frac{4}{9} \cdot \frac{9}{4} l = \frac{4}{9} \cdot 9$$

$$l = 4$$

Jay should make the sandbox 4 feet long.

8. −8

The only numbers that the problem wants you to check are −8, −6, and −4. Substitute each one into the inequality $-2x + 3 > 15$ to find out which ones make the inequality true.

Test for −8	Test for −6	Test for −4
$-2 \cdot -8 + 3 > 15$	$-2 \cdot -6 + 3 > 15$	$-2 \cdot -4 + 3 > 15$
$16 + 3 > 15$	$12 + 3 > 15$	$8 + 3 > 15$
$19 > 15$ is true	$15 > 15$ is false	$11 > 15$ is false

The number from the set $\{-8, -6, -4\}$ that makes the inequality $-2x + 3 > 15$ true is −8.

9. **B**

$$-2x + 16 \geq 28$$
$$\underline{-16 = -16}$$

$$-2x \geq 12$$

Be Careful! When multiplying both sides of an inequality by a negative number, reverse the inequality sign.

$$-\frac{1}{2} \cdot -2x \leq -\frac{1}{2} \cdot 12$$
$$x \leq -6$$

10. **C** To begin the problem, it is best to move the $10 - x$ to the left side and the 12 to the right side. Remember if 12 is more than $10 - x$, then $10 - x$ must be smaller than 12.

$$12 > 10 - x$$
$$10 - x < 12$$
$$\underline{-10 \quad = -10}$$
$$-x < 2$$

To isolate the x, you must multiply both sides by -1.

$$-1 \cdot -x > -1 \cdot 2$$
$$x > -2$$

11. $\dfrac{90 + 96 + 86 + P}{4} \geq 92$

To find the average of scores, add all of the numbers and then divide by how many numbers were added together. Sue has earned 90, 96, and 86 points so far. She has to take one more test, so there will be four tests in all. She wants to have an average of 92 points or more than 92 points on the tests.

$$\frac{\text{Test 1} + \text{Test 2} + \text{Test 3} + \text{Test 4}}{4} \geq 92$$

You want to find the number of points (P) that Sue must earn on her last test.

$$\frac{90 + 96 + 86 + P}{4} \geq 92$$

12. **C** *28 subtracted from a number* means that you are taking a number and subtracting 28 from it, so you must write $x - 28$, not $28 - x$.
Is less than 18 is written < 18.
28 subtracted from a number is less than 18 translates to $x - 28 < 18$.

13. $x < -10$ or $x \geq 5$

These must be solved separately and kept separated. The parts cannot be joined together.

$$5x - 13 > 9x + 27$$
$$\underline{-9x \qquad = -9x}$$
$$-4x \qquad - 13 > 27$$
$$\underline{+13 = +13}$$
$$-4x > 40$$

$$-\frac{1}{4} \cdot -4x < -\frac{1}{4} \cdot 40$$

$$x < -10$$

$$5(2x + 6) \geq 80$$
$$10x + 30 \geq 80$$
$$\underline{-30 = -30}$$
$$10x \geq 50$$

$$\frac{1}{10} \cdot 10x \geq \frac{1}{10} \cdot 50$$

$$x \geq 5$$

So the answer is $x < -10$ or $x \geq 5$.

14. **C** Solve $-12 < 7 - x < 15$ either by separating it into $7 - x > -12$ and $7 - x < 15$ and solving each part separately or by solving $-12 < 7 - x < 15$ all at once.

$$7 - x > -12 \qquad\qquad 7 - x < 15$$
$$\underline{-7 \quad = -7} \qquad\qquad \underline{-7 \quad = -7}$$
$$-x > -19 \qquad\qquad -x < 8$$
$$-1 \cdot -x < -1 \cdot -19 \qquad -1 \cdot -x > -1 \cdot 8$$
$$x < 19 \qquad\qquad x > -8$$

Putting the two inequalities together, you get $19 > x > -8$.

Solve $-12 < 7 - x < 15$ all at once.

$$-12 < 7 - x < 15$$
$$\underline{-7 = -7 \quad = -7}$$
$$-19 < -x < 8$$
$$-1 \cdot -19 > -1 \cdot -x > -1 \cdot 8$$
$$19 > x > -8$$

Solutions to Practice Problems

1. $G = 14$

$$28 = 2G$$
$$2G = 28$$

$$\frac{1}{2} \cdot 2G = \frac{1}{2} \cdot 28$$

$$G = 14$$

2. $R = 12$

$$14 = 2 + R$$
$$2 + R = 14$$
$$\underline{-2 \quad\quad = -2}$$
$$R = 12$$

3. $E = -6$

$$-14 = E - 8$$
$$E - 8 = -14$$
$$\underline{+8 = +8}$$
$$E = -6$$

4. $A = -27$

$$\frac{A}{-3} = 9$$

$$-\frac{1}{3}A = 9$$

$$-3 \cdot -\frac{1}{3}A = -3 \cdot 9$$

$$A = -27$$

5. $s = 4$

$$8s - 16 = 24 - 2s$$
$$\underline{+2s \quad\quad = \quad\quad +2s}$$
$$10s - 16 = 24$$
$$\underline{+16 = +16}$$
$$10s = 40$$

$$\frac{1}{10} \cdot 10s = \frac{1}{10} \cdot 40$$

$$s = 4$$

6. $p = -4$

$$15 + p - 12 = 4p + 15$$
$$p \;\; + 3 = 4p + 15$$
$$\underline{-4p \quad\quad = -4p}$$
$$-3p \;\; + 3 = 15$$
$$\underline{-3 = -3}$$
$$-3p = 12$$

$$-\frac{1}{3} \cdot -3p = -\frac{1}{3} \cdot 12$$

$$p = -4$$

7. $e = -5$

$$17 - e = 12 - 2e$$
$$+2e = \qquad +2e$$
$$e + 17 = 12$$
$$\underline{-17} = \underline{-17}$$
$$e = -5$$

8. $I = 1$

$$18 = 2 + 2(3I + 5)$$
$$2 + 2(3I + 5) = 18$$
$$2 + 6I + 10 = 18$$
$$6I + 12 = 18$$
$$\underline{-12} = \underline{-12}$$
$$6I = 6$$

$$\frac{1}{6} \cdot 6I = \frac{1}{6} \cdot 6$$

$$I = 1$$

9. $E = 5$

$$11 - 3E = -4(6 - E)$$
$$11 - 3E = -24 + 4E$$
$$\underline{-4E} = \qquad \underline{-4E}$$
$$-7E + 11 = -24$$
$$\underline{-11} = \underline{-11}$$
$$-7E = -35$$

$$-\frac{1}{7} \cdot -7E = -\frac{1}{7} \cdot -35$$

$$E = 5$$

10. $e = \dfrac{I \cdot ERA}{9}$

$$ERA = \frac{9 \cdot e}{I}$$

$$I \cdot ERA = I \cdot \frac{9 \cdot e}{I}$$

$$I \cdot ERA = 9e$$

$$\frac{1}{9} \cdot I \cdot ERA = \frac{1}{9} \cdot 9e$$

$$\frac{I \cdot ERA}{9} = e$$

$$e = \frac{I \cdot ERA}{9}$$

11. −2 and −3

Test each number in the inequality.

Test for −1
$$4x - 3 \geq 5x - 1$$
$$4(-1) - 3 \geq 5(-1) - 1$$
$$-4 - 3 \geq -5 - 1$$
$$-7 \geq -6, \text{ which is false.}$$

Test for −2
$$4x - 3 \geq 5x - 1$$
$$4(-2) - 3 \geq 5(-2) - 1$$
$$-8 - 3 \geq -10 - 1$$
$$-11 \geq 11, \text{ which is true.}$$

Test for −3
$$4x - 3 \geq 5x - 1$$
$$4(-3) - 3 \geq 5(-3) - 1$$
$$-12 - 3 \geq -15 - 1$$
$$-15 \geq -16, \text{ which is true.}$$

The only numbers from the set {−1, −2, −3} that make the inequality true are −2 and −3.

12. Nev R. Right is wrong.

Numbers that satisfy the inequality −8 < x are all of the numbers that are larger than −8. Any number that is greater than −8 would work. Examples include −7, −6, −5, −4, −3, −2, −1, 0, 1, and 2.

13. x > −8

$$-14 - 3x < 2x + 26$$
$$\underline{-2x = -2x}$$
$$-5x - 14 < 26$$
$$\underline{+14 = +14}$$
$$-5x < 40$$

$$-\frac{1}{5} \cdot -5x > -\frac{1}{5} \cdot 40$$

$$x > -8$$

14. p ≥ 12

$$\frac{p}{-3} + 5 \leq 1$$

$$\underline{-5 = -5}$$

$$\frac{p}{-3} \leq -4$$

$$-3 \cdot \frac{p}{-3} \geq -3 \cdot -4$$

$$p \geq 12$$

15. $r \le 17$

$$5(r + 3) \ge 7r - 19$$
$$5r + 15 \ge 7r - 19$$
$$\underline{-7r \qquad = -7r}$$
$$-2r + 15 \ge -19$$
$$\underline{-15 = -15}$$
$$-2r \ge -34$$

$$-\frac{1}{2} \cdot -2r \le -\frac{1}{2} \cdot -34$$
$$r \le 17$$

16. $t < -16$

$$18 < -\frac{3}{4}t + 6$$

$$-\frac{3}{4}t + 6 > 18$$

$$\underline{-6 = -6}$$

$$-\frac{3}{4}t > 12$$

$$-\frac{4}{3} \cdot -\frac{3}{4}t < -\frac{4}{3} \cdot 12$$
$$t < -16$$

17. $-15 < x \le -7$ or $-7 \ge x > -15$

Solve the inequality by separating it into two parts, $5 - 3x \ge 26$ and $5 - 3x < 50$, and then solving each of those parts.

$$5 - 3x \ge 26 \qquad\qquad 5 - 3x < 50$$
$$\underline{-5 \quad = -5} \qquad\qquad \underline{-5 \quad = -5}$$
$$-3x \ge 21 \qquad\qquad -3x < 45$$

$$-\frac{1}{3} \cdot -3x \le -\frac{1}{3} \cdot 21 \qquad\qquad -\frac{1}{3} \cdot -3x > -\frac{1}{3} \cdot 45$$

$$x \le -7 \qquad\qquad\qquad x > -15$$

Then, $x \le -7$ and $x > -15$ can be combined into $-15 < x \le -7$ or $-7 \ge x > -15$.

The problem can be solved together.

$$26 \le 5 - 3x < 50$$
$$\underline{-5 = -5 = \quad -5}$$
$$21 \le -3x < 45$$

$$-\frac{1}{3} \cdot 21 \ge -\frac{1}{3} \cdot -3x > -\frac{1}{3} \cdot 45$$

$$-7 \ge x > -15$$

18. $x < 4$ or $x > 6$

This problem must be solved separately and kept apart in the answer.

$$5 > 2x - 3$$
$$2x - 3 < 5$$
$$\underline{+3 = +3}$$
$$2x < 8$$
$$\frac{1}{2} \cdot 2x < \frac{1}{2} \cdot 8$$
$$x < 4$$

$$-26 > 4 - 5x$$
$$4 - 5x < -26$$
$$\underline{-4 \quad = -4}$$
$$-5x < -30$$
$$-\frac{1}{5} \cdot -5x > -\frac{1}{5} \cdot 30$$
$$x > 6$$

19. $-3 - F = 18$

A number, F, subtracted from –3 means that you are going to take the number F away from –3. That requires you to subtract –3 minus F.
Is 18 means equals 18.
A number, F, subtracted from –3 is 18 translates to $-3 - F = 18$.

20. $N + 2N = 27$

The phrase *the number, N, increased by* means a number N added to.
Twice that number means two times that number N or $2N$.
Would give 27 means equal to 27.
Combined, *a number, N, such that the number increased by twice that number would give 27* translates into $N + 2N = 27$.

21. $8.26 \left(6\frac{1}{2} \right) + T = 73.19$

After reading the problem several times, determine what you are trying to find. The problem asks how much Kreigh earned in tips (T). You are given the variable to use, T. Kreigh earns \$8.26 for every hour that he works plus he earns tips. Altogether he earned \$73.19. Translate the word problem into an equation by thinking of what parts go together, as shown in the diagram below.

$$\boxed{\begin{array}{c}\$8.26 \\ \text{for every hour} \\ \text{of work}\end{array}} \cdot \boxed{\begin{array}{c}6\frac{1}{2} \\ \text{hours} \\ \text{of work}\end{array}} + \boxed{\text{Tips } T} = \boxed{\begin{array}{c}\$73.19 \\ \text{Total} \\ \text{earnings}\end{array}}$$

The diagram translates into the equation $8.26 \cdot \left(6\frac{1}{2} \right) + T = 73.19$.

22. 171.9464 square centimeters

Look at the formulas on the reference sheet. A baseball is in the shape of a sphere. The problem wants to know the amount of leather it would take to cover the surface of a baseball (surface area). You do not want to know what amount goes inside the baseball (volume). The formula for the surface area of a sphere is $SA = 4\pi r^2$.

You need the radius, but the problem gave the diameter. Remember that the diameter is twice the radius. If the diameter is 7.4 centimeters, then the radius is only half of that distance, 3.7 centimeters.

Put the radius into the surface area formula.

$$SA = 4\pi r^2$$
$$SA = 4 \cdot 3.14 \cdot 3.7^2$$
$$SA = 4 \cdot 3.14 \cdot 13.69$$
$$SA = 171.9464 \text{ square centimeters}$$

23. $3 \cdot .52 + .06G \leq 2.90$

The problem wants you to find the number of gummy bears that you can purchase. You know that you are buying three candy bars that cost $.52 for each candy bar. Gummy bears cost $.06 each. You only have $2.90 to spend. The diagram below summarizes the ideas of the problem.

The diagram translates into $3 \cdot .52 + .06G \leq 2.90$.

$$3 \cdot .52 + .06G \leq 2.90$$
$$1.56 + .06G \leq 2.90$$
$$\underline{-1.56 \qquad = -1.56}$$
$$.06G \leq 1.34$$

$$\frac{1}{.06} \cdot .06G \leq -\frac{1}{.06} \cdot 1.34$$

$$G \leq 22.333$$

Since the number of gummy bears must be smaller than 22.3, you can only buy 22 gummy bears. If you buy more than 22 gummy bears, you will spend more than $2.90.

24. $7,300 + 1,250B \geq 15,400$

The problem wants to know how many bake sales are necessary to earn at least $15,400 for the sound system. The school can earn more than $15,400. Each bake sale earns $1,250. The school has already earned $7,300. These ideas are represented in the diagram below.

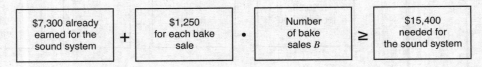

The diagram translated into an equation gives you $7,300 + 1,250B \geq 15,400$.

Chapter 4	# Algebra 1: Operations with Real Numbers

INDIANA'S ACADEMIC MATHEMATICS STANDARDS ADDRESSED:

A1.1.1 Compare real number expressions.

A1.1.2 Simplify square roots using factors.

A1.1.3 Understand and use the distributive, associative, and commutative properties.

A1.1.4 Use the law of exponents for rational exponents.

A1.1.5 Use dimensional (unit) analysis to organize conversions and computations.

A1.9.3 Use the properties of the real number system and the order of operations to justify the steps of simplifying functions and solving equations.

8.5.1 Convert common measurements for length, area, volume, weight, capacity, and time to equivalent measurements within the same system.

REAL NUMBERS

A rational number can be written as a fraction. An irrational number cannot be written as a fraction. An irrational number is a nonrepeating, nonterminating decimal. Reread Chapter 2 for a more complete explanation of the difference between rational and irrational numbers.

Rational numbers combined with irrational numbers are real numbers. Real numbers are every number that you know at this time. Real numbers include fractions, all kinds of decimals, square roots of positive numbers, negative square roots of positive numbers, and π. Numbers that are not real numbers appear in the Algebra 2 curriculum.

Real number expressions involve several real numbers combined by operations. Comparing or evaluating real number expressions involves finding approximations for any irrational numbers. Calculators are not allowed on these problems.

To compare $\sqrt{25} - \sqrt{16}$ and $\sqrt{25-16}$, evaluate each expression separately.

$$\sqrt{25} - \sqrt{16} = 5 - 4 = 1$$
$$\sqrt{25-16} = \sqrt{9} = 3$$

Now compare the answers to the two expressions.

$$\sqrt{25} - \sqrt{16} < \sqrt{25-16}$$

Exercises

1. Which is larger, $\sqrt{17} + \sqrt{9}$ or $\sqrt{17+9}$?

2. Which of the following is the best approximation for $10 + \sqrt{35}$?

    A. 15.9
    B. 7.9
    C. 17.9
    D. 27.9

Solutions to Exercises

1. $\sqrt{17} + \sqrt{9}$ is larger.

$\sqrt{17}$ is a little larger than 4. $\sqrt{17} + \sqrt{9}$ is a little more than 4 plus $\sqrt{9} = 3$. So $\sqrt{17} + \sqrt{9}$ is a little more than 7.

$\sqrt{17+9} = \sqrt{26}$, which is a little more than $\sqrt{25} = 5$.

2. **A** 15.9

$\sqrt{35}$ is a little less than $\sqrt{36} = 6$. $\sqrt{35} \approx 5.9$.

$$10 + \sqrt{35} \approx 10 + 5.9 = 15.9$$

Practice Problems

1. Which is the best approximation for $2^3 + \sqrt{48}$?

   A. 12.9
   B. 14.9
   C. 29.9
   D. 31.9

2. Place the expressions $\sqrt{26} + \sqrt{9}$, $\sqrt{26+9}$, $\sqrt{33+4}$, and $\sqrt{33} + \sqrt{4}$ in order from smallest value to largest value.

(Solutions are found at the end of the chapter.)

ORDER OF OPERATIONS

Suppose you need to find the value of $4 + 6 \cdot 2$ without using a calculator. Should you add first or multiply first? Is the answer 20 or 16? Everyone must get the same answer to problems like this, so there is a specific order in which problems are worked. This specific order is called the order of operations.

> First, do all calculations inside of grouping symbols. Grouping symbols include parentheses and brackets.
> Second, evaluate all of the exponents or powers.
> Third, do division and multiplication as they are written from left to right.
> Finally, do subtraction and addition as they are written from left to right.

In the problem $4 + 6 \cdot 2$, multiplication must be done before addition. $4 + 6 \cdot 2$ is changed to $4 + 12 = 16$.

Example
Evaluate $(4 + 6) \cdot 2$. In this case, do the calculations inside of the grouping symbols first. The first step is underlined.

$$\underline{(4 + 6)} \cdot 2 = (10) \cdot 2 = 20$$

Example

Find the value of $(2^4 - 4) \div 2 \cdot 3$. First, do the work inside the parentheses. Underlining the steps in order helps keep your work organized.

$$(\underline{2^4} - 4) \div 2 \cdot 3$$

$$(\underline{16 - 4}) \div 2 \cdot 3$$

$$(12) \div 2 \cdot 3$$

Do division and multiplication as they appear from left to right.

$$\underline{(12) \div 2} \cdot 3$$

$$6 \cdot 3 = 18$$

Exercises

1. Evaluate: $15 + 6 \cdot 2 \div 3 + 8 \cdot 4$

2. Simplify: $15 - 2 + 6$

3. Find the value of $24 \div 2 \cdot 3$.

Solutions to Exercises

1. 51

$$15 + \underline{6 \cdot 2} \div 3 + 8 \cdot 4$$

$$15 + \underline{12 \div 3} + 8 \cdot 4$$

$$15 + 4 + \underline{8 \cdot 4}$$

$$\underline{15 + 4} + 32$$

$$19 + 32$$

$$51$$

2. 19

$$\underline{15 - 2} + 6$$

$$13 + 6$$

$$19$$

3. 36

$$\underline{24 \div 2} \cdot 3$$

$$12 \cdot 3$$

$$36$$

Practice Problems

3. Evaluate: $10 + 16 \div 2 \cdot 4 - 8$

4. Simplify: $18 - 12 \div 3 + 1$

(Solutions are found at the end of the chapter.)

SIMPLIFYING SQUARE ROOTS

Square roots can be multiplied.

$$\sqrt{2} \cdot \sqrt{5} = \sqrt{10}$$

Some problems can be simplified once the square roots are multiplied.

$$\sqrt{2} \cdot \sqrt{18} = \sqrt{36} = 6$$

Example

To simplify $\sqrt{3} \cdot \sqrt{27}$, multiply the square roots. Then simplify your answer.

$$\sqrt{3} \cdot \sqrt{27} = \sqrt{81} = 9$$

Square roots can be simplified by removing factors that are perfect squares.

A perfect square is an integer that is a square of another integer. 289 is a perfect square because $17^2 = 289$.

$\sqrt{28}$ breaks apart into $\sqrt{4} \cdot \sqrt{7}$. Then $\sqrt{4}$ is simplified to 2.

$$\sqrt{28} = \sqrt{4} \cdot \sqrt{7} = 2 \cdot \sqrt{7} = 2\sqrt{7}$$

$2\sqrt{7}$ is considered to be the simplest form of $\sqrt{28}$.

To simplify $\sqrt{63}$, think of the factors of 63 that are also in the list of perfect squares. You can break down $\sqrt{63} = \sqrt{3} \cdot \sqrt{21}$, but neither 3 nor 21 are perfect squares. Think of other factors of 63. $\sqrt{63} = \sqrt{9} \cdot \sqrt{7}$ and 9 is a perfect square.

$$\sqrt{63} = \sqrt{9} \cdot \sqrt{7} = 3 \cdot \sqrt{7} = 3\sqrt{7}$$

Exercises

1. Simplify: $\sqrt{24}$

2. Simplify: $\sqrt{300}$

Solutions to Exercises

1. Pairs of factors of 24 include $1 \cdot 24$, $2 \cdot 12$, $3 \cdot 8$, and $4 \cdot 6$. The only pair of factors with a perfect square is $4 \cdot 6$.

$$\sqrt{24} = \sqrt{4} \cdot \sqrt{6} = 2 \cdot \sqrt{6} = 2\sqrt{6}$$

2. $\sqrt{300} = \sqrt{100} \cdot \sqrt{3} = 10 \cdot \sqrt{3} = 10\sqrt{3}$

Practice Problems

5. Simplify: $\sqrt{75}$

6. Change $\sqrt{72}$ into its simplest form.

(Solutions are found at the end of the chapter.)

COMMUTATIVE, ASSOCIATIVE, AND DISTRIBUTIVE PROPERTIES

> **Definition:** Commutative Property for Addition: $a + b = b + a$

> **Definition:** Commutative Property for Multiplication: $a \cdot b = b \cdot a$

The commutative property states that you can add or multiply numbers in any order. You will get the same answer whether you multiply $5 \cdot 6$ or $6 \cdot 5$.

There is no commutative property for subtraction or division. You do not get the same answer when you subtract $6 - 5$ or $5 - 6$.

> **Definition:** Associative Property for Addition: $(a + b) + c = a + (b + c)$

> **Definition:** Associative Property for Multiplication: $(a \cdot b) \cdot c = a \cdot (b \cdot c)$

The associative property states that you can group numbers in any way when you are adding or multiplying. You will get the same answer whether you multiply $(2 \cdot 3) \cdot 4$ or $2 \cdot (3 \cdot 4)$.

There is no associative property for subtraction or division. You do not get the same answer when you divide $(24 \div 4) \div 2$ or $24 \div (4 \div 2)$.

The commutative and associative properties are used when simplifying algebraic expressions like $(5x + 3) + (2x + -7)$. All of the operations in the problem are addition, so the numbers can be grouped together in any order. In this case, group $5x$ and $2x$ together.

$$(5x + 3) + (2x + -7)$$
$$\underline{5x + 2x} + \underline{3 + -7}$$
$$7x + -4$$

Problems that involve subtraction or division cannot be simplified using commutative or associative properties until they are changed into problems involving only addition and multiplication. For example, $(7x - 4) + (2x - 5)$ must be changed to $(7x + -4) + (2x + -5)$.

Now, combine the distributive property with the commutative and associative properties to simplify more complicated problems. To simplify $5(2x - 7) - 3(4x - 5)$, first change the subtraction problems into addition problems.

$$5(2x - 7) - 3(4x - 5) = 5(2x + -7) + -3(4x + -5)$$

Use the distributive property. Be careful with the negative signs.

$$5(2x + -7) + -3(4x + -5) = 5 \cdot 2x + 5 \cdot -7 + -3 \cdot 4x + -3 \cdot -5$$
$$= 10x + -35 + -12x + 15$$

Use the commutative and associative properties to regroup and simplify. You do not need to rewrite the problem when you regroup. Just underline the terms that can be added.

$$\underline{10x} + \underline{-35} + \underline{-12x} + \underline{15} = -2x + -20$$

Exercises

1. Simplify: $4(2x + 7) + 5x$

2. Simplify: $2(5x + 3) - (2x - 1)$

3. When simplifying $5x + (3 + 2x)$, what property justifies changing the problem into $5x + (2x + 3)$?

Solutions to Exercises

1. Use the distributive property first.

$$4(2x + 7) + 5x = 4 \cdot 2x + 4 \cdot 7 + 5x$$
$$= 8x + 28 + 5x$$

Use the commutative and associative properties to regroup and simplify.

$$\underline{8x} + 28 + \underline{5x} = 13x + 28$$

2. The original problem must be rewritten or thought of as $2(5x + 3) - 1(2x - 1)$.

> **Remember:**
> $(2x - 1)$ is the same as $1(2x - 1)$.

Use the distributive property after changing the subtraction problems to addition problems.

$$2(5x + 3) - 1(2x - 1) = 2(5x + 3) + -1(2x + -1)$$
$$= 2 \cdot 5x + 2 \cdot 3 + -1 \cdot 2x + -1 \cdot -1$$
$$= 10x + 6 + -2x + 1$$

Use the associative and commutative properties to regroup.

$$\underline{10x} + \underline{6} + \underline{-2x} + \underline{1} = 8x + 7$$

3. The problem is changed by switching $3 + 2x$ into $2x + 3$. The commutative property says that you can switch the order of the terms.

Practice Problems

7. Simplify: $(4w + -7) + (-6w + 9)$

8. Simplify: $5(2c - 6) - (3c - 9)$

9. Name the property used to change from the first step to the second step in the figure below.

$$(7 + 5x) + 12x$$

⇓

$$7 + (5x + 12x)$$

(Solutions are found at the end of the chapter.)

RADICALS OTHER THAN SQUARE ROOTS

The square root of 25, $\sqrt{25}$, means what number multiplied by itself gives you 25. We know that $\sqrt{25} = 5$. You are asking for the square root, so you could write $\sqrt[2]{25} = 5$. The 2 is called the index of the radical. Square roots are used so frequently that no one writes the 2.

If you want to know what number multiplied by itself three times gives you 64, you would write the cube root of 64, $\sqrt[3]{64}$. The 3 is the index of the radical. The 3 tells you how many times the number needs to be multiplied by itself.

$$\sqrt[3]{64} = 4$$

To be able to simplify a radical with an index other than 2, you need to be familiar with higher powers of numbers, as shown in the table below.

Higher Powers			
$2^3 = 2 \cdot 2 \cdot 2 = 8$	$2^4 = 2 \cdot 2 \cdot 2 \cdot 2 = 16$	$2^5 = 2 \cdot 2 \cdot 2 \cdot 2 \cdot 2 \cdot = 32$	$2^6 = 2 \cdot 2 \cdot 2 \cdot 2 \cdot 2 \cdot 2 \cdot = 64$
$3^3 = 3 \cdot 3 \cdot 3 = 27$	$3^4 = 3 \cdot 3 \cdot 3 \cdot 3 = 81$	$3^5 = 3 \cdot 3 \cdot 3 \cdot 3 \cdot 3 = 243$	
$4^3 = 4 \cdot 4 \cdot 4 = 64$	$4^4 = 4 \cdot 4 \cdot 4 \cdot 4 = 256$		
$5^3 = 5 \cdot 5 \cdot 5 = 125$	$5^4 = 5 \cdot 5 \cdot 5 \cdot 5 = 625$		
$6^3 = 6 \cdot 6 \cdot 6 = 216$			
$7^3 = 7 \cdot 7 \cdot 7 = 343$			

Example

$\sqrt[5]{32}$ asks for the number when multiplied by itself five times will give you 32. Using the table, you find that $\sqrt[5]{32} = 2$.

Exercises

1. Compute: $\sqrt[3]{8}$

2. Find: $\sqrt[4]{625}$

Solutions to Exercises

1. $\sqrt[3]{8}$ asks for the number when multiplied by itself three times gives you 8. The answer is 2.

2. Think of a number when multiplied by itself four times gives you 625. The number is 5.

RATIONAL EXPONENTS

Rational exponents are exponents that are fractions.

Definition: $b^{\frac{x}{y}} = \sqrt[y]{b^x} = \left(\sqrt[y]{b}\right)^x$ for any positive number b.

Using the definition, $27^{\frac{2}{3}}$ means $\sqrt[3]{27^2}$ or $\left(\sqrt[3]{27}\right)^2$. Choose the meaning that is easier to do because you cannot use a calculator on these problems. It is easier to do $\left(\sqrt[3]{27}\right)^2$. First find $\sqrt[3]{27}$ and then square your answer.

$$\left(\sqrt[3]{27}\right)^2 = (3)^2 = 9$$

Example

To compute $16^{\frac{3}{2}}$ without using a calculator, rewrite the problem using the definition of rational exponents.

$$16^{\frac{3}{2}} = \sqrt[2]{16^3} \text{ or } 16^{\frac{3}{2}} = \left(\sqrt[2]{16}\right)^3$$

Use $16^{\frac{3}{2}} = \left(\sqrt[2]{16}\right)^3$ because it is easy to find $\sqrt[2]{16}$.

$$16^{\frac{3}{2}} = \left(\sqrt[2]{16}\right)^3 = (4)^3 = 64$$

Exercises

1. Without using a calculator, find $32^{\frac{3}{5}}$.

2. Compute $125^{\frac{2}{3}}$ without using a calculator.

Solutions to Exercises

1. Use the definition of rational exponents to rewrite $32^{\frac{3}{5}}$ as $\left(\sqrt[5]{32}\right)^3$. Then find $\sqrt[5]{32}$ and take that answer to the third power.

$$32^{\frac{3}{5}} = \left(\sqrt[5]{32}\right)^3 = (2)^3 = 8$$

2. Rewrite $125^{\frac{2}{3}}$ as $\left(\sqrt[3]{125}\right)^2$. Find $\sqrt[3]{125}$ and square the answer.

$$125^{\frac{2}{3}} = \left(\sqrt[3]{125}\right)^2 = (5)^2 = 25$$

Practice Problems

10. Compute $81^{\frac{3}{4}}$ without using a calculator.

(Solution is found at the end of the chapter.)

DIMENSIONAL OR UNIT ANALYSIS

In mathematics, 50 miles per hour can be written as the fraction $\dfrac{50 \text{ miles}}{\text{hour}}$. A heartbeat of

34 times per minute can be written as $\dfrac{34 \text{ times}}{\text{minute}}$ or $\dfrac{34 \text{ heartbeats}}{\text{minute}}$. Fractions like these can be used to change dimensions from miles per hour to feet per second or from heartbeats per minute to heartbeats per hour.

We will review some work with fractions so that we can use the same process when changing dimensions.

> **Remember:** In Chapter 1, $\dfrac{29}{8} \cdot \dfrac{2}{3}$, the 8 in the denominator of the first fraction can be reduced with the 2 in the numerator of the second fraction by dividing both numbers by 2.
>
> $$\frac{29}{8} \cdot \frac{2}{3} = \frac{29}{\underset{4}{\cancel{8}}} \cdot \frac{\overset{1}{\cancel{2}}}{3}$$

When simplifying problems with unit labels, cancel the labels in the same way as the numbers are cancelled.

$$\frac{34 \text{ heartbeats}}{\cancel{\text{minute}}} \cdot \frac{60 \, \cancel{\text{minutes}}}{1 \text{ hour}}$$

The word "minute" in the denominator of the first fraction can be cancelled with the word "minutes" in the numerator of the second fraction.

> **Remember:** In Chapter 1, $\dfrac{12}{12} = 1$.

When changing dimensions, $\dfrac{60 \text{ seconds}}{1 \text{ minute}}$ is the same thing as 1 because 60 seconds and 1 minute are equal. Even though they do not look exactly the same, they mean the same.

Combining these two ideas with fractions allows us to change from 50 miles per hour,

$\dfrac{50 \text{ miles}}{\text{hour}}$, into feet per hour, $\dfrac{\text{feet}}{\text{hour}}$. Begin with $\dfrac{50 \text{ miles}}{\text{hour}}$ and multiply it by 1, written as

$\dfrac{5{,}280 \text{ feet}}{1 \text{ mile}}$.

$$\frac{50 \text{ miles}}{\text{hour}} \cdot \frac{5{,}280 \text{ feet}}{1 \text{ mile}}$$

The "miles" in the numerator of the first fraction cancels with the "mile" in the denominator of the second fraction. The answer will contain only $\dfrac{\text{feet}}{\text{hour}}$.

$$\frac{50 \text{ \sout{miles}}}{\text{hour}} \cdot \frac{5{,}280 \text{ feet}}{1 \text{ \sout{mile}}} = \frac{50 \cdot 5{,}280 \text{ feet}}{\text{hour}} = \frac{264{,}000 \text{ feet}}{\text{hour}}$$

So, 50 miles per hour is the same as 264,000 feet per hour.

Example

A cheetah ran 300 feet in 2.92 seconds. What was the cheetah's speed in miles per hour?

First, write the cheetah's speed as a fraction, $\dfrac{300 \text{ feet}}{2.92 \text{ seconds}}$. Multiply this fraction by 1,

written to show the relationship between feet and miles, either $\dfrac{5{,}280 \text{ feet}}{1 \text{ mile}}$ or $\dfrac{1 \text{ mile}}{5{,}280 \text{ feet}}$.

Choose $\dfrac{1 \text{ mile}}{5{,}280 \text{ feet}}$ so that the "feet" in the first fraction will cancel with the "feet" in the

second fraction.

$$\frac{300 \text{ \sout{feet}}}{2.92 \text{ seconds}} \cdot \frac{1 \text{ mile}}{5{,}280 \text{ \sout{feet}}}$$

Now the labels in the fractions are $\dfrac{\text{miles}}{\text{second}}$. Multiply the answer so far by 1, written to

show the relationship between seconds and hours, either $\dfrac{60 \text{ seconds}}{1 \text{ minute}} \cdot \dfrac{60 \text{ minutes}}{1 \text{ hour}}$ or

$\dfrac{1 \text{ minute}}{60 \text{ seconds}} \cdot \dfrac{1 \text{ hour}}{60 \text{ minutes}}$.

Be careful! To change from seconds to hours, first change from seconds to minutes and then from minutes to hours. There are 60 seconds in 1 minute, so that fraction

can be written as $\dfrac{60 \text{ seconds}}{1 \text{ minute}}$ or $\dfrac{1 \text{ minute}}{60 \text{ seconds}}$. There are 60 minutes in 1 hour, so that

fraction can be written as $\dfrac{60 \text{ minutes}}{1 \text{ hour}}$ or $\dfrac{1 \text{ hour}}{60 \text{ minutes}}$.

Remember you need for the "seconds" in the problem to cancel with the "seconds" in what

you are multiplying by. Use $\dfrac{60 \text{ seconds}}{1 \text{ minute}} \cdot \dfrac{60 \text{ minutes}}{1 \text{ hour}}$.

$$\frac{300 \text{ \sout{feet}}}{2.92 \text{ \sout{seconds}}} \cdot \frac{1 \text{ mile}}{5{,}280 \text{ \sout{feet}}} \cdot \frac{60 \text{ \sout{seconds}}}{1 \text{ \sout{minute}}} \cdot \frac{60 \text{ \sout{minutes}}}{1 \text{ hour}}$$

Use your calculator to multiply the numerators and the denominators.

$$\frac{300 \text{ \sout{feet}}}{2.92 \text{ \sout{seconds}}} \cdot \frac{1 \text{ mile}}{5{,}280 \text{ \sout{feet}}} \cdot \frac{60 \text{ \sout{seconds}}}{1 \text{ \sout{minute}}} \cdot \frac{60 \text{ \sout{minutes}}}{1 \text{ hour}} = \frac{1{,}080{,}000 \text{ miles}}{15{,}417.6 \text{ hour}}$$

Use your calculator to divide and get 70.05 miles per hour. A cheetah can run 70.05 miles per hour for short periods of time.

Exercises

1. To find the number of cars stuck in a traffic jam on a 13-mile stretch of a 3-lane

   highway, simplify: $3 \text{ lanes} \cdot \dfrac{13 \text{ miles}}{1 \text{ lane}} \cdot \dfrac{5,280 \text{ feet}}{1 \text{ mile}} \cdot \dfrac{1 \text{ vehicle}}{30 \text{ feet}}$.

2. Land near downtown Chicago costs $1,500,000 per acre. Change $1,500,000 per acre into dollars per square foot.

Solutions to Exercises

1. Cancel the labels. Remember that $3 \text{ lanes} = \dfrac{3 \text{ lanes}}{1}$.

   $$\dfrac{3 \cancel{\text{ lanes}}}{1} \cdot \dfrac{13 \cancel{\text{ miles}}}{1 \cancel{\text{ lane}}} \cdot \dfrac{5,280 \cancel{\text{ feet}}}{1 \cancel{\text{ mile}}} \cdot \dfrac{1 \text{ vehicle}}{30 \cancel{\text{ feet}}} = \dfrac{3 \cdot 13 \cdot 5,280}{30} \text{ vehicles}$$

   6,864 vehicles are stuck in the traffic jam.

2. Write $1,500,000 per acre as a fraction, $\dfrac{1,500,000 \text{ dollars}}{\text{acre}}$. Use the reference sheet to find the number of square feet in 1 acre, 1 acre = 43,560 square feet. Write the relationship between square feet and acres as a fraction that can be used with $\dfrac{1,500,000 \text{ dollars}}{\text{acre}}$.

   $$\dfrac{1,500,000 \text{ dollars}}{\cancel{\text{acre}}} \cdot \dfrac{1 \cancel{\text{ acre}}}{43,560 \text{ square feet}} = \dfrac{1,500,000}{43,560} \times \text{dollars per square foot}$$

   $$= 34.44 \text{ dollars per square foot}$$

Practice Problems

11. A sloth, a slow-moving mammal, travels 0.15 miles in 1 hour. Convert this speed to feet per minute.

12. At your sister's lemonade stand, she sells 5 pints of lemonade every 6 minutes. If she continues selling lemonade at the same pace, how many quarts of lemonade will your sister sell per hour?

SAMPLE GQE QUESTIONS FOR CHAPTER 4

 1. Which of the following has the numbers listed in order from largest to smallest?

A. $\sqrt{25+9}$, $\sqrt{25-9}$, $\sqrt{25} + \sqrt{9}$, $\sqrt{25} - \sqrt{9}$

B. $\sqrt{25} + \sqrt{9}$, $\sqrt{25+9}$, $\sqrt{25-9}$, $\sqrt{25} - \sqrt{9}$

C. $\sqrt{25+9}$, $\sqrt{25} + \sqrt{9}$, $\sqrt{25-9}$, $\sqrt{25} - \sqrt{9}$

D. $\sqrt{25} + \sqrt{9}$, $\sqrt{25-9}$, $\sqrt{25+9}$, $\sqrt{25} - \sqrt{9}$

 2. In which of the following is $\sqrt{48}$ written in simplest form?

A. $4\sqrt{12}$

B. $2\sqrt{12}$

C. $16\sqrt{3}$

D. $4\sqrt{3}$

 3. Evaluate: $16 - 10 \div 2 \cdot 5$

A. 15

B. –9

C. $\dfrac{6}{10} = \dfrac{3}{5}$

D. 55

 4. Which of the following is the same as $40 \div 2 \cdot 5 + 5 \cdot 3$?

A. $40 \div 10 + 5 \cdot 3$

B. $40 \div 20 \cdot 3$

C. $20 \cdot 5 + 5 \cdot 3$

D. $20 \cdot 10 \cdot 3$

 5. The peregrine falcon has a record diving speed of 168 miles per hour. Change this speed to feet per minute.

 6. Which of the following is the same as $25^{\frac{1}{2}} + 9^{\frac{3}{2}}$?

A. 26

B. 32

C. 34

D. 36

7. In the figure below, look at the change made from the first equation to the second equation.

$$7 + 2x + 3 = 24$$

$$\Downarrow$$

$$2x + 7 + 3 = 24$$

Which property justifies the change?

A. Addition Property of Equality

B. Distributive Property

C. Associative Property of Addition

D. Commutative Property of Addition

SOLUTIONS TO SAMPLE GQE QUESTIONS FOR CHAPTER 4

Answer Key

1. **B**
2. **D**
3. **B**

4. **C**
5. 14,784 feet per minute

6. **B**
7. **D**

Answers Explained

1. **B** $\sqrt{25} + \sqrt{9}$, $\sqrt{25+9}$, $\sqrt{25-9}$, $\sqrt{25} - \sqrt{9}$

$$\sqrt{25} + \sqrt{9} = 5 + 3 = 8$$

$$\sqrt{25+9} = \sqrt{34} \approx 5.8$$

$$\sqrt{25-9} = \sqrt{16} = 4$$

$$\sqrt{25} - \sqrt{9} = 5 - 3 = 2$$

2. **D** $4\sqrt{3}$. Think of the pairs of factors of 48, $2 \cdot 24$, $3 \cdot 16$, $4 \cdot 12$, and $6 \cdot 8$. Both $3 \cdot 16$ and $4 \cdot 12$ have a factor that is a perfect square. Choose the largest perfect square.

$$\sqrt{48} = \sqrt{16} \cdot \sqrt{3} = 4\sqrt{3}$$

$2\sqrt{12}$ is correct but it is not simplified completely. Always check if the number under the radical sign has any perfect square factors.

$$2\sqrt{12} = 2 \cdot \sqrt{4} \cdot \sqrt{3} = 2 \cdot 2 \cdot \sqrt{3} = 4\sqrt{3}$$

3. **B** -9

$$16 - \underline{10 \div 2} \cdot 5$$
$$16 - \underline{5 \cdot 5}$$
$$16 - 25$$
$$-9$$

4. **C** $20 \cdot 5 + 5 \cdot 3$

$$\underline{40 \div 2} \cdot 5 + 5 \cdot 3$$
$$20 \cdot 5 + 5 \cdot 3$$

5. 14,784 feet per minute

Write 168 miles per hour as a fraction, $\dfrac{168 \text{ miles}}{\text{hour}}$. There are 5,280 feet in one mile

and 60 minutes in 1 hour. Rewrite both of these as fractions equal to 1 so that the labels will cancel with those in the original fraction.

$$\frac{168 \;\cancel{\text{miles}}}{\cancel{\text{hour}}} \cdot \frac{5,280 \text{ feet}}{1 \;\cancel{\text{mile}}} \cdot \frac{1 \;\cancel{\text{hour}}}{60 \text{ minutes}} = 14,784 \text{ feet per minute}$$

6. **B** 32

$$25^{\frac{1}{2}} = \sqrt[2]{25^1} = \sqrt{25} = 5$$

$$9^{\frac{3}{2}} = \left(\sqrt[2]{9}\right)^3 = \left(\sqrt{9}\right)^3 = (3)^3 = 27$$

$$25^{\frac{1}{2}} + 9^{\frac{3}{2}} = 5 + 27 = 32$$

7. **D** Commutative Property
The order of the terms was switched. $7 + 2x$ was switched to $2x + 7$.

Solutions to Practice Problems

1. **B** 14.9
$2^3 = 2 \cdot 2 \cdot 2 = 8$. $\sqrt{48}$ is a little less than $\sqrt{49} = 7$.

$$2^3 + \sqrt{48} \approx 8 + 6.9 = 14.9$$

2. $\sqrt{26+9}$, $\sqrt{33+4}$, $\sqrt{33} + \sqrt{4}$, $\sqrt{26} + \sqrt{9}$

$$\sqrt{26+9} = \sqrt{35} \approx 5.9$$
$$\sqrt{33+4} = \sqrt{37} \approx 6.1$$
$$\sqrt{33} + \sqrt{4} \approx 5.7 + 2 = 7.7$$
$$\sqrt{26} + \sqrt{9} \approx 5.1 + 3 = 8.1$$

3. 34

$$10 + \underline{16 \div 2} \cdot 4 - 8$$
$$10 + \underline{8 \cdot 4} - 8$$
$$\underline{10 + 32} - 8$$
$$42 - 8$$
$$34$$

4. 15

$$18 - \underline{12 \div 3} + 1$$
$$\underline{18 - 4} + 1$$
$$14 + 1$$
$$15$$

5. $5\sqrt{3}$
Use a pair that includes a perfect square.

$$\sqrt{75} = \sqrt{3} \cdot \sqrt{25} = 5 \cdot \sqrt{3} = 5\sqrt{3}$$

6. $6\sqrt{2}$
Many students immediately think $9 \cdot 8 = 72$. There are two pairs of factors of 72 that include perfect squares, $9 \cdot 8$ and $2 \cdot 36$. Always use the factors with the largest perfect square.

$$\sqrt{72} = \sqrt{2} \cdot \sqrt{36} = 6 \cdot \sqrt{2} = 6\sqrt{2}$$

Using the other pair of factors increases the number of steps to do the problem.

$$\sqrt{72} = \sqrt{9} \cdot \sqrt{8} = 3 \cdot \sqrt{8} = 3\sqrt{8}$$

But $\sqrt{8}$ can be broken into $\sqrt{4} \cdot \sqrt{2}$ and then simplified.

$$3\sqrt{8} = 3 \cdot \sqrt{4} \cdot \sqrt{2} = 3 \cdot 2 \cdot \sqrt{2} = 6 \cdot \sqrt{2} = 6\sqrt{2}$$

Always look at your final answer to be sure that you have simplified completely.

7. $-2w + 2$

The associative and commutative properties state that the order and grouping does not matter when working solely with addition.

$$(\underline{4w} + \underline{-7}) + (\underline{-6w} + \underline{9}) = -2w + 2$$

8. $7c - 21$

Think of $5(2c - 6) - (3c - 9)$ as $5(2c - 6) - 1(3c - 9)$.

$$5(2c - 6) - 1(3c - 9) = 5(2c + -6) + -1(3c + -9)$$

$$5(2c + -6) + -1(3c + -9) = 5 \cdot 2c + 5 \cdot -6 + -1 \cdot 3c + -1 \cdot -9$$

$$= 10c + -30 + -3c + 9$$

$$\underline{10c} + \underline{-30} + \underline{-3c} + \underline{9} = 7c + -21$$

9. The associative property of addition was used.

10. 27

$$81^{\frac{3}{4}} = \left(\sqrt[4]{81}\right)^3 = (3)^3 = 27$$

11. 13.2 feet per minute

Write .15 mile in 1 hour as a fraction, $\dfrac{.15 \text{ mile}}{\text{hour}}$. Rewrite 1 to show the relationship between miles and feet, $\dfrac{5,280 \text{ feet}}{1 \text{ mile}}$. Rewrite 1 to show the relationship between hours and minutes, $\dfrac{1 \text{ hour}}{60 \text{ minutes}}$.

$$\frac{.15 \ \cancel{\text{miles}}}{\cancel{\text{hour}}} \cdot \frac{5,280 \text{ feet}}{1 \ \cancel{\text{mile}}} \cdot \frac{1 \ \cancel{\text{hour}}}{60 \text{ minutes}}$$

$$= \frac{792 \text{ feet}}{60 \text{ minutes}} = 13.2 \text{ feet per minute}$$

12. 25 quarts per hour

Write 5 pints of lemonade every 6 minutes as a fraction, $\dfrac{5 \text{ pints}}{6 \text{ minutes}}$. Use your reference sheet to rewrite 1 showing the relationship between pints and quarts, $\dfrac{1 \text{ quart}}{2 \text{ pints}}$. Write 1 showing the relationship between hours and minutes, $\dfrac{60 \text{ minutes}}{1 \text{ hour}}$.

$$\frac{5 \ \cancel{\text{pints}}}{6 \ \cancel{\text{minutes}}} \cdot \frac{1 \text{ quart}}{2 \ \cancel{\text{pints}}} \cdot \frac{60 \ \cancel{\text{minutes}}}{1 \text{ hour}} = 25 \frac{\text{quarts}}{\text{hour}}$$

GEOMETRY DEFINITIONS

There are lots of geometric terms that must be part of your vocabulary for the GQE. As these terms are reviewed, try to develop a way to remember their meanings. The first set of words to be defined will be important in polygons, a two-dimensional shape with straight sides, and with some three-dimensional shapes.

Definition: An altitude is a line segment drawn from the vertex perpendicular to the opposite side.

A vertex of a shape is a corner of the shape. Perpendicular means that it intersects at a right angle (90° angle). An altitude is a line drawn from a corner to the opposite side so that it meets the opposite side at a 90° angle. Altitude in flying means the height of the airplane. Altitude in geometry means the height of the shape.

Altitude $\overline{CD}$ is drawn for each triangle shown below.

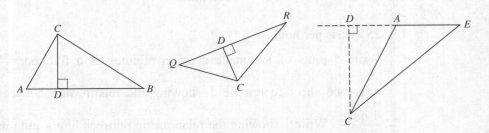

In $\triangle ACE$, side $\overline{AE}$ had to be extended so that altitude $\overline{CD}$ could be drawn. Altitude $\overline{CD}$ is actually outside of $\triangle ACE$.

Some three-dimensional shapes have altitudes. Altitude $\overline{CD}$ is drawn in the cone shown below.

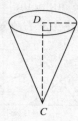

Definition: A diagonal is a line segment drawn from one vertex to a nonadjacent vertex.

Nonadjacent means not beside. A diagonal is a line segment drawn from one corner of a shape to a corner that is not beside the first corner. Walking diagonally across a field means walking from one corner to the opposite corner.

In the polygons shown below, all of the diagonals from vertex A are drawn.

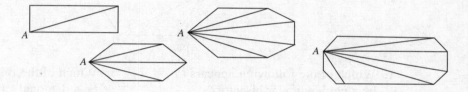

A diagonal is always inside the polygon.

Definition: An angle bisector is a line, line segment, or ray that splits the angle in half.

To bisect means to cut into two equal parts. To bisect an angle means to cut the angle into two equal parts. If an angle measures 50°, then the angle bisector splits the angle into two 25° angles.

Angle bisector $\overrightarrow{AB}$ is drawn in each shape shown below.

Definition: A perpendicular bisector of a line segment is a line, line segment, or ray that intersects the midpoint of a line segment at a 90° angle.

The two words "perpendicular bisector" describe the line. First, the line must intersect the line segment at a 90° angle. Second, the line must intersect at the point in the middle of the line segment, the midpoint.

A perpendicular bisector $\overline{HE}$ is drawn in both shapes shown below.

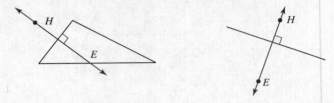

Exercises

Use the hexagon shown below to answer questions 1–4.

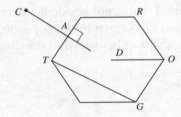

1. Which of the following appears to be a perpendicular bisector?

   A. $\overline{CA}$
   B. $\overline{DO}$
   C. $\overline{TG}$
   D. There is no perpendicular bisector drawn.

2. Which of the following appears to be an altitude?

   A. $\overline{CA}$
   B. $\overline{DO}$
   C. $\overline{TG}$
   D. There is no altitude drawn.

3. Which of the following appears to be a diagonal?

   A. $\overline{CA}$
   B. $\overline{DO}$
   C. $\overline{TG}$
   D. There is no diagonal drawn.

4. Which of the following appears to be an angle bisector?

   A. $\overline{CA}$
   B. $\overline{DO}$
   C. $\overline{TG}$
   D. There is no angle bisector drawn.

Solutions to Exercises

1. **A** $\overline{CA}$
   $\overline{CA}$ intersects the line segment at its midpoint and is perpendicular to the line segment.

2. **D** There is no altitude drawn.
   None of the line segments goes from a vertex to the opposite side.

3. **C** $\overline{TG}$
   $\overline{TG}$ goes from vertex T to vertex G. Vertices T and G are not beside each other.

4. **B** $\overline{DO}$
   $\overline{DO}$ cuts $\angle ROG$ into two equal angles, $\angle DOR$ and $\angle DOG$.

The second set of words to be defined deals with the angles and line segments associated with circles.

> **Definition:** A radius of a circle is a line segment joining the center of the circle to any point on the circle.

In a circle there are many radii (plural of radius). All of the radii of one circle are the same length. Several circles are shown below with radii drawn on the circles.

  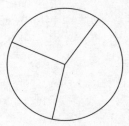

> **Definition:** A diameter of a circle is the line segment joining two points of the circle that also contains the center of the circle.

In a circle there are many diameters. All of the diameters within a circle are the same length. Several circles are shown below with one diameter drawn on each.

  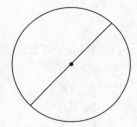

> **Definition:** A central angle in a circle is an angle formed by joining two points on a circle to the center of the circle.

The word "central angle" implies an angle at the center. A central angle $\angle HAT$ is drawn in each circle shown below.

  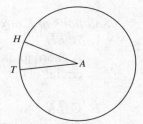

> **Definition:** A chord is a line segment joining two points on a circle.

A chord joins any two points on a circle. If a chord contains the center of a circle, then it is also called a diameter. There are several chords drawn in each circle shown below.

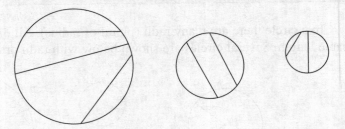

Exercises

Use the following figure to answer questions 1–4.

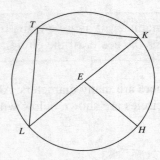

1. Which of the following is not a chord?

    A. $\overline{TK}$
    B. $\overline{EH}$
    C. $\overline{LK}$
    D. $\overline{TL}$

2. Which of the following is a diameter?

    A. $\overline{TK}$
    B. $\overline{EH}$
    C. $\overline{LK}$
    D. $\overline{TL}$

3. Which of the following is a central angle?

    A. $\angle TKL$
    B. $\angle TLK$
    C. $\angle LTK$
    D. $\angle KEH$

4. Which of the following is a radius?

    A. $\overline{TK}$
    B. $\overline{EH}$
    C. $\overline{LK}$
    D. $\overline{TL}$

Solutions to Exercises

1. **B** $\overline{EH}$

    A chord joins any two points on a circle. Point E is the center of the circle, not on the circle.

2. **C** $\overline{LK}$

    A diameter contains two points on the circle and also contains the center of the circle. $\overline{LK}$ contains the points L and K that are on the circle. $\overline{LK}$ also contains point E, the center of the circle.

3. **D** ∠*KEH*

A central angle must have its vertex at the center of the circle. In ∠*KEH*, *E* is the vertex.

4. **B** $\overline{EH}$

A radius joins the center of the circle to any point on the circle. Point *E* is the center of the circle and point *H* is on the circle.

Practice Problems

Use the following figure to answer questions 1–4.

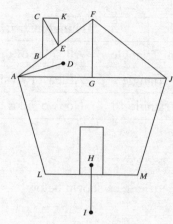

1. Which of the following appears to be an angle bisector?

   A. $\overline{AD}$
   B. $\overline{AF}$
   C. $\overline{HI}$
   D. $\overline{FJ}$

2. Which of the following appears to be a diagonal?

   A. $\overline{AD}$
   B. $\overline{FG}$
   C. $\overline{HI}$
   D. $\overline{CE}$

3. Which of the following appears to be a perpendicular bisector?

   A. $\overline{AD}$
   B. $\overline{AF}$
   C. $\overline{HI}$
   D. $\overline{CE}$

4. Which of the following appears to be an altitude?

   A. $\overline{AD}$
   B. $\overline{FG}$
   C. $\overline{HI}$
   D. $\overline{CE}$

(Solutions are found at the end of the chapter.)

TWO-DIMENSIONAL FIGURES

A plane is a flat surface that extends indefinitely, while space is three-dimensional. A rectangle lies on a plane, while a cone is in space. A triangle lies on a plane, while a sphere is in space. The reference sheet for the GQE has shapes that are on a plane and figures in space.

    Triangles are polygons with three sides. There are two different ways to classify triangles, by their angles and by their sides. Triangle classifications by angles are listed in the table below.

Triangle Classification by Angles	
Acute Triangle	All angles in the triangle are less than 90°.
Obtuse Triangle	One angle in the triangle is more than 90°.
Right Triangle	One angle in the triangle is exactly 90°.
Equiangular Triangle	All angles in the triangle are equal in measure.

Triangle classifications by sides are given in the table below.

Triangle Classification by Sides	
Equilateral Triangle	All sides of the triangle are equal.
Isosceles Triangle	At least two sides of the triangle are the same length.
Scalene Triangle	No two sides of the triangle are the same length.

Exercises

1. Look at the triangles shown below.

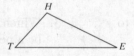

   Classify each triangle according to its angles and its sides.

2. Can a triangle be both scalene and isosceles? Why or why not?

Solutions to Exercises

1. △*THE* is an obtuse scalene triangle because it has one obtuse angle, *H*, and all the sides are different lengths.
   △*CAT* is a right isosceles triangle because $\overline{CA}$ and $\overline{CT}$ are congruent and ∠*C* is a right angle. Remember that congruent means the same size and the same shape.
   △*RAN* is an acute equilateral triangle because all of the sides are the same length and all of the angles are acute.

2. No.
   A triangle cannot have all three sides of different lengths at the same time that two sides are the same length.

   Quadrilaterals, four-sided polygons, are the next group of shapes on the GQE reference sheet. The four kinds of quadrilaterals shown on the reference sheet are parallelogram, rectangle, square, and trapezoid. A parallelogram is a quadrilateral with both pairs of opposite sides parallel, as shown below.

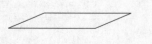

The last parallelogram shown above is a rhombus, a special kind of parallelogram where all of the sides are congruent.

A rectangle is a parallelogram with four right angles as shown below.

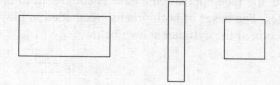

The last rectangle shown above is a square, a special kind of rectangle where all of the sides are congruent.

A trapezoid is a quadrilateral with exactly one pair of opposite sides parallel as shown below.

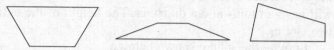

The parallel sides in a trapezoid are called the bases.

Exercises

1. Which quadrilaterals (rectangle, trapezoid, parallelogram, square, and rhombus) can have both pairs of opposite sides congruent?

2. Is a square always a rhombus? Why or why not? Is a rhombus always a square? Why or why not?

Solutions to Exercises

1. rectangle, parallelogram, square, and rhombus
   A trapezoid can have only one pair of opposite sides congruent.

2. A square is always a rhombus, but a rhombus is not always a square.
   A square is a special kind of rhombus with four right angles. A rhombus may or may not be a square, depending on whether all of the angles are right angles.

THREE-DIMENSIONAL GEOMETRIC OBJECTS

Look at the rectangular prism shown below.

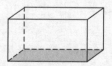

A rectangular prism is a solid shape where all of the faces are rectangles. Think of a rectangular prism as a shoebox. Each flat side is called a face. Each corner is called a vertex (plural is vertices). Notice that opposite faces are congruent rectangles. There are six faces on a rectangular prism. The base of the rectangular prism is shaded gray.

Look at the general prisms shown below.

General prisms are named by the shape of their base. The base of the first prism is a triangle, so the prism is called a triangular prism. The base of the second prism is a trapezoid, so the prism is called a trapezoidal prism. Notice that in each of the prisms shown above, the face opposite the base is congruent to the base. All of the other faces are rectangles. The bases of both prisms are shaded in gray.

Look at the cylinder shown below.

Think of a cylinder as a can of vegetables. The base of a cylinder is a circle, shaded in gray. Notice there is a circle congruent to the base on the top of the cylinder. The base will have a radius and a diameter. The height of the cylinder is the distance between the two circles.

Look at the sphere shown below.

Think of a sphere as a ball. A sphere has a radius and a diameter, but no faces. If you cut a sphere in half, you would have half of a sphere or a hemisphere. The bottom of the hemisphere is a circle, called a great circle. This great circle has a radius that is considered the radius of the sphere.

Look at the right circular cone shown below.

Think of a cone as an ice cream cone without the ice cream. The base, the gray shaded part, is a circle. The base will have a radius that is considered the radius of the cone. The height of the cone is the line segment from the top of the cone perpendicular to the base of the cone.

Look at the regular pyramid shown below.

Think of a pyramid as the Pyramids in Egypt. The base, the gray shaded part, is a square. In other regular pyramids, the base could be a pentagon or a hexagon with all sides the same length and all angles the same measure. The faces other than the base are all congruent triangles. The height of the pyramid is the line segment from the top of the pyramid perpendicular to the base of the pyramid.

Exercises

1. Identify each three-dimensional shape.

A. 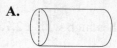   B.   C.   D.

2. Identify the three-dimensional shapes that compose the irregular shapes.

A.   B.   C.

Solutions to Exercises

1. **A.** A cylinder
   B. A triangular prism
   C. A cone
   D. A sphere

2. **A.** A hemisphere is put on each end of a cylinder.
   B. Half of a sphere *or* a hemisphere
   C. A cylinder with a cone on top

Practice Problems

5. Look at the triangles shown below.

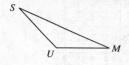 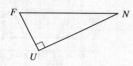

Classify each triangle by its angles and by its sides.

6. Is it possible to draw an obtuse triangle that has angles that measure 40° and 50°? Why or why not?

7. Look at the shape drawn below.

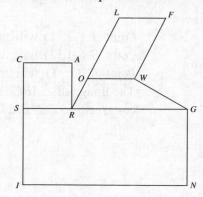

The shape has been divided into four quadrilaterals. State whether each statement is true or false. If false, change the underlined word or phrase to make the statement true.

A. Quadrilateral *CARS* is <u>a square and a rectangle</u>.
B. Quadrilateral *SING* is a <u>rhombus</u>.
C. Quadrilateral *GROW* is a <u>parallelogram</u>.
D. Quadrilateral *WOLF* is a <u>rectangle</u>.

8. Identify the three-dimensional figure that would best represent the object described.
   A. a basketball
   B. a swimming pool that is 50 meters long and 25 meters wide and is filled to a depth of 6 feet
   C. a 12 fluid ounce can of soda that is 5 inches high with a 2.75-inch diameter
   D. a wooden block in a toy set

(Solutions are found at the end of the chapter.)

TRANSFORMATIONS: TRANSLATIONS, ROTATIONS, REFLECTIONS, AND DILATIONS

Translations, rotations, and reflections are changes that are made in the location of a shape. The size of the shape does not change. The moved shape is congruent to the original shape. The area, perimeter, or circumference of the shape is the same before and after the move is made.

Translations

A translation slides each point of a shape in one direction. $\triangle ABC$ is shown in the figure below.

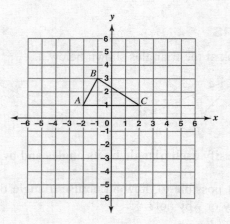

To translate $\triangle ABC$ four units down, move each of the points down four units, as shown below.

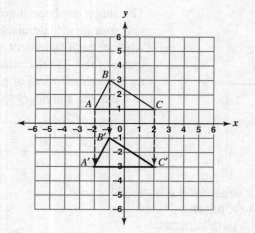

Point A (−2, 1) will move to A' (−2, −3).
Point B (−1, 3) will move to B' (−1, −1).
Point C (2, 1) will move to C' (2, −3).
The image of $\triangle ABC$ is $\triangle A'B'C'$. $\triangle ABC$ is congruent to $\triangle A'B'C'$.

One translation can be followed by another translation. The quadrilateral shown below is translated three units up and five units to the left.

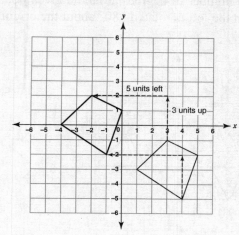

Each one of the points is moved three units up and five units to the left. The two quadrilaterals are congruent, the same shape and the same size.

Exercise

Use the coordinate plane below to translate the triangle two units down and six units to the right.

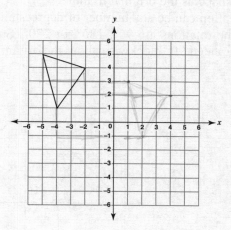

Solution to Exercise

Move each point down two units and right six units.

Rotations

Rotations spin all points a certain number of degrees around a given point. On the coordinate plane below, the triangle on the left is rotated 180° about the origin (0, 0).

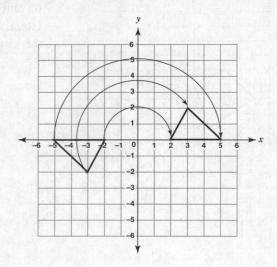

The arrows show how the points were moved 180° clockwise. To understand how a rotation works, trace the triangle on a separate sheet of paper. Place your traced triangle on top of the original triangle. Put your pencil point on the origin and turn the paper so it is upside down. This is a rotation of 180°. As you can see, the rotated triangle is the same size and shape as the original triangle.

A rotation can be any number of degrees either clockwise or counterclockwise. Most of the time rotations are 90°, 180°, or 270°. Suppose the angle on the coordinate plane below is rotated 90° clockwise about the origin.

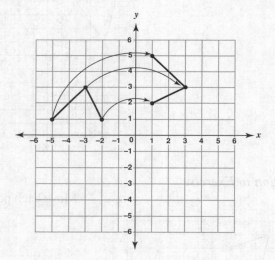

The easiest way to see how this rotation works is to trace the angle on a sheet of paper. Place the angle on top of the original angle. Put your pencil point on the origin and spin the paper 90° clockwise.

Some rotations are around a point that is not at the origin. The quadrilateral on the coordinate plane below is rotated 90° counterclockwise about point A.

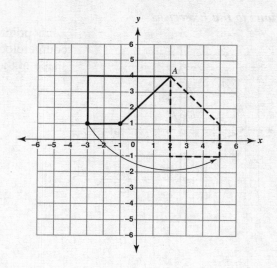

Each point of the quadrilateral is rotated 90° counterclockwise with point *A* as the center of the rotation.

Exercises

1. Look at the shape on the coordinate plane below.

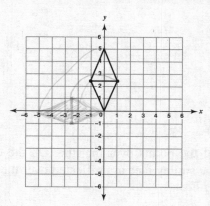

On the same coordinate plane, draw the rotation of the shape 90° counterclockwise about the origin.

2. Look at the shape on the coordinate plane below.

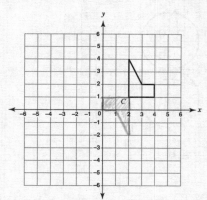

On the same coordinate plane, draw the rotation of the shape 180° about point *C*.

Solutions to the Exercises

1.

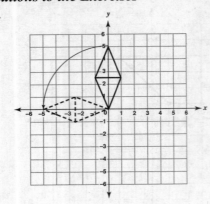

Each point in the shape must be rotated 90° counterclockwise about the origin. Spin the shape in the direction of the dotted arrow.

2.

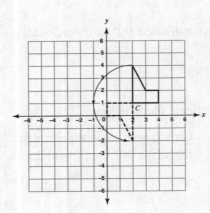

Each point of the shape must be rotated 180° around point *C*. This means that you spin the shape in the direction of the dotted arrow, first 90° and then another 90° to make the full 180°.

Reflections

A reflection across a line folds all of the points over the line. On the coordinate plane below, all of the points of the solid line smiley face are reflected over the *x*-axis to get the dotted line smiley face.

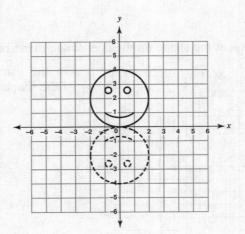

Any point above the *x*-axis in the solid line smiley face is below the *x*-axis the same distance in the dotted line smiley face. The two smiley faces are congruent to each other; the exact same size and same shape. To see how a reflection works, trace the solid line smiley face and the *x*-axis on a separate sheet of paper. Turn the traced smiley face upside down, placing the traced *x*-axis on top of the *x*-axis. This is a reflection over the *x*-axis.

The solid line smiley face on the coordinate plane below is reflected across the *y*-axis to give the dotted line smiley face.

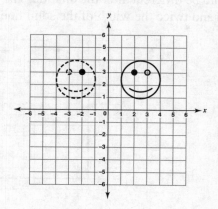

As you can see, the dotted line smiley face is congruent to the solid line smiley face.

Exercise

Look at the triangle on the coordinate plane below.

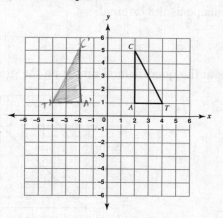

On the same coordinate plane, draw the reflection of the triangle across the *y*-axis. Be sure to label the new locations of points *C*, *A*, and *T* as *C′*, *A′*, and *T′*.

Solution to Exercise

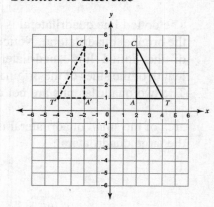

To reflect points across the *y*-axis, fold them across the *y*-axis and make them exactly the same distance away from the *y*-axis. The image points are labeled *C′*, *A′*, and *T′*. The triangle and its image are congruent.

Dilations

Dilations are changes that are made to the size of a shape. The image of a shape can be twice as large as the original shape or half as large as the original shape. The area of the shape would be different after the dilation. The dotted line rectangle shown below is twice the length and twice the width of the solid line rectangle.

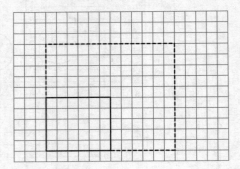

The rectangles are not congruent. Look at the area of each rectangle. The area of the small rectangle is 30. The area of the large rectangle is 120.

The large rectangle is twice the length and twice the width of the small rectangle, but the area of the large rectangle is four times the area of the small rectangle. Dilation does not create a rectangle the same size as the original rectangle but it does create a rectangle the same shape as the original.

Exercise

1. Look at the quadrilateral drawn on the grid below.

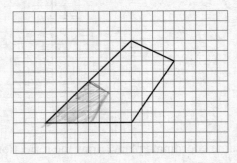

On the same grid, draw a quadrilateral that is half the size of the original quadrilateral.

Solution to Exercise

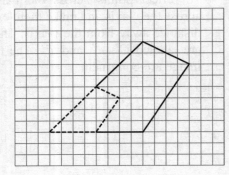

The dotted line quadrilateral is half the size of the original quadrilateral. Notice that each side of the dotted line quadrilateral is half the length of the sides of the original quadrilateral. The two quadrilaterals are not congruent. The area of the small quadrilateral is $\frac{1}{4}$ the size of the large quadrilateral.

Practice Problems

9. Look at the triangle shown on the coordinate plane below.

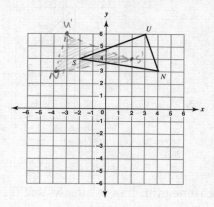

On the same coordinate plane, draw the reflection of the triangle over the y-axis. Label the new locations of points S, U, and N as S', U', and N'. Is $\triangle SUN$ congruent to $\triangle S'U'N'$? Yes

10. Look at the quadrilateral shown on the coordinate plane below.

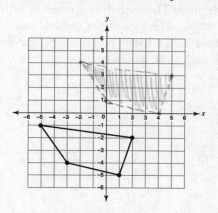

On the same coordinate plane, draw the translation of the quadrilateral five spaces up and three spaces to the right. Are the two quadrilaterals congruent? Yes

11. Look at the letter E on the coordinate plane below.

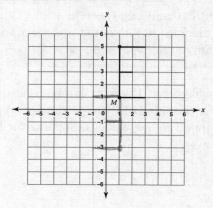

On the same coordinate plane, draw the letter E rotated 180° about point M. Are the two letters congruent?

Yes

12. Look at the letter M on the grid below.

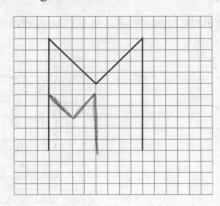

On the same grid, draw a letter M that is half the size of the original letter M. Are the two letters congruent? No

(Solutions are at the end of the chapter.)

PYTHAGOREAN THEOREM

The Pythagorean Theorem only applies to right triangles. A right triangle is a triangle with one right angle. The sides of a right triangle have special names. The side opposite the right angle is called the hypotenuse, c. The two sides that make the right angle are called the legs, a and b, as shown below.

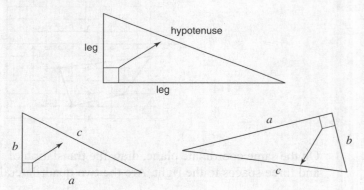

Definition: Pythagorean Theorem
In a right triangle the sum of the squares of the lengths of the legs is equal to the square of the length of the hypotenuse. $a^2 + b^2 = c^2$

The Pythagorean Theorem can be used to find the missing side when two sides are given. The easiest kind of Pythagorean Theorem problem involves finding the hypotenuse when you are given the two legs. Suppose the legs in a right triangle measure 5 and 12, as shown below.

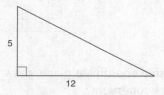

To find the length of the hypotenuse, use the Pythagorean Theorem.

$$5^2 + 12^2 = c^2$$
$$25 + 144 = c^2$$
$$169 = c^2$$
$$c = \sqrt{169} = 13$$

Sometimes problems are written only in words. What is the length of the diagonal of a square whose sides measure 8 centimeters? Always sketch a picture to help you understand the situation.

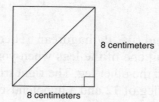

A square has sides that are 8 centimeters long, as shown above. Remember a square has four right angles and all of the sides of a square are the same length. The lengths of the sides of the square are the legs of the right triangle. The question is asking you to find the length of the hypotenuse. Use the Pythagorean Theorem.

$$8^2 + 8^2 = c^2$$
$$64 + 64 = c^2$$
$$128 = c^2$$
$$c = \sqrt{128} \approx 11.31$$

Use your calculator to get an approximation for $\sqrt{128}$.

Exercises

1. Look at the right triangle shown below.

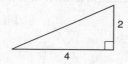

Find the length of the missing side.

2. A clock has a minute hand that is 4 inches long and an hour hand that is 3 inches long. What is the distance between the ends of the hands at 9 o'clock?

Solutions to Exercises

1. The Pythagorean Theorem can be used to find the hypotenuse because you are given the two legs.

$$2^2 + 4^2 = c^2$$
$$4 + 16 = c^2$$
$$20 = c^2$$
$$c = \sqrt{20} \approx 4.47$$

2.

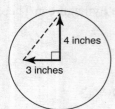

The figure to the left shows clock hands at 9:00 with the minute hand 4 inches long and the hour hand 3 inches long. At 9:00, the hands will meet at a right angle, so the Pythagorean Theorem can be used.

$$3^2 + 4^2 = c^2$$
$$9 + 16 = c^2$$
$$25 = c^2$$
$$c = 5$$

The other kind of Pythagorean Theorem problem involves finding one of the legs when you are given the hypotenuse and the other leg. The right triangle shown at the right has a leg of 12 and a hypotenuse of 15.

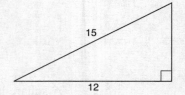

Use the Pythagorean Theorem to set up the problem, but this time you know c and you do not know one of the legs. It does not matter which letter you use as the missing letter when it is one of the legs.

$$a^2 + 12^2 = 15^2$$
$$a^2 + 144 = 225$$
$$a^2 = 81$$
$$a = 9$$

The missing leg is 9 units long.

Exercises

1. Look at the triangle shown below.

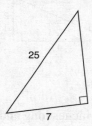

Find the length of the missing side.

2. The diagonal brace on a rectangular gate is 2 meters long. The height of the gate is 1 meter. How wide is the gate?

Solutions to Exercises

1. Set up the equation using the Pythagorean Theorem.

$$7^2 + b^2 = 25^2$$
$$49 + b^2 = 625$$
$$b^2 = 576$$
$$b = 24$$

When you have $b^2 = 576$, you may not know the answer. Use your calculator to find $\sqrt{576}$.

2. The figure below shows the problem information in a sketch.

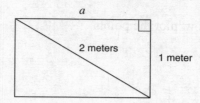

The gate is rectangular, so all of the angles are right angles. There is a triangle formed with the width that you do not know, the height that is 1 meter, and the hypotenuse (diagonal) that is 2 meters. Use the Pythagorean Theorem to find the width.

$$a^2 + 1^2 = 2^2$$
$$a^2 + 1 = 4$$
$$a^2 = 3$$
$$a = \sqrt{3} \approx 1.73$$

Practice Problems

13. Oscar is planning to paint his house. To make the ladder safe, the foot of the 17-foot-ladder should be placed 4 feet from the house, as shown below.

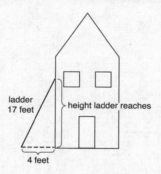

How far up the side of the house will the ladder reach?

14. Lucas walked 62 yards due north, then 30 yards due east, as shown in the diagram below.

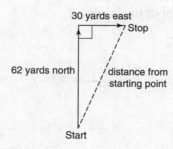

How far is Lucas from his starting point?

SAMPLE GQE QUESTIONS FOR CHAPTER 5

1. A rectangular garden is 6 meters wide and has a diagonal measuring 10 meters. Find the perimeter of the garden.

2. On the coordinate plane below, plot the points (−1, 1), (1, 3), and (4, 3).

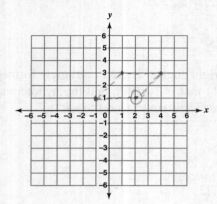

Now plot a fourth point so that the four points form a parallelogram.

3. Look at the triangle on the coordinate plane below.

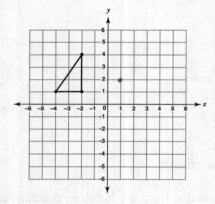

On the same coordinate plane, draw a rotation of the triangle 90° clockwise about the origin.

4. Look at the trapezoid on the coordinate plane below.

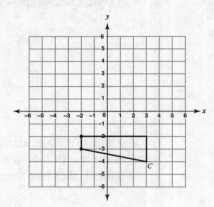

On the same coordinate plane, draw a reflection of the trapezoid over the x-axis and label the reflection of C as C′. Then rotate the new trapezoid 90° counterclockwise about point C′.

5. Look at the obtuse triangle shown below.

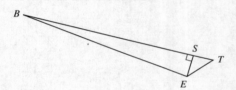

Which of the following statements is correct?

A. $\overline{ES}$ is a diagonal of $\triangle BET$.
B. $\overline{ES}$ is a perpendicular bisector of $\overline{BT}$.
C. $\overline{ES}$ is an altitude of $\triangle BET$.
D. $\overline{ES}$ is an angle bisector of $\angle BET$.

6. Look at the circle shown below.

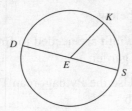

Which of the following statements is not correct?

A. $\overline{DS}$ is a diameter of circle E.
B. $\overline{DS}$ is a chord of circle E.
C. $\overline{EK}$ is a radius of circle E.
D. $\overline{EK}$ is a chord of circle E.

7. Look at the isosceles triangle on the coordinate plane below.

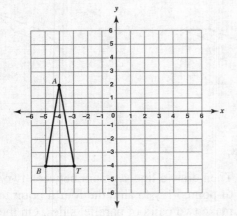

On the same coordinate plane, draw a translation of the isosceles triangle three units up and seven units to the right. Be sure to label the new locations of B, A, and T as B', A', and T'.

8. Look at the triangle shown below.

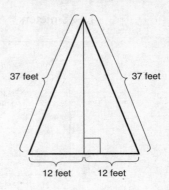

What is the length of the altitude of the triangle?

Answers Explained

1. **28 meters.** A drawing of the garden is shown below.

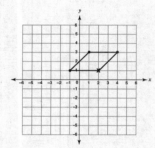

The length and width are needed to find the perimeter of the garden.

The width is 6. Use the Pythagorean Theorem to find the length.

$$6^2 + b^2 = 10^2$$
$$36 + b^2 = 100$$
$$b^2 = 64$$
$$b = 8$$

The length of the rectangular garden is 8 meters. Find the perimeter using the formula $P = 2l + 2w$.

$$P = 2(8) + 2(6) = 16 + 12 = 28$$

2.   or 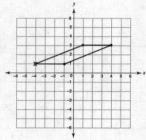 or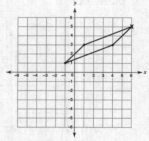

Graph the three points as shown below.

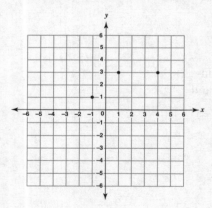

Then draw a line segment between two pairs of points. Try to find the fourth point to make two pairs of parallel sides. On the first coordinate plane, the fourth point is (2, 1). On the second coordinate plane the fourth point is (−4, 1). On the third coordinate plane, the fourth point is (6, 5).

3.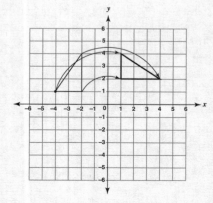

The bold triangle shown is the original triangle rotated 90° clockwise about the origin. The arrows are drawn to help you see the 90° rotation.

4.

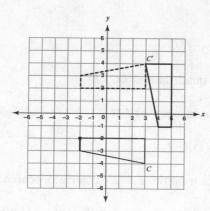

The final answer is the bold trapezoid. First, the trapezoid was reflected over the *x*-axis, as shown with the dotted line trapezoid. Then the dotted line trapezoid is rotated 90° counterclockwise about point *C′*.

5. **C** $\overline{ES}$ is an altitude of $\triangle BET$. In choice A, a triangle does not have any diagonals. In choice B, $\overline{ES}$ is perpendicular to $\overline{BT}$. $\overline{ES}$ does not bisect $\overline{BT}$. In choice C, $\overline{ES}$ is drawn from the vertex perpendicular to the opposite side, an altitude. In choice D, $\overline{ES}$ does not cut $\angle BET$ into two equal angles.

6. **D** $\overline{EK}$ is a chord of circle *E* is not correct. In choice A, $\overline{DS}$ is a diameter because it connects two points of the circle through the center *E*. In choice B, $\overline{DS}$ is a chord because it connects two points on the circle. A diameter is a special kind of chord. In choice C, $\overline{EK}$ is a radius because it connects the center of the circle to a point on the circle. In choice D, $\overline{EK}$ is not a chord because it does not connect two points on the circle.

7.

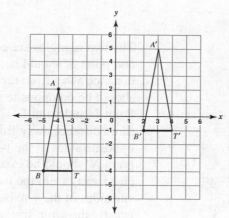

Move each point up three units and right seven units. Point *B* (−5, −4) moves to *B′* (2, −1). Point *A* (−4, 2) moves to *A′* (3, 5). Point *T* (−3, −4) moves to *T′* (4, −1).

8. 35 feet

The altitude in the isosceles triangle forms a right triangle, as shown below.

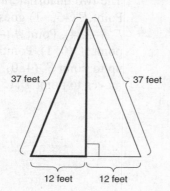

37 feet   37 feet

12 feet   12 feet

Use the Pythagorean Theorem to find the altitude, which is the leg that is missing in the right triangle.

$$12^2 + b^2 = 37^2$$
$$144 + b^2 = 1{,}369$$
$$b^2 = 1{,}225$$
$$b = 35$$

Solutions to Practice Problems

1. **A** $\overline{AD}$. $\angle GAB$ is bisected by $\overline{AD}$.

2. **D** $\overline{CE}$. $\overline{CE}$ is a diagonal in quadrilateral $BCKE$.

3. **C** $\overline{HI}$. $\overline{HI}$ is a perpendicular bisector of $\overline{LM}$.

4. **B** $\overline{FG}$. $\overline{FG}$ is an altitude in $\triangle AFJ$.

5. $\triangle SUM$ is an obtuse, isosceles triangle. $\triangle FUN$ is a right, scalene triangle.

6. No. Remember that the sum of the measures of the angles in a triangle is 180°. The sum of the measures of two of the angles of a triangle is 40° + 50° = 90°. That leaves 90° left for the measure of the third angle of the triangle. A triangle that has one angle exactly 90° is called a right triangle.

7. Choice A is true.
   Choice B is false. Quadrilateral $SING$ is a parallelogram and a rectangle.
   Choice C is false. Quadrilateral $GROW$ is a trapezoid.
   Choice D is false. Quadrilateral $WOLF$ is a parallelogram.

8. A. a sphere
   B. a rectangular prism
   C. a cylinder
   D. a rectangular prism

9.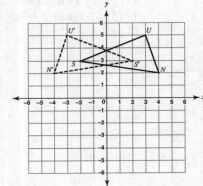

   The two triangles are congruent. Point S and point S' should be exactly the same distance away from the y-axis. They should be on opposite sides of the y-axis. Point U and point U' should be exactly the same distance away from the y-axis and on opposite sides of the y-axis. Point N and point N' should be exactly the same distance away from the y-axis and on opposite sides of the y-axis. The triangles are exactly the same size and shape, so they are congruent.

10.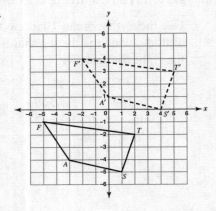

    The two quadrilaterals are congruent. Point F (−5, −1) goes up to point F' (−2, 4). Point A (−3, −4) slides to point A' (0, 1). Point S (1, −5) slides up to point S' (4, 0). Point T (2, −2) moves to point T' (5, 3).

11.

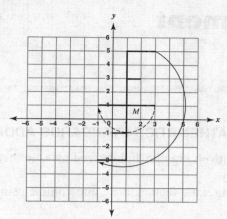

The two letters are congruent.

When you rotate a shape 180°, it does not matter whether you go clockwise or counterclockwise. Place your pencil point on letter M. Think of spinning the letter E halfway around without moving point M. Point M stays in exactly the same place while all of the other points spin around. The letters are still exactly the same size and shape.

12.

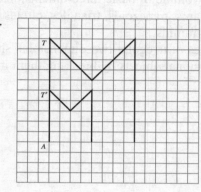

The two letters are not congruent.

Take each line segment of the large letter M and make the small letter M half the size. $\overline{AT}$ is 10 units long, so line segment AT' should be 5 units long. The rest of the line segments should be done exactly the same way, half the length of the large M.

13. $\sqrt{273} \approx 16.5$ feet

The ladder, the ground, and the house form a right triangle. Use the Pythagorean Theorem to find the height the ladder reaches.

$$4^2 + b^2 = 17^2$$
$$16 + b^2 = 289$$
$$b^2 = 273$$
$$b = \sqrt{273} \approx 16.5$$

14. $\sqrt{4744} \approx 68.88$ yards

Use the Pythagorean Theorem to find the hypotenuse.

$$62^2 + 30^2 = c^2$$
$$3844 + 900 = c^2$$
$$4744 = c^2$$
$$c = \sqrt{4744} \approx 68.87$$

Chapter 6 | **Measurement**

==

```
INDIANA'S ACADEMIC MATHEMATICS STANDARDS ADDRESSED:

8.5.2  Solve simple problems involving rate and derived measurements for attributes
       such as velocity and density.
8.5.3  Solve problems involving scale factors, area, and volume using ratio and
       proportion.
8.5.4  Use formulas for finding the perimeter and area of basic two-dimensional
       shapes and the surface area and volume of basic three-dimensional shapes,
       including rectangles, parallelograms, trapezoids, triangles, circles, prisms,
       cylinders, spheres, cones, and pyramids.
8.5.5  Estimate and compute the area of irregular two-dimensional shapes and the
       volume of irregular three-dimensional objects by breaking them into more basic
       geometric objects.
```

DISTANCE FORMULA

On the reference sheet the distance formula is given, $d = rt$, where d represents the distance traveled, r represents the rate at which you are traveling, and t represents the time traveled. The distance formula can be applied to distances driving a car or swimming laps in a pool. When driving a car, the distance is in miles or kilometers, the rate is in miles per hour or kilometers per hour, and the time is in hours. When swimming laps in a pool, the distance is in laps, the rate is in laps per minute, and the time is in minutes. Many story problems use the distance formula to find a missing piece of information.

Example
A truck is traveling 70 miles per hour. How long (in hours) will it take for the truck to travel 168 miles?

You are given the distance, 168 miles, and the rate, 70 miles per hour. Find the time it takes. Use the distance formula.

$$d = r \cdot t$$
$$168 = 70 \cdot t$$
$$t = 2.4$$

It would take the truck 2.4 hours.

Exercise

1. Who is going faster—an athlete running 100 meters in 9.09 seconds or a bicyclist riding 760 meters in 64.5 seconds?

Solution to Exercise

Calculate the rate of speed for the athlete.

$$d = r \cdot t$$
$$100 = r \cdot 9.09$$
$$r \approx 11.00 \text{ meters per second}$$

Calculate the rate of speed for the bicyclist.

$$d = r \cdot t$$
$$760 = r \cdot 64.5$$
$$r \approx 11.78 \text{ meters per second}$$

The bicyclist travels 11.78 meters in 1 second while the athlete runs 11.00 meters in 1 second. The bicyclist is going faster.

Practice Problem

1. On the indoor track, Spee Dee sprints at an average rate of 1.25 laps per minute. How many minutes will it take Spee Dee to sprint 40 laps?

(Solution is found at the end of the chapter.)

PROPORTIONS

Definition: A proportion is two ratios that are equal to each other.

Example: $\dfrac{4}{8} = \dfrac{3}{6}$

In the proportion $\dfrac{a}{b} = \dfrac{c}{d}$, a and d are called the extremes while b and c are called the means.

In a proportion, the product of the means is equal to the product of the extremes. Using the proportion $\dfrac{4}{8} = \dfrac{3}{6}$, the product of the means is $3 \cdot 8 = 24$ and the product of the extremes is $4 \cdot 6 = 24$.

"Solve a proportion" means to find the missing number or numbers. In the proportion $\dfrac{5}{x} = \dfrac{4}{8}$, the property of proportions can be used to find the value of x. The product of the means equals the product of the extremes.

$$4 \cdot x = 5 \cdot 8$$
$$4x = 40$$
$$x = 10$$

Exercise

Solve: $\dfrac{7}{5} = \dfrac{x}{9}$

Solution to Exercise

1. Use the property of proportions. The product of the means equals the product of the extremes.

$$5x = 63$$

$$x = \frac{63}{5} = 12\frac{3}{5}$$

SCALE FACTORS

A map is a visual representation of a geographical area. Maps are drawn to a scale, the ratio of the size on the map to the actual size. For instance, if the scale of the map is 1 inch represents 50 miles, then 2 inches would represent 100 miles and 225 miles would be represented by 4.5 inches.

> **Remember:** Figures that are the same shape (proportional sides) but not necessarily the same size are called similar figures.

Architects use scale drawings to represent a house or a shopping mall. If the scale on an architectural drawing is $\dfrac{1}{4}$ inch represents 1 foot, then 2 inches represent 8 feet, and 14 feet would be represented by $3\dfrac{1}{2}$ inches. The architectural drawing of a building is similar to the actual building.

Example

You are making a scale drawing of your dream house. You have chosen the scale, 1 inch represents 2 feet. The living room of your dream house is rectangular, 23 feet by 20 feet. What are the dimensions of the scale drawing of your living room? The dimensions of a room are the length and the width of the room.

$$\frac{\text{length of scale drawing}}{\text{length of actual room}} = \frac{1 \text{ inch}}{2 \text{ feet}} = \frac{x \text{ inches}}{23 \text{ feet}}$$

$$\frac{1}{2} = \frac{x}{23}$$

$$2x = 23$$
$$x = 11.5$$

So the scale drawing of the living room should be 11.5 inches long. Use the same process to find the width of the scale drawing of the living room.

$$\frac{1}{2} = \frac{x}{20}$$
$$2x = 20$$
$$x = 10$$

The scale drawing of the living room should be 10 inches wide. The dimensions of the scale drawing of the living room are 11.5 inches by 10 inches.

Using the dimensions of the living room of your dream house, find the ratio of the area of the scale drawing to the area of the actual living room.
Area of the scale drawing of the living room is 115 square inches.

$$A = l \cdot w$$
$$A = 11.5 \cdot 10$$
$$A = 115$$

Area of the actual living room is 460 square feet.

$$A = 23 \cdot 20$$
$$A = 460$$

The ratio of the area of the scale drawing of the living room to the area of the actual living room is not the same as the ratio of the length of the scale drawing of the living room to the length of the actual living room.

$$\frac{\text{length of scale drawing}}{\text{length of actual room}} = \frac{1 \text{ inch}}{2 \text{ feet}} = \frac{1}{2}$$

$$\frac{\text{area of the scale drawing}}{\text{area of the living room}} = \frac{115}{460} = \frac{1}{4}$$

> The ratio of the areas of similar figures is the square of the ratio of the sides.
>
> $$\frac{\text{Area of figure A}}{\text{Area of figure B}} = \left(\frac{\text{side of figure A}}{\text{corresponding side of figure B}} \right)^2$$

A similar rule can be made when comparing the volumes of similar objects.

> $$\frac{\text{Volume of object A}}{\text{Volume of object B}} = \left(\frac{\text{side of object A}}{\text{corresponding side of object B}} \right)^3$$

Example

A model of a sculpture is made using the scale 2 inches represent 3 feet. It takes 160 cubic inches of water to fill the sculpture model. How many cubic feet of water will it take to fill the actual sculpture? You do not need to know the shape of the sculpture.

$$\frac{\text{Volume of model}}{\text{Volume of sculpture}} = \left(\frac{\text{length of model}}{\text{corresponding length of sculpture}}\right)^3$$

$$\frac{160 \text{ cubic inches}}{\text{Volume of sculpture}} = \left(\frac{2 \text{ inches}}{3 \text{ feet}}\right)^3$$

$$\frac{160 \text{ cubic inches}}{V} = \frac{8 \text{ cubic inches}}{27 \text{ cubic feet}}$$

$$8V = 4{,}320$$
$$V = 540$$

You would need 540 cubic feet of water to fill the sculpture.

Exercises

1. The scale used on a map is 2 centimeters represent 5 kilometers. If the map distance between Mathville and Algebra City is 15 centimeters, what is the actual distance between Mathville and Algebra City?

2. The drawing of a playground uses the scale 2 centimeters represent 5 meters. The area of the drawing of the playground is 48 square centimeters. How many square meters of turf grass will be needed to cover the actual playground?

Solutions to Exercises

1. $\dfrac{\text{distance on the map}}{\text{actual distance}} = \dfrac{2 \text{ centimeters}}{5 \text{ kilometers}} = \dfrac{15 \text{ centimeters}}{x \text{ kilometers}}$

$$\frac{2}{5} = \frac{15}{x}$$
$$2x = 75$$
$$x = 37.5$$

2. The ratio of the areas of similar figures is the square of the ratio of the sides.

$$\frac{\text{Area of drawing of playground}}{\text{Area of playground}} = \left(\frac{2 \text{ centimeters}}{5 \text{ meters}}\right)^3$$

$$\frac{48}{A} = \left(\frac{2}{5}\right)^2$$

$$\frac{48}{A} = \frac{4}{25}$$

$$4A = 1{,}200$$
$$A = 300$$

Practice Problems

2. A model of a hot air balloon uses the scale 2 centimeters represent 7 meters. The model of the hot air balloon needed 56 cubic centimeters of a helium mixture. How much helium mixture would the real hot air balloon need?

3. The letters on the scale drawing of a billboard are 3.5 centimeters high. The letters on the actual billboard are 4.2 meters high. The area of the scale drawing of the billboard is 1,450 square centimeters. What is the area of the real billboard?

(Solutions are found at the end of the chapter.)

TWO-DIMENSIONAL SHAPES: PERIMETER AND CIRCUMFERENCE

Perimeter of a shape is the length of its boundary. To find the perimeter, add the lengths of the sides. In a circle, the distance around the circle is called the circumference, not the perimeter. A formula for circumference of a circle is given on the reference sheet.

Example
Find the perimeter of the shape shown below.

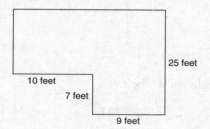

The lengths of all of the sides are not given, but there are enough lengths to be able to find the perimeter. The horizontal missing side must be the same length as the other two horizontal sides combined, 10 feet + 9 feet = 19 feet.

The vertical missing side must be the same length as the difference between the other two vertical sides, 25 feet − 7 feet = 18 feet.

Now you can find the perimeter of the shape.

$$P = 19 + 25 + 9 + 7 + 10 + 18$$
$$P = 88 \text{ feet}$$

Example
Find the perimeter of the trapezoid shown below.

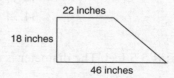

The lengths of three of the sides of the trapezoid are given, so you must find the length of the fourth side. The figure below shows the trapezoid separated in such a way that a right triangle is formed with sides 18, 24, and c. Use the Pythagorean Theorem to find the missing side, c.

$$18^2 + 24^2 = c^2$$
$$c^2 = 900$$
$$c = 30$$

$$P = 22 + 30 + 46 + 18$$
$$P = 116 \text{ inches}$$

Example

Find the circumference of the circle shown below.

20 yards

The circle has a diameter of 20. The diameter is twice the radius, so the radius of the circle shown is 10 yards.

$$C = 2 \cdot 3.14 \cdot 10$$
$$C = 62.8 \text{ yards}$$

Exercises

1. If the circumference of a circle is 34.54 feet, what is the radius of the circle? What is the diameter of the circle?

2. Find the distance around the shape shown below.

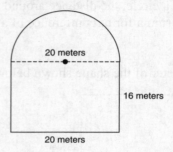

20 meters

16 meters

20 meters

3. Find the perimeter of the triangle shown below.

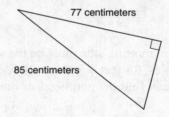

77 centimeters

85 centimeters

Solutions to Exercises

1. Use the formula for circumference of a circle and solve for *r*.

$$34.54 = 2 \cdot 3.14 \cdot r$$
$$34.54 = 6.28r$$
$$r = 5.5$$

The radius is 5.5 feet. The diameter is twice the radius, 11 feet.

2. The bottom of the shape is a rectangle with sides 16 meters and 20 meters. The top of the shape is half of a circle. The diameter of the circle is 20 meters, so the radius of the circle is 10 meters. Find the circumference of the entire circle and then take half of that amount.

$$C = 2 \cdot 3.14 \cdot 10$$
$$C = 62.8 \text{ meters}$$

Then, the circumference of half of the circle is half of 62.8, 31.4 meters.

$$\text{Distance around} = 16 + 20 + 16 + 31.4$$
$$\text{Distance around} = 83.4 \text{ meters}$$

3. Before finding the perimeter, use the Pythagorean Theorem to find the missing side.

$$a^2 + 77^2 = 85^2$$
$$a^2 + 5{,}929 = 7{,}225$$
$$a^2 = 1{,}296$$
$$a = 36$$

Now find the perimeter.

$$P = 77 + 85 + 36$$
$$P = 198 \text{ centimeters}$$

Practice Problem

4. Find the distance around the shape shown below.

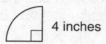 4 inches

(Solution is found at the end of the chapter.)

TWO-DIMENSIONAL SHAPES: AREA

The area of a shape is the amount of region that the shape encloses. If you want to put a fence around a field, then you are finding the perimeter of the field. If you want to paint a wall, then you are finding the area of the wall.

The units for perimeter or circumference are distances of feet, centimeters, or miles. The units for area are square feet, square centimeters, or square miles.

The base and the height are needed for the area of a triangle. In each triangle shown below, the base is $\overline{AB}$ and the height is $\overline{CD}$.

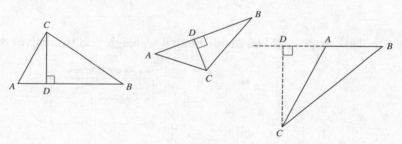

The height or the altitude of a triangle is drawn from a vertex of the triangle to the opposite side. The side that the height is drawn to is called the base.

The bases and the height are needed for the area of a trapezoid. The height is the distance between the two parallel sides or bases. In each trapezoid *MATH*, shown below, the bases are $\overline{MA}$ and $\overline{TH}$.

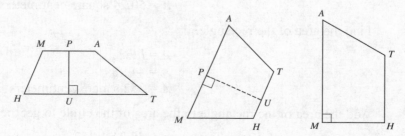

In the first two trapezoids, the height is $\overline{UP}$. In the third trapezoid, the height is $\overline{MH}$.

The base and the height are needed to find the area of a parallelogram. Find the area of the parallelogram shown below.

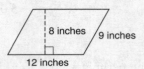

The base is 12 inches and the height is 8 inches. The 9 inches is given to confuse you.

$$A = b \cdot h$$
$$A = 12 \cdot 8$$
$$A = 96 \text{ square inches}$$

The radius is needed to find the area of a circle. Find the area of the circle shown below.

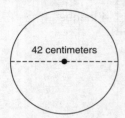

The diameter is 42 centimeters, so the radius is 21 centimeters.

$$A = \pi \cdot r^2$$
$$A = 3.14 \cdot 21^2$$
$$A = 3.14 \cdot 441$$
$$A = 1{,}384.74 \text{ square centimeters}$$

Example

Find the area of the shape shown below.

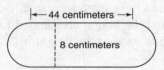

This shape could be divided into a rectangle and two halves of a circle, as shown below.

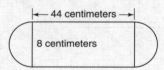

Put the two halves of the circle together to make an entire circle with a diameter of 8 centimeters. Find the area of a circle with radius 4 centimeters.

$$A = \pi \cdot r^2$$
$$A = 3.14 \cdot 4^2$$
$$A = 3.14 \cdot 16$$
$$A = 50.24 \text{ square centimeters}$$

Find the area of the rectangle.

$$A = l \cdot w$$
$$A = 44 \cdot 8$$
$$A = 352 \text{ square centimeters}$$

Add the area of the rectangle to the area of the circle to get the total area.

$$\text{Total Area} = 50.24 + 352$$
$$\text{Total Area} = 402.24 \text{ square centimeters}$$

Exercises

1. Find the area of the circle shown below.

34 miles

2. Look at the diagram below.

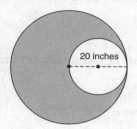

20 inches

The radius of the large circle is the same as the diameter of the small circle. Find the area of the shaded portion of the circle.

3. If the circumference of a circle is 169.56 feet, find the radius, the diameter, and the area of the circle.

Solutions to Exercises

1. The problem gives the diameter of the circle as 34 miles. Radius is half of the diameter, so the radius is 17 miles.

$$A = \pi \cdot r^2$$
$$A = 3.14 \cdot 17^2$$
$$A = 3.14 \cdot 289$$
$$A = 907.46 \text{ square miles}$$

2. The area of the shaded portion of the circle is the area of the large circle minus the area of the small circle. The large circle has a radius of 20 inches.

$$A = \pi \cdot r^2$$
$$A = 3.14 \cdot 20^2$$
$$A = 3.14 \cdot 400$$
$$A = 1,256 \text{ square inches}$$

The small circle has a radius of 10 inches.

$$A = \pi \cdot r^2$$
$$A = 3.14 \cdot 10^2$$
$$A = 3.14 \cdot 100$$
$$A = 314 \text{ square inches}$$

Shaded Area = Area of the Large Circle – Area of the Small Circle
Shaded Area = 1,256 – 314
Shaded Area = 942 square inches

3. Use the formula for circumference of a circle to find the radius of the circle.

$$C = 2 \cdot \pi \cdot r$$
$$169.56 = 2 \cdot 3.14 \cdot r$$
$$6.28r = 169.56$$
$$r = 27 \text{ feet}$$

Once you have the radius, then find the diameter, 54 feet. Then, use the formula for the area of a circle.

$$A = \pi \cdot r^2$$
$$A = 3.14 \cdot 27^2$$
$$A = 3.14 \cdot 729$$
$$A = 2,289.06 \text{ square feet}$$

Practice Problems

5. The area of the trapezoid shown below is 150 square centimeters.

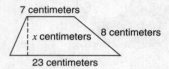

Find the height of the trapezoid.

6. Which is the bigger slice of pie: one fourth of a 7-inch diameter pie, or one eighth of a 10-inch diameter pie?

7. Find the area of the shape shown below.

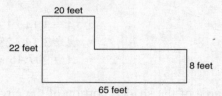

(Solutions are found at the end of the chapter.)

THREE-DIMENSIONAL GEOMETRIC FIGURES: SURFACE AREA

The surface area of a basketball is the amount of leather it would take to cover the ball. The volume of a basketball is the amount of air it would take to fill the ball. The surface area of a rectangular prism is the amount of paper it would take to cover the box. The volume of a rectangular prism is the amount you could put inside the box. The units for surface area are square inches, square feet, or square miles. The units for volume are cubic inches, cubic feet, or cubic miles.

To find the surface area of a rectangular prism, you need the length, the width, and the height. To find the surface area of a general prism, you need the sum of the areas of the faces. The faces are the flat surfaces that make up the prism. In a rectangular prism the faces are all rectangles. In a general prism the two faces that are parallel and congruent are called the bases. The other faces of a general prism are parallelograms.

Example

Find the surface area of the triangular prism shown below.

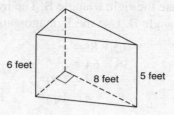

The surface area of the triangular prism is the sum of the areas of the faces. The faces of the triangular prism are shown in the figures below.

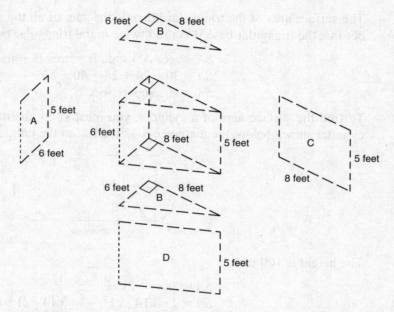

Shape A is a rectangle with length 5 feet and width 6 feet. The shape looks like a parallelogram in the three-dimensional drawing, but it is a rectangle.

$$A = 5 \cdot 6$$
$$A = 30 \text{ square feet}$$

Shape B is the base in the shape of a right triangle. Notice that there are two congruent bases, so there will be two areas that are the same.

$$A = \frac{1}{2} \cdot 6 \cdot 8$$

$$A = 24 \text{ square feet}$$

Shape C is a rectangle with length 8 feet and width 5 feet. The shape looks like a parallelogram in the three-dimensional drawing, but it is a rectangle.

$$A = 8 \cdot 5$$
$$A = 40 \text{ square feet}$$

Shape D is a rectangle with width 5 feet but an unknown length. The shape looks like a parallelogram in the three-dimensional drawing, but it is a rectangle. To find the length of rectangle D, use the right triangle B. The hypotenuse of right triangle B is the same as the length of rectangle D. Use the Pythagorean Theorem to find the hypotenuse.

$$6^2 + 8^2 = c^2$$
$$36 + 64 = c^2$$
$$c^2 = 100$$
$$c = 10$$

Now find the area of rectangle D.

$$A = 10 \cdot 5$$
$$A = 50 \text{ square feet}$$

The surface area of the triangular prism is the sum of all the areas of the bases. Remember that the triangular base B is used twice in the triangular pyramid.

$$SA = \text{area A} + \text{area B} + \text{area B} + \text{area C} + \text{area D}$$
$$SA = 30 + 24 + 24 + 40 + 50$$
$$SA = 168 \text{ square feet}$$

To find the surface area of a cylinder, you must know the radius and the height. In the cylinder shown below, the diameter is 42 inches, so the radius is 21 inches.

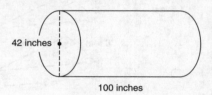

42 inches

100 inches

The height is 100 inches.

$$SA = 2\pi r^2 + 2\pi rh$$
$$SA = 2 \cdot 3.14 \cdot 21^2 + 2 \cdot 3.14 \cdot 21 \cdot 100$$
$$SA = 2 \cdot 3.14 \cdot 441 + 2 \cdot 3.14 \cdot 21 \cdot 100$$
$$SA = 2{,}769.48 + 13{,}188$$
$$SA = 15{,}957.48 \text{ square inches}$$

To find the surface area of a sphere, you need the radius. In the sphere shown below, the radius is 9 inches.

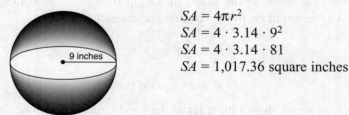

9 inches

$$SA = 4\pi r^2$$
$$SA = 4 \cdot 3.14 \cdot 9^2$$
$$SA = 4 \cdot 3.14 \cdot 81$$
$$SA = 1{,}017.36 \text{ square inches}$$

Exercises

1. Find the surface area of the hemisphere shown below.

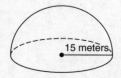

15 meters

2. A cylindrical storage tank has a diameter of 9 meters and a height of 30 meters, as shown below.

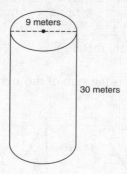

The owner needs to paint the sides and the top of the tank, but not the base of the tank. What is the area to be painted?

3. What is the surface area of the trapezoidal prism shown below?

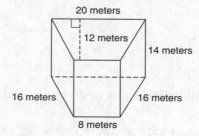

Solutions to Exercises

1. To find the surface area of the hemisphere, add half of the surface area of the sphere and the area of the base on which the hemisphere sits.

$$SA \text{ of sphere} = 4 \cdot 3.14 \cdot 15^2$$
$$SA \text{ of sphere} = 4 \cdot 3.14 \cdot 225$$
$$SA \text{ of sphere} = 2{,}826 \text{ square meters}$$

The base on which the hemisphere sits is a circle with radius 15 meters.

$$A = \pi \cdot 15^2$$
$$A = 3.14 \cdot 225$$
$$A = 706.5 \text{ square meters}$$

Add half of the surface area of the sphere to the area of the circle to get the total surface area of the hemisphere.

$$SA = \frac{1}{2} \text{ surface area of sphere} + \text{area of circle}$$

$$SA = \frac{1}{2} \cdot 2{,}826 + 706.5$$

$$SA = 2{,}119.5 \text{ square meters}$$

2. The area to be painted on the tank is the surface area of the cylinder minus the area of one of the bases. The formula for the surface area of a cylinder is given on the reference sheet. Since you do not need the area of one of the bases, the formula will change to add only one of the bases.

$$SA = 1\pi r^2 + 2\pi r h$$

Remember if the diameter is 9 meters, then the radius is 4.5 meters.

$$SA = 1 \cdot 3.14 \cdot 4.5^2 + 2 \cdot 3.14 \cdot 4.5 \cdot 30$$
$$SA = 3.14 \cdot 20.25 + 2 \cdot 3.14 \cdot 4.5 \cdot 30$$
$$SA = 63.585 + 847.8$$
$$SA = 911.385 \text{ square meters}$$

3. The surface area of a trapezoidal prism is the sum of the areas of the faces. The faces are shown in the figures below.

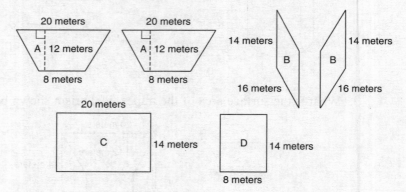

Trapezoid A has bases of 20 meters and 8 meters with a height of 12 meters.

$$\text{Area of trapezoid A} = \frac{1}{2}(8 + 20) \cdot 12$$

$$\text{Area of trapezoid A} = \frac{1}{2}(28) \cdot 12$$

Area of trapezoid A = 168 square meters

Rectangle B has length 14 meters and width 16 meters.

Area of rectangle B = $l \cdot w$
Area of rectangle B = $14 \cdot 16$
Area of rectangle B = 224 square meters

Rectangle C has length 20 meters and width 14 meters.

Area of rectangle C = $20 \cdot 14$
Area of rectangle C = 280 square meters

Rectangle D has length 14 meters and width 8 meters.

Area of rectangle D = $14 \cdot 8$
Area of rectangle D = 112 square meters

$SA = 2 \cdot \text{trapezoid A} + 2 \cdot \text{area rectangle B} + \text{area rectangle C} + \text{area rectangle D}$
$SA = 2 \cdot 168 + 2 \cdot 224 + 280 + 112$
$SA = 336 + 448 + 280 + 112$
$SA = 1{,}176 \text{ square meters}$

Practice Problems

8. Rosy wants to paint the inside of the lidless box shown below.

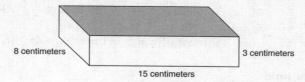

How many square centimeters of surface will Rosy paint?

9. Look at the cylindrical peg shown below.

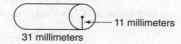

Sam Tailspin wants to paint the peg for his model airplane. What is the surface area in square millimeters of the peg?

(Solutions are found at the end of the chapter.)

THREE-DIMENSIONAL GEOMETRIC FIGURES: VOLUME

The volume of a three-dimensional figure is the amount of space enclosed by the figure. The volume of a cylindrical can is the amount of space inside the can.

To find the volume of a pyramid, you need the area of the base and the height. The Great Pyramid of Giza in Egypt has a square base about 756 feet long. Its height, when first built, was 481 feet. Use the formula for the area of a rectangle to find the area of the square.

$$\text{Area of square base} = 756 \cdot 756$$
$$\text{Area of square base} = 571{,}536 \text{ square feet}$$

Use the formula for the volume of a regular pyramid.

$$V = \frac{1}{3} \cdot 571{,}536 \cdot 481$$

$$V = 91{,}636{,}272 \text{ cubic feet}$$

Example
A spherical water tank is half full of water. The diameter of the tank is 12 feet. To find the amount of water in the tank, find half of the volume of the sphere.

$$\text{Volume of sphere} = \frac{4}{3}\pi r^3$$

Remember to use the radius in the formula, not the diameter.

$$\text{Volume of sphere} = \frac{4}{3} \cdot 3.14 \cdot 6^3$$

$$\text{Volume of sphere} = \frac{4}{3} \cdot 3.14 \cdot 216$$

$$\text{Volume of sphere} = 904.32 \text{ cubic feet}$$

The tank is only half filled with water. Take half of the volume of the sphere to find the amount of water in the tank.

$$\text{Amount of water in the tank} = \frac{1}{2} \cdot 904.32$$

$$\text{Amount of water in the tank} = 452.16 \text{ cubic feet}$$

Example

An 11-ounce can of corn is $3\frac{1}{2}$ inches high with a 3-inch diameter. To find the amount of corn in the can, you need to find the volume of the can. The shape of a can is a cylinder.

$$V = \pi r^2 h$$

Remember to use the radius of the can, not the diameter.

$$V = 3.14 \cdot 1.5^2 \cdot 3.5$$
$$V = 3.14 \cdot 2.25 \cdot 3.5$$
$$V = 24.7275 \text{ cubic inches}$$

Example

Find the volume of the cone shown below.

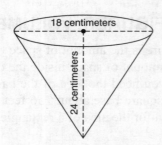

$$\text{Volume of cone} = \frac{1}{3}\pi r^2 h$$

$$\text{Volume of cone} = \frac{1}{3} \cdot 3.14 \cdot 9^2 \cdot 24$$

$$\text{Volume of cone} = \frac{1}{3} \cdot 3.14 \cdot 81 \cdot 24$$

$$\text{Volume of cone} = 2{,}034.72 \text{ cubic centimeters}$$

Exercises

1. A cylindrical pipe at an oil refinery is 200 feet long. The diameter of the pipe is 6 inches. How many cubic feet of oil can this pipe hold?

2. The base of a general prism has an area of 25 square meters. The prism is 4.2 meters high. Find the volume of the prism.

3. Find the volume of the cone shown below.

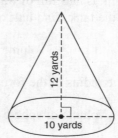

Solutions to Exercises

1. The amount of oil the pipe can hold is the volume of the cylindrical pipe. The diameter of the pipe is 6 inches so the radius of the pipe is 3 inches. All of the units must be the same when calculating the volume. Change the radius of the pipe, 3 inches, into $\frac{3}{12} = \frac{1}{4} = .25$ of a foot.

$$V = 3.14 \cdot .25^2 \cdot 200$$
$$V = 3.14 \cdot .0625 \cdot 200$$
$$V = 39.25 \text{ cubic feet}$$

2. Volume of prism = *Bh*. *B* represents the area of the base of the prism.

$$\text{Volume of prism} = 25 \cdot 4.2$$
$$\text{Volume of prism} = 105 \text{ cubic meters}$$

3. Volume of cone = $\frac{1}{3}\pi r^2 h$. Remember to use the radius of the cone, not the diameter.

$$\text{Volume of cone} = \frac{1}{3} \cdot 3.14 \cdot 5^2 \cdot 12$$

$$\text{Volume of cone} = \frac{1}{3} \cdot 3.14 \cdot 25 \cdot 12$$

$$\text{Volume of cone} = 314 \text{ cubic yards}$$

Practice Problems

10. A conical ice cream cone at Jen and Berry's has a radius of 6 centimeters and a height of 18 centimeters. A spherical scoop of ice cream has a diameter of 12 centimeters. If the ice cream melted inside the cone, would the cone overflow? Ignore the volume change with the phase shift from solid to liquid.

11. A swimming pool with a rectangular base is 50 meters long and 25 meters wide. It is filled to a height of 1.5 meters. How many cubic meters of water must be added to fill the pool to a height of 2 meters?

12. A swimming pool with a circular base has a diameter of 15 feet and a depth of 6 feet. The water level is 1.5 feet below the rim of the pool. Find the volume of water in the pool.

13. Look at the square pyramid shown below.

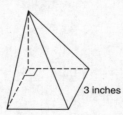

3 inches

The height of the square pyramid is 5 inches. Find the volume of the pyramid.

(Solutions are found at the end of the chapter.)

SAMPLE GQE QUESTIONS FOR CHAPTER 6

1. To resurface a football field, workers dug down 6 inches and removed all of that dirt. How many cubic feet of dirt are necessary to fill the field to its original depth? A football field is a rectangle, 65 yards wide and 120 yards long.

2. The speed of sound at sea level is 340.3 meters per second. The speed of sound 20,000 meters above sea level is 295.1 meters per second. How much faster will sound travel a distance of 10,000 meters at sea level than at 20,000 meters above sea level? Round your answer to the nearest tenth of a second.

3. A model of a statue is constructed using the scale of 2 centimeters represent 3 decimeters. If the volume of the scale model statue is 20 cubic centimeters, what is the volume of the actual statue?

4. One base of a trapezoid is 12 inches long and the height is 16 inches long. If the area is 248 square inches, find the length of the other base.

   A. 19 inches
   B. 25 inches
   C. 152 inches
   D. 484 inches

5. Phen Cit is going to put a fence around his backyard. A diagram of the backyard is shown below.

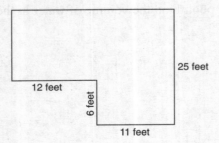

How many square feet are in Phen Cit's backyard?

   A. 96 square feet
   B. 300 square feet
   C. 347 square feet
   D. 503 square feet

6. Find the area of the shape shown below.

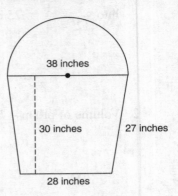

Show All Work

7. A basketball has a diameter of 24 centimeters. Find the amount of leather needed to cover the basketball.

8. Find the volume of the cone shown below.

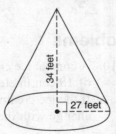

Show All Work

9. A mouse found a piece of cheese that is shaped like a prism with a triangular base. The height of the cheese is 2 inches. The area of the base of the prism is 3.5 square inches. Find the volume of the piece of cheese.

Show All Work

10. Find the volume of the pyramid shown below.

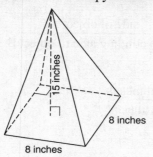

15 inches

8 inches

8 inches

Show All Work

 11. Find the volume of the hemisphere shown below.

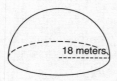

18 meters

Show All Work

 12. The Math Teachers Retirement Home has just installed a circular fountain 8 meters in diameter. The United Math Teachers Organization has requested that a 1.5-meter wide path be paved around the fountain. If paving costs the city $10 per square meter, find the cost of the paved ring around the fountain.

Show All Work

Answers Explained

1. **35,100 cubic feet.** Your answer is to be given in cubic feet. Change all of the units into feet. The width of the football field is 195 feet. The length of the football field is 360 feet. The depth of the resurfacing is .5 feet.

   The amount of dirt necessary to fill the field is the volume of a rectangular prism with length 360 feet, width 195 feet, and height .5 feet.

   $$\text{Volume} = 360 \cdot 195 \cdot .5$$
   $$\text{Volume} = 35,100 \text{ cubic feet}$$

2. **4.5 seconds.** Use the distance formula to find the time it takes for sound to travel a distance of 10,000 meters at sea level.

   $$d = rt$$
   $$10,000 = 340.3 \cdot t$$
   $$t \approx 29.386 \text{ seconds}$$

   Use the distance formula to find the time it takes for sound to travel a distance of 10,000 meters when it is 20,000 meters above sea level.

   $$10,000 = 295.1 \cdot t$$
   $$t \approx 33.887 \text{ seconds}$$

   The difference is $33.887 - 29.386 = 4.501$ seconds.

3. 81 cubic decimeters. The ratio of the volumes of similar figures is not the same as the ratio of the sides.

$$\frac{\text{Volume of object A}}{\text{Volume of object B}} = \left(\frac{\text{side of object A}}{\text{corresponding side of object B}}\right)^3$$

$$\frac{24 \text{ cubic centimeters}}{\text{Volume of actual statue}} = \left(\frac{2 \text{ centimeters}}{3 \text{ decimeters}}\right)^3$$

$$\frac{24}{V} = \frac{8}{27}$$
$$8V = 648$$
$$V = 81 \text{ cubic decimeters}$$

4. **A** 19 inches

$$248 = \frac{1}{2}(12 + b) \cdot 16$$

$$248 = 8(12 + b)$$
$$248 = 8b + 96$$
$$248 = 8b + 96$$
$$8b = 152$$
$$b = 19 \text{ inches}$$

5. **D** 503 square feet. Divide the shape into rectangles, as shown below.

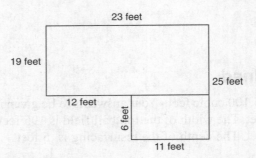

Total Area = Area of large rectangle + Area of small rectangle
Total Area = $23 \cdot 19 + 11 \cdot 6$
Total Area = $437 + 66$
Total Area = 503 square feet

6. 1,556.77 square inches

$$\text{Area of trapezoid} = \frac{1}{2}(38 + 28) \cdot 30$$

$$\text{Area of trapezoid} = \frac{1}{2}(66) \cdot 30$$

$$\text{Area of trapezoid} = 990 \text{ square inches}$$

$$\text{Area of circle} = 3.14 \cdot 19^2$$
$$\text{Area of circle} = 3.14 \cdot 361$$
$$\text{Area of circle} = 1,133.54 \text{ square inches}$$

$$\text{Total Area} = \text{Area of trapezoid} + \frac{1}{2} \text{ Area of circle}$$

$$\text{Total Area} = 990 + \frac{1}{2} \cdot 1{,}133.54$$

$$\text{Total Area} = 1{,}556.77 \text{ square inches}$$

7. 1,808.64 square centimeters. The amount of leather to cover the basketball is the surface area of the sphere with diameter 24 centimeters. The radius of the sphere is 12 centimeters.

$$SA = 4 \cdot 3.14 \cdot 12^2$$
$$SA = 4 \cdot 3.14 \cdot 144$$
$$SA = 1{,}808.64 \text{ square centimeters}$$

8. 25,942.68 cubic feet

$$V = \frac{1}{3} \cdot 3.14 \cdot 27^2 \cdot 34$$

$$V = \frac{1}{3} \cdot 3.14 \cdot 729 \cdot 34$$

$$V = 25{,}942.68 \text{ cubic feet}$$

9. 7 cubic inches

$$V = Bh$$
$$V = 3.5 \cdot 2$$
$$V = 7 \text{ cubic inches}$$

10. 320 cubic inches

$$\text{Volume of a regular pyramid} = \frac{1}{3} Bh$$

B represents the area of the base of the pyramid. The base of the pyramid is a square with side of 8 inches. To find the area of the base, square the length of the side.

$$\text{Volume of a regular pyramid} = \frac{1}{3} \cdot 8^2 \cdot 15$$

$$\text{Volume of a regular pyramid} = \frac{1}{3} \cdot 64 \cdot 15$$

$$\text{Volume of a regular pyramid} = 320 \text{ cubic inches}$$

11. 12,208.32 cubic meters

$$V = \frac{4}{3} \cdot 3.14 \cdot 18^3$$

$$V = \frac{4}{3} \cdot 3.14 \cdot 5{,}832$$

$$V = 24{,}416.64 \text{ cubic meters}$$

Half of the sphere would have half of the volume = 12,208.32 cubic meters.

12. **$447.45** The diagram below shows the fountain and the path around it.

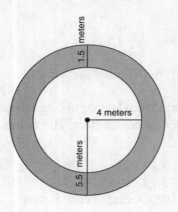

Since the diameter of the fountain is 8 meters, the radius of the fountain is 4 meters. The radius of the large circle is 5.5 meters because it combines the radius of the fountain and the width of the path. The shaded part of the diagram is the path around the fountain. To find the area of the path, subtract the area of the large circle minus the area of the small circle.

Area of large circle = $3.14 \cdot 5.5^2$
Area of large circle = $3.14 \cdot 30.25$
Area of large circle = 94.985 square meters

Area of small circle = $3.14 \cdot 4^2$
Area of small circle = $3.14 \cdot 16$
Area of small circle = 50.24 square meters

Area of path = Area of large circle − Area of small circle
Area of path = 94.985 − 50.24
Area of path = 44.745

Each square meter of paving costs $10. Multiply the number of square meters by $10 to get the total cost of $447.45.

Solutions to Practice Problems

1. **32 minutes.** Use the distance formula putting 40 in for the distance and 1.25 in for the rate.

$$d = r \cdot t$$
$$40 = 1.25 \cdot t$$
$$t = 32$$

2. **2,401 cubic meters.** When measuring the amount needed to fill a three-dimensional object, volumes are used.

$$\frac{\text{Volume of model hot air balloon}}{\text{Volume of actual hot air balloon}} = \left(\frac{2 \text{ centimeters}}{7 \text{ meters}} \right)^3$$

$$\frac{56 \text{ cubic centimeters}}{\text{Volume of actual hot air balloon}} = \frac{8 \text{ cubic centimeters}}{343 \text{ cubic meters}}$$

$$\frac{56}{V} = \frac{8}{343}$$
$$8V = 19,208$$
$$V = 2,401$$

3. 2,088 square meters

$$\frac{\text{Area of drawing of billboard}}{\text{Area of billboard}} = \left(\frac{3.5 \text{ centimeters}}{4.2 \text{ meters}}\right)^2$$

$$\frac{1450}{A} = \left(\frac{3.5}{4.2}\right)^2$$

$$\frac{1450}{A} = \frac{12.25}{17.64}$$

$$12.25A = 25,578$$

$$A = 2,088$$

4. 14.28 inches
   The shape is one-fourth of a circle, as shown below.

Find the circumference of the circle and then take one-fourth of the circumference.

$$C = 2 \cdot 3.14 \cdot 4$$
$$C = 25.12 \text{ inches}$$

One-fourth of the circumference would be 6.28 inches. To find the distance around, add the two radii and the part of the circumference.

$$\text{Distance around} = 6.28 + 4 + 4$$
$$\text{Distance around} = 14.28 \text{ inches}$$

5. 10 centimeters

$$150 = \frac{1}{2}(23 + 7) \cdot h$$

$$150 = \frac{1}{2}(30) \cdot h$$

$$150 = 15 \cdot h$$
$$h = 10 \text{ centimeters}$$

6. One eighth of a 10-inch diameter pie is bigger.
   The bigger piece of pie would have the bigger area. Find the area of each piece of pie and then compare them. Remember to use the radius not the diameter when finding the area.

$$\text{Area of } \frac{1}{4} \text{ of 7-inch diameter pie} = \frac{1}{4} \cdot 3.14 \cdot 3.5^2$$

$$\text{Area of } \frac{1}{4} \text{ of 7-inch diameter pie} = \frac{1}{4} \cdot 3.14 \cdot 12.25$$

$$A = 9.61625 \text{ square inches}$$

$$\text{Area of } \frac{1}{8} \text{ of 10-inch diameter pie} = \frac{1}{8} \cdot 3.14 \cdot 5^2$$

$$\text{Area of } \frac{1}{8} \text{ of 10-inch diameter pie} = \frac{1}{8} \cdot 3.14 \cdot 25$$

$$A = 9.8125 \text{ square inches}$$

7. 800 square feet
    Break down this shape into two rectangles, as shown in either figure below.

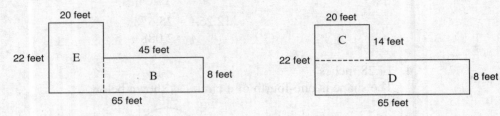

Using the first figure, add the area of rectangle E and the area of rectangle B.

$$\text{Area of rectangle E} = 22 \cdot 20$$
$$\text{Area of rectangle E} = 440 \text{ square feet}$$

$$\text{Area of rectangle B} = 45 \cdot 8$$
$$\text{Area of rectangle B} = 360 \text{ square feet}$$

The total area is the sum of the area of rectangle E and the area of rectangle B.

$$\text{Total area} = 440 + 360$$
$$\text{Total area} = 800 \text{ square feet}$$

Using the second figure, add the area of rectangle C and the area of rectangle D.

$$\text{Area of rectangle C} = 20 \cdot 14$$
$$\text{Area of rectangle C} = 280 \text{ square feet}$$

$$\text{Area of rectangle D} = 65 \cdot 8$$
$$\text{Area of rectangle D} = 520 \text{ square feet}$$

The total area is the sum of the area of rectangle C and the area of rectangle D.

$$\text{Total Area} = 280 + 520$$
$$\text{Total Area} = 800 \text{ square feet}$$

8. 258 square centimeters. Rosy will paint five faces, two rectangles with length 8 centimeters and width 3 centimeters, two rectangles with length 15 centimeters and width 3 centimeters, and one rectangle with length 8 centimeters and width 15 centimeters.

$$SA = 2 \cdot 8 \cdot 3 + 2 \cdot 15 \cdot 3 + 1 \cdot 8 \cdot 15$$
$$SA = 48 + 90 + 120$$
$$SA = 258 \text{ square centimeters}$$

9. 2901.36 square millimeters

$$SA = 2\pi r^2 + 2\pi rh$$
$$SA = 2 \cdot 3.14 \cdot 11^2 + 2 \cdot 3.14 \cdot 11 \cdot 31$$
$$SA = 2 \cdot 3.14 \cdot 121 + 2 \cdot 3.14 \cdot 11 \cdot 31$$
$$SA = 759.88 + 2,141.48$$
$$SA = 2,901.36 \text{ square millimeters}$$

10. The ice cream cone would overflow with ice cream.

$$\text{Volume of ice cream cone} = \frac{1}{3} \cdot 3.14 \cdot 6^2 \cdot 18$$

$$\text{Volume of ice cream cone} = \frac{1}{3} \cdot 3.14 \cdot 36 \cdot 18$$

$$\text{Volume of ice cream cone} = 678.24 \text{ cubic centimeters}$$

$$\text{Volume of scoop of ice cream} = \frac{4}{3} \cdot 3.14 \cdot 6^3$$

$$\text{Volume of scoop of ice cream} = \frac{4}{3} \cdot 3.14 \cdot 216$$

$$\text{Volume of scoop of ice cream} = 904.32 \text{ cubic centimeters}$$

The volume of the scoop of ice cream is more than the volume of the ice cream cone, so the cone would overflow.

11. 625 cubic meters of water. The amount of water to be added will be the volume of a rectangular prism that is 50 meters long, 25 meters wide, and .5 meter high. It is .5 meter high because you want to increase the height from 1.5 meters to 2 meters.

$$\text{Volume of water} = 50 \cdot 25 \cdot .5$$
$$\text{Volume of water} = 625 \text{ cubic meters}$$

12. 794.8125 cubic feet. The shape of the swimming pool is a cylinder with a radius of 7.5 feet and a height of 6 feet. The water level is 1.5 feet below the rim of the pool, so the height of the water is 6 feet − 1.5 feet = 4.5 feet.

$$V = 3.14 \cdot 7.5^2 \cdot 4.5$$
$$V = 3.14 \cdot 56.25 \cdot 4.5$$
$$V = 794.8125 \text{ cubic feet}$$

13. 15 cubic inches

$$\text{Volume of pyramid} = \frac{1}{3} Bh$$

B is the area of the base. The base is a square with a side of 3 inches, so the area of the base is 9 square inches.

$$\text{Volume of pyramid} = \frac{1}{3} \cdot 9 \cdot 5$$

$$\text{Volume of pyramid} = 15 \text{ cubic inches}$$

Chapter 7 | Data Analysis

INDIANA'S ACADEMIC MATHEMATICS STANDARDS ADDRESSED:

8.6.1. Identify claims based on statistical data and, in simple cases, evaluate the reasonableness of the claims. Design a study to investigate the claim.

8.6.2 Identify different methods of selecting samples, analyzing the strengths and weaknesses of each method, and the possible bias in a sample or display.

8.6.3 Understand the meaning of, and be able to identify or compute the minimum, the lower quartile, the median, the upper quartile, the interquartile range, and the maximum of a data set.

8.6.4 Analyze, interpret, and display single- and two-variable data in appropriate bar, line, and circle graphs; stem-and-leaf plots and box-and-whisker plots and explain which types of display are appropriate for various data sets.

SURVEYS AND SAMPLE POPULATIONS

A survey is a statistical study of a small group of people called a sample population. The results of surveys are used to make generalizations about larger groups of people. Surveys are used to predict the winner of an election before all of the results have been counted. The designer of a survey must choose the sample population very carefully or the results of the survey will be invalid.

Example

A survey is taken to determine what kind of music people like best. Cray Zee took the survey at the entrance to an orchestra concert. The results of the survey showed that more people enjoy classical music than any other kind of music. Cray Zee's survey results are biased because he did not select a representative sample. Sum Art took the survey at a supermarket while people waited in the checkout line. Sum Art's results are totally different because her sample was made up of a broader range of people.

Exercise

Spay Cee opened a men's shoe store. She needed to determine the sizes of shoes to put in her store. She used the shoe sizes of the players for the Indiana Pacers basketball team as a basis for her purchases of men's shoes. Spay Cee was surprised to find that she had not purchased shoes that were the correct size for her shoppers. Explain how Spay Cee's sample choice created biased results.

Solution to Exercise

The Indiana Pacers' basketball team is not a representative sample of men's shoe size. The Pacers' sample is not likely to include people who are short and is likely to include mainly tall people.

Practice Problems

1. Carol is conducting a survey to determine what kind of dessert people like best. She conducts her first survey outside Jen and Berry's Ice Cream Store. Explain how Carol's choice of location could produce a biased sample.

(Solution is found at the end of the chapter.)

STATISTICAL CLAIMS

A statement that is based upon statistics is called a statistical claim. Some people misuse statistics and make unrealistic claims from very little information.

Amy surveyed 10 people at her high school and found that 9 of them love Mama Joan's Pizza. Amy claims that over 90 percent of the 50,000 people in her town love Mama Joan's Pizza. This claim is not reasonable because Amy only spoke with 10 high school students and there are 50,000 people in her town.

The number and types of stuffed animals at U R Toy Store is shown in the bar graph below.

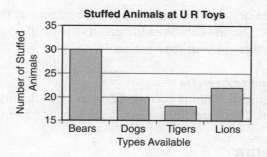

By looking at the bars on the graph, Suzy claims that there are three times as many bears as there are dogs. The bar graph is misleading because the scale on the vertical axis starts at 15 instead of 0. Notice the difference in the same information shown in the bar graph below.

Using this bar graph, Suzy would not have made the incorrect claim.

Exercise

The bar graph below shows the results of a survey in which juniors and seniors were asked if they were in choir.

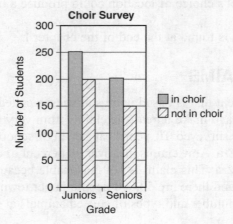

The seniors claimed that a greater percentage of the seniors participate in choir than the juniors. Use the data from the graph to explain whether the claim is correct. Include both percentages in your explanation.

Solution to Exercise

The seniors' claim is correct. There are 250 juniors in choir out of 450 juniors, about 56 percent. There are 200 seniors in choir out of 350 seniors, about 57 percent.

Practice Problems

2. The bar graph below shows the results of a survey in which teenagers, adults, and senior citizens were asked if they can or cannot drive a car.

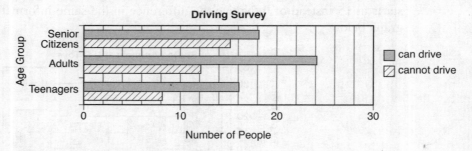

Classify each claim as true or false. Include numbers to justify your reasoning.

A. A larger percentage of adults can drive a car than teenagers.
B. More than 50 percent of the senior citizens drive a car.

(Solutions are found at the end of the chapter.)

LINE GRAPHS AND CIRCLE GRAPHS

Data plotted as points on a graph can be connected to form a line graph. The line makes it easy to see a trend over time.

The life expectancy of a baby is given in the line graph below.

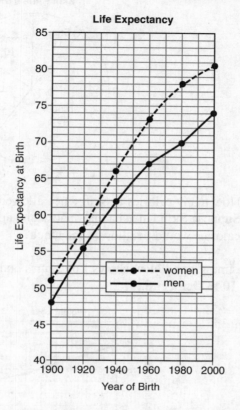

Lots of information is available on a line graph. Life expectancy of both men and women is increasing. Women's life expectancy is increasing more quickly than men's life expectancy. The life expectancy of a baby boy born in 1960 is about 67 years. The life expectancy of a baby girl born in 1960 is about 73 years.

Data is graphed as parts of a whole on a circle graph. The entire circle represents 100 percent or all of the data. The different parts of the circle are called sectors.

Ree Derr's 120 books are classified, as shown in the circle graph below.

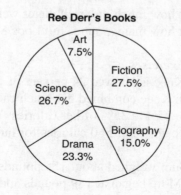

Lots of information is available from a circle graph. Art books compose 7.5 percent of Ree Derr's 120 books or 9 books. Science and drama books compose about 50 percent of all of her books.

Exercises

1. The circle graph below shows the number of calories Skih Knee allows for each meal on her diet of 1,600 calories.

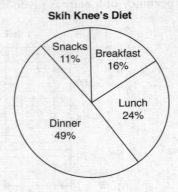

Skih Knee's Diet

A. How many calories is Skih Knee allowed for lunch?

B. Suppose Skih Knee had no snacks and applied those calories to lunch. How many calories would be allowed for lunch?

2. The line graph below shows Lou Sum's and Loos More's weight loss from September 10 to October 8.

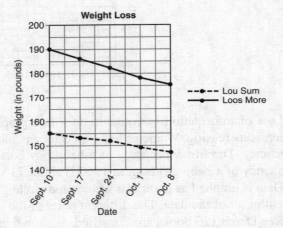

A. About how much did each dieter weigh on September 10?

B. About how many pounds did Loos More lose over the time shown?

Solutions to Exercises

1. A. Skih Knee is allowed 24 percent of 1,600 or 384 calories for lunch.

 B. If Skih Knee combined her lunch calories and her snack calories, she would have 11% + 24% = 35% of her calories for lunch. Skih Knee would eat 35 percent of 1,600 calories or 560 calories for lunch.

2. A. Lou Sum weighed about 155 pounds and Loos More weighed about 189 pounds.

 B. Loos More began at 189 pounds and ended at 175 pounds, so he lost 14 pounds.

Practice Problems

3. The results of a high school survey concerning chocolate are shown below.

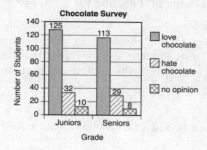

A. What percent of juniors love chocolate?
B. What percentage of all juniors and seniors love chocolate?

4. Jane has been keeping track of her dog Spot's weight for the first 18 months of his life, as shown below.

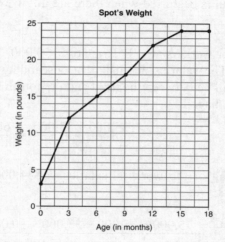

A. How much did Spot weigh at 6 months?
B. For which 3-month time period did Spot's weight increase the most? By how much did it increase?

5. The results of a survey concerning the number of movies people saw last year are given in the circle graph below.

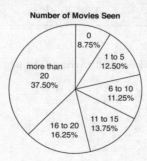

A. What percent of the people surveyed say that they saw five or fewer movies last year?
B. If 2,000 people were surveyed, how many said they saw more than 20 movies last year?

(Solutions are found at the end of the chapter.)

MEAN, MEDIAN, AND MODE

People who collect data need a way to describe the data using a number. The mean, median, and mode all describe the data, giving three different kinds of information.

> **Definition:** The mean is the average of all the numbers. To find the mean, add all of the numbers and divide by how many numbers were added.

The mean is best to use when there are no extreme values.

> **Definition:** The median is the middle number when the numbers are listed in order. If there are two numbers in the middle, then average the two middle numbers.

The median is best used when there are a few extreme measures in the data.

Example

Salaries for workers at U R Toys are $150,000, $75,000, $35,000, $33,000, $30,000, and $18,000. The salaries are listed in descending order. Since there are six salaries, the median will be between the third and the fourth salaries. Average $35,000 and $33,000 to get the median.

$$\frac{35,000 + 33,000}{2} = \frac{68,000}{2} = 34,000$$

The median of the workers' salaries is $34,000. The mean of the workers' salaries is $56,833.33.

$$\frac{150,000 + 75,000 + 35,000 + 33,000 + 30,000 + 18,000}{6} = \frac{341,000}{6} = 56,833.33$$

The median is a better description of the workers' salaries due to the influence of the high pay of two workers.

> **Definition:** The mode is the number that occurs most frequently. There may be no mode, one mode, or several modes.

The mode is best used when you want to describe what happens most.

Example

The following table gives the shoe sizes of all of the shoes sold on Monday at Footsy Shoe Store.

Shoes Sizes Sold on Monday				
$7\frac{1}{2}$	7	$7\frac{1}{2}$	5	7
9	10	8	$8\frac{1}{2}$	$5\frac{1}{2}$
6	7	7	$7\frac{1}{2}$	6
6	7	7	5	$7\frac{1}{2}$

The mode of the list of shoe sizes is 7 because six pairs of shoes size 7 were sold. The median and the mean would have little meaning when dealing with shoe sizes.

Exercise

The speeds of the fastest turtles at the World Championship Turtle Race are listed in the table below.

Speeds of Leatherback Turtles				
21.9	22.1	21.7	21.4	21.8
21.7	21.6	22.3	21.5	22.1

Find the mean, median, and mode of the turtles' speeds.

Solution to Exercise

To find the mean, add all of the times and divide that answer by 10.

$$\frac{\text{sum of all times}}{10} = \frac{218.1}{10} = 21.81$$

To find the median, list the numbers in numerical order and find the middle one.

21.4, 21.5, 21.6, 21.7, 21.7, 21.8, 21.9, 22.1, 22.1, 22.3

Since 21.7 and 21.8 are in the middle, add them together and divide by 2.

$$\frac{21.7 + 21.8}{2} = \frac{43.5}{2} = 21.75$$

The mode is the number that occurs most frequently. There are 2 modes because both 21.7 and 22.1 occur twice.

Practice Problem

6. Five crates are loaded onto an empty delivery truck. The weight of each crate is listed in the table below.

Weight of Crates (in pounds)					
34.8	39.8	39.8	40.2	46.1	50.1

Find the mean, median, and mode of the weights.

(Solution is found at the end of the chapter.)

QUARTILES AND THE INTERQUARTILE RANGE

The median is the middle of a set of numbers listed in order. The median is shown in the list of numbers below.

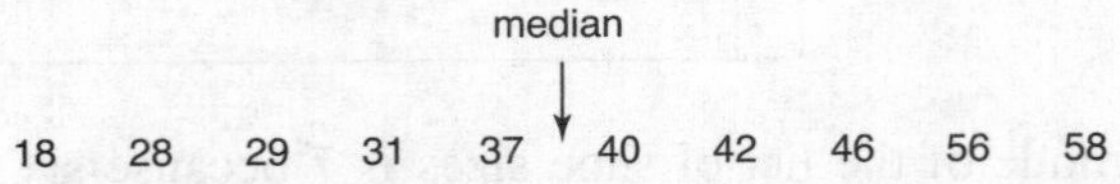

median
↓
18   28   29   31   37   40   42   46   56   58

The median is halfway between 37 and 40, $38\frac{1}{2}$. There are 5 numbers above the median and 5 numbers below the median.

Definition: The lower quartile is the middle or median of the bottom half of the list of numbers.

Definition: The upper quartile is the middle or the median of the top half of the list of numbers.

Definition: The interquartile range is the difference between the upper and the lower quartiles.

The median, the upper quartile, and the lower quartile are all shown below.

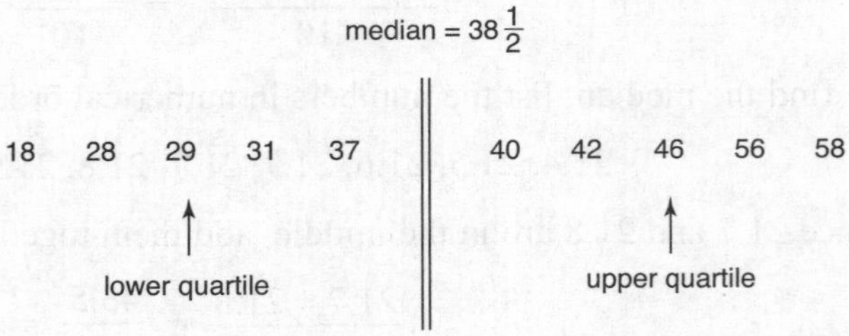

median = $38\frac{1}{2}$

18   28   29   31   37 | 40   42   46   56   58

↑   ↑
lower quartile   upper quartile

The lower quartile is 29. The upper quartile is 46. The interquartile range is $46 - 29 = 17$. The minimum value is 18. The maximum value is 58.

Example

The scores on a Business Test are given below.

Scores on Business Test									
84	89	76	65	74	93	82	68	76	94
73	85	89	91	74	63	83	80	80	70

Before finding the median or quartiles, the numbers must be placed in numerical order, as shown below.

Scores on Business Test

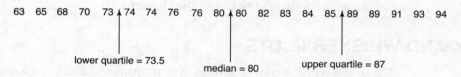

The median is 80. There will be 10 numbers above the median and 10 numbers below the median. The lower quartile is 73.5. In the bottom half of the numbers, there will be 5 numbers above the lower quartile and 5 numbers below the lower quartile. The upper quartile is 87. In the top half of the numbers, there will be 5 numbers above the upper quartile and 5 numbers below the upper quartile. The interquartile range is 87 – 73.5 = 13.5.

Exercise

Carol's bowling scores from last month are listed in ascending order.

105, 121, 123, 124, 133, 134, 135, 136, 137, 138, 138, 139, 141, 153, 174, 188

What are the lower quartile, the median, and the upper quartile of all of the bowling scores? What is the interquartile range?

Solution to Exercise

Carol's Bowling Scores

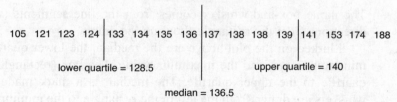

Since there are 16 scores in the list, the middle number will be halfway between the eighth and the ninth numbers, 136 and 137. The median is 136.5. Since there are eight scores in the bottom half, the lower quartile, the median of the bottom half of the scores will be halfway between the third and the fourth numbers, 124 and 133. The lower quartile is 128.5. Since there are eight scores in the top half, the upper quartile will be halfway between the third and the fourth numbers of the top half, 139 and 141. The upper quartile is 140. The interquartile range is 140 – 128.5 = 11.5.

Practice Problems

Mr. Math T. Cher collected the average monthly temperatures in Mathville for a year.

65 66 66 66 68 69 70 71 75 76 77 78

7. What are the lower quartile, the median, and the upper quartile of the temperatures? What is the interquartile range?

8. What percent of the temperatures are below the median of 69.5?

9. What percent of the temperatures are between the lower quartile and the upper quartile?

(Solutions are found at the end of the chapter.)

BOX-AND-WHISKER PLOTS

A box-and-whisker plot is another way to display data. A box-and-whisker plot is effective when you want to show how the data is distributed around the median.

The heights of the students in Mr. Math T. Cher's class are listed:

75 72 71 71 70 69 69 68 67 65 65 65 63 62 61

The median is 68, the lower quartile is 65, and the upper quartile is 71.
The box-and-whisker plot displaying this data is shown below.

Heights of Students in Mr. Math T. Cher's Class

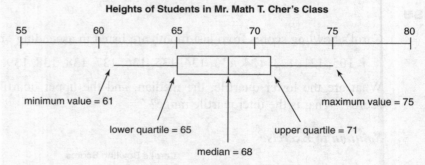

The name box-and-whisker comes from the line segments (whiskers) that extend both directions from the rectangle (box) on the graph.

Marked on the plot above are the median, the lower quartile, the upper quartile, the minimum value, and the maximum value. The box (rectangle) is drawn from the lower quartile to the upper quartile. The median is a mark made inside the rectangle. The whiskers are drawn from the left of the rectangle to the minimum value and from the right of the rectangle to the maximum value.

Example

Hamerin Homer played first base for the Mathville team for 13 years. During each year of his career, he hit a lot of home runs as listed below.

12 14 20 21 25 29 30 30 31 32 32 39 46

The median of the home runs is 30, the lower quartile is 20.5, and the upper quartile is 32.

To make a box-and-whisker plot, draw a number line showing numbers from the minimum value to the maximum value. Below the number line, draw a rectangle from the lower quartile to the upper quartile. Place a vertical line in the rectangle to mark the median. Draw a whisker from the left side of the rectangle to the minimum value. Draw the other whisker from the right side of the rectangle to the maximum value. The box-and-whisker plot should look like the one shown below.

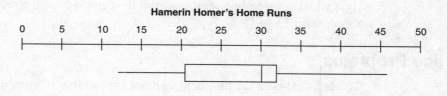

Hamerin Homer's Home Runs

Exercises

1. Test scores from two different classes are shown in the box-and-whisker plot below.

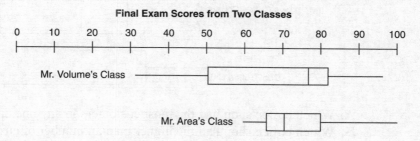

Final Exam Scores from Two Classes

    A. Which class had the higher median score?

    B. What was the lower quartile in Mr. Volume's class?

    C. In which class were the scores grouped more closely around its median?

2. Make a box-and-whisker plot for the data given below.

Points Scored by the Players on the Mathville High School Basketball Team											
0	0	3	4	4	5	7	8	12	14	15	18

Solutions to Exercises

1. A. Mr. Volume's class had a higher median score.

    B. The lower quartile in Mr. Volume's class was 50.

    C. Mr. Area's class was more closely grouped around the median.

2.

Points Scored by Players on the Mathville High School Basketball Team

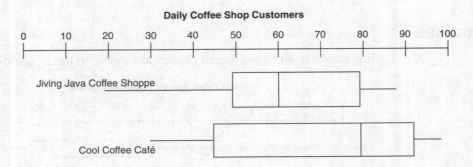

The median is 6, the lower quartile is $3\frac{1}{2}$, and the upper quartile is 13. The rectangle goes from the lower quartile to the upper quartile, from $3\frac{1}{2}$ to 13. The median, 6, is a vertical line segment drawn inside the rectangle. The left whisker is the line segment from the left side of the rectangle to the minimum value, from $3\frac{1}{2}$ to 0. The right whisker is the line segment from the right side of the rectangle to the maximum value, from 13 to 18.

Practice Problems

10. The box-and-whisker plot below shows the number of customers in the coffee shops in Mathville.

Daily Coffee Shop Customers

Jiving Java Coffee Shoppe

Cool Coffee Café

A. Which coffee shop had the most customers in any one day?
B. Which coffee shop had the higher median number of customers?
C. What is the lower quartile at the Jiving Java Coffee Shoppe?

11. Make a box-and-whisker plot for the information given below.

Mathville Baseball Team Batting Averages								
.123	.144	.156	.255	.257	.263	.274	.292	.298

What is the interquartile range?

(Solutions are found at the end of the chapter.)

STEM-AND-LEAF PLOTS

A stem-and-leaf plot is a way to organize data and display the data. A stem-and-leaf plot is helpful when analyzing data for clusters and gaps.

Remember the problem concerning Hamerin Homer's home runs during his 13-year baseball career.

12 14 20 21 25 29 30 30 31 32 32 39 46

The stem-and-leaf plot below displays the same data.

Hamerin Homer's Home Runs

stem	leaf
1	2 4
2	0 1 5 9
3	0 0 1 2 2 9
4	6

Key
2 | 0 = 20

The name stem-and-leaf comes from separating the data into the first part of the number, called the stem, and the part of the number that is left, called the leaf.

Marked on the figure below are the stems, the leaves, the largest value, the smallest value, and the key.

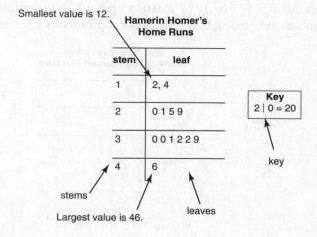

Smallest value is 12.

Hamerin Homer's Home Runs

stem	leaf
1	2, 4
2	0 1 5 9
3	0 0 1 2 2 9
4	6

Key
2 | 0 = 20

key

stems

Largest value is 46.

leaves

Example

The data given below are the cholesterol readings of the teachers at Mathville High School.

Cholesterol Levels of Mathville High School Teachers									
146	177	185	186	188	192	197	199	200	201
208	208	208	209	215	216	232	234	234	259

The stems for this data will be the hundreds and tens columns, while the leaves will be the units digits. Think of the number 146 as 14│6 and 177 as 17│7. The stems will go from 14 to 25. The beginning of the stem-and-leaf plot will look like the figure below.

**Cholesterol Levels of
Mathville High School Teachers**

Stem	Leaf
14	
15	
16	
17	
18	
19	
20	
21	
22	
23	
24	
25	

The stems are all listed in the first column. Now list all of the leaves in the appropriate second column. The number 146 or 14│6 means to put a 6 in the row beside the 14. The next number 177 or 17│7 means to put a 7 in the column beside the 17. Once all of the numbers are placed, the stem-and-leaf plot looks like the figure below.

**Cholesterol Levels of
Mathville High School Teachers**

Stem	Leaf
14	6
15	
16	
17	7
18	5 6 8
19	2 7 9
20	0 1 8 8 8 9
21	5 6
22	
23	2 4 4
24	
25	9

Key
14│6 = 146

Remember to title your stem-and-leaf plot and to provide a key for the plot.

Exercises

1. The stem-and-leaf plot below shows the number of points scored during all of the Mathville High School's girls' basketball games.

Number of Points Scored by Mathville High School Girls Basketball Team

Stem	Leaves
4	8 9 9
5	2 3 7 9
6	1 2 2 5 7 9
7	1 1 3 3 4 5 8 9
8	1 2 3 5 5 5
9	
10	2

Key
4|8 = 48

A. How many games are represented?
B. What was the highest number of points scored by the team?
C. What is the mode?

2. Mathville lies near a fault in Earth's crust. The following data are the magnitudes of the earthquakes over the past 20 years.

Magnitude of Earthquakes in Mathville

2.5	2.6	2.8	3.1	3.2	3.3	3.3	3.8	3.9	4.1
4.1	4.1	4.3	4.5	4.7	4.8	5.0	5.1	5.1	5.3

Make a stem-and-leaf plot for the data.

Solutions to Exercises

1. A. There are 28 games represented. Even though there is a stem of 9 there are no leaves. There were no scores in the nineties.
   B. The highest number of points scored by the team was 102.
   C. The mode was 85.

2. **Magnitude of Earthquakes in Mathville**

Stem	Leaves
2	5 6 8
3	1 2 3 3 8 9
4	1 1 1 3 5 7 8
5	0 1 1 3

Key
2|5 = 2.5

Practice Problems

12. Coach Hoops kept track of the percent of free throws one of his players, Mack Shack, made each game of the season. The results are shown below.

Percent of Free Throws Made									
85	45	66	72	46	55	59	56	46	43
24	64	47	43	41	79	53	60	39	57

Organize the data in a stem-and-leaf plot. Be sure to include a title and a key.

A. In how many games did Mack Shack score on at least 60 percent of his free throws?

B. What is the median percent of scoring free throws that Mack Shack made per game?

13. The numbers of games won by each team in major league baseball in 2005 are displayed in the back-to-back stem-and-leaf plot below.

Number of Games Won

American League Key 6|5 = 56	American League		National League	National League Key 6|7 = 67
		6	5	
	9 7	6	7 7	
	9 4 1	7	1 3 5 7 9	
	8 3 0	8	1 1 2 3 3 8 9	
	9 5 5 5 3	9	0	
		10	0	

The number of wins for the teams in the American League is listed on the left with the National League listed on the right.

A. What is the fewest number of games won by any National League team?

B. What is the mode for the number of games won by any National League team?

(Solutions are found at the end of the chapter.)

SAMPLE GQE QUESTIONS FOR CHAPTER 7

1. Mr. Math T. Cher kept track of the number of competitions that took place each week at Mathville High School. He displayed his results in the line graph shown below.

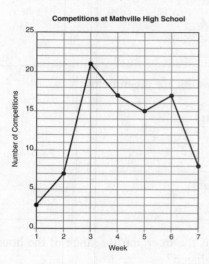

Which interval has the largest decrease in competitions?

A. week 2 to week 3
B. week 3 to week 4
C. week 4 to week 5
D. week 6 to week 7

2. Anna Variable surveyed the juniors and seniors at Mathville High School to determine which they liked better, geometry or algebra. Her results are shown below.

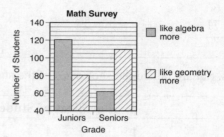

Anna Variable stated that twice as many juniors liked algebra more than geometry because the algebra bar was twice as tall as the geometry bar. Use the data from the graph to explain why Anna's claim is incorrect and how the bar graph could be changed to more accurately depict the data. Be sure to include data in your explanation.

3. Mr. Oval is going to conduct a survey of adults on developing a dog park. Which of the following methods of sampling would provide the most unbiased results?

A. survey customers leaving a pet store
B. survey workers at a humane society fundraiser
C. survey people whose phone numbers end in 3
D. survey shoppers in the pet aisle at the grocery store

4. The heights of students in Mr. Triangle's math class are displayed below

Heights of Mr. Triangle's Students (in inches)																
60	60	61	62	62	63	64	65	66	66	67	68	70	72	73	75	77

What is the upper quartile of the heights?

A. 70
B. 66
C. 71
D. 77

5. Ms. Square tallied the shoe sizes of the students in her class. The results are given below.

Shoe Sizes of Students in Ms. Square's Class										
4	5	$5\frac{1}{2}$	6	$6\frac{1}{2}$	7	8	$8\frac{1}{2}$	10	12	$12\frac{1}{2}$

Make a box-and-whisker plot of the data.

6. Students in Ms. Drive's class were asked the distance from home to school. The results of the survey are given in the stem-and-leaf plot below.

Distance from Home to Mathville High School

0	5 6 8 9
1	0 2 5 7 8
2	1 2 2 3
3	5 6
4	1 2 5
5	
6	7 8 9 9
7	
8	0 1 2 3 5
9	0 1 1
10	5

Key
6 | 7 = 6.7

A. How many students travel more than 7 miles to school?
B. How many students travel less than 2 miles to school?

7. What is the interquartile range of the hourly wages listed?

$6.50	$6.53	$6.57	$6.83
$6.91	$6.98	$7.46	$7.55

A. $6.55
B. $7.22
C. $.67
D. $.37

8. A survey was taken of favorite board games played by 500 Mathville residents. The results are shown in the circle graph below.

Favorite Mathville Games

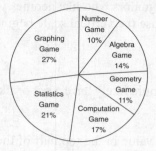

How many people surveyed chose the Algebra game as their favorite board game?

A. 14
B. 35
C. 70
D. There is not enough information given.

9. Mr. Zup R. Teech tallied his students scores on their first quiz. The results are shown below.

Scores on the Quiz				
50	46	43	39	35
48	45	42	39	33
48	45	40	37	31
48	44	40	35	30

An absent student took the quiz and received a 50.

A. Explain how the score of the absent student will affect the mean of the scores.
B. Explain how the score of the absent student will affect the median of the scores.
C. Explain how the score of the absent student will affect the mode of the scores.

10. The Biology class found the height, in feet, of 20 buildings in Mathville.

123 154 134 117   106 145 176 188 145 193
118 130 140 153   127 120 149 137 195 191

Use the data to fill in the stem-and-leaf plot below.

Building Heights

Stem	Leaf

Key
12 | 3 = 123

Answers Explained

1. **D** Week 6 to week 7
   The competitions decreased from 17 to 8.

2. There are 120 juniors who like algebra more and 80 juniors who like geometry more. The graph does not show a fair representation because the vertical scale begins at 40 instead of 0. Make the vertical scale begin at 0.

3. **C** Survey people whose phone numbers end in 3
   All of the other choices involve people who would more likely have a pet or like pets.

4. **C** 71
   The median is 66. The upper quartile is the middle value in the top half of the numbers. There are 8 numbers above the median so the middle value is halfway between 70 and 72.

5.

Shoe Sizes

The median is 7, the lower quartile is $5\frac{1}{2}$, and the upper quartile is 10. The minimum value is 4 and the maximum value is $12\frac{1}{2}$.

6.
   A. Nine students travel more than 7 miles.
   B. Nine students travel less than 2 miles.

7. **C** $.67
   The median of the hourly wages is $6.87. The upper quartile is $7.22 and the lower quartile is $6.55. The interquartile range is $7.22 – $6.55 = $.67.

8. **C** 70
   The Algebra game won 14 percent of the 500 people surveyed.

$$.14 \cdot 500 = 70$$

9. A. The mean will be raised. If a top score is added to the data, the mean will be raised. If a bottom score is added to the data, the mean will be lowered. It is not necessary to calculate the mean.
   B. The median will be raised. The median is halfway between the tenth and eleventh number, so it is 41. If a number is added above the median, then the score will move to the eleventh number that is now higher. If a number is added below the median, then the score will move to the eleventh number that is now lower.
   C. The mode will not be affected. Since the mode is 48 and it happens three times, the extra 50 will not affect the mode. The mode could have changed if the additional score would have made a number repeat three times.

10.

Building Heights

Stem	Leaf
10	6
11	7 8
12	0 3 7
13	0 4 7
14	0 5 5 9
15	3 4
16	
17	6
18	8
19	1 3 5

Key
12 | 3 = 123

Solutions to Practice Problems

1. The people at an ice cream parlor are not a representative sample. The parlor sample is not likely to include people who dislike ice cream and is likely to include people who love ice cream.

2. A. False. There are 24 out of 36 adults who drive a car, about 67 percent. There are 16 out of 24 teenagers who drive a car, about 67 percent. Adults and teenagers have the same percentages.
   B. True. There are 18 out of 33 senior citizens who drive a car, about 55 percent.

3. A. There are 125 juniors who love chocolate. There were 167 juniors surveyed. The percent of juniors who love chocolate is about 74.9.
   B. The percentage of juniors and seniors who love chocolate is 238 out of 317, about 75.1 percent.

4. A. Spot weighed 15 pounds at 6 months.
   B. Spot's weight increased the most from birth to 3 months. His weight increased 9 pounds.

5. A. Combine the percent who saw 0 movies and the percent who saw between 1 and 5 movies to get 8.75% + 12.50% = 21.25%.
   B. The number of people who saw more than 20 movies last year is 37.5 percent of 2,000, about 750 people.

6. Mean is 41.8. Median is 40. Mode is 39.8.

$$\text{Mean} = \frac{\text{sum of all weights}}{6} = \frac{250.8}{6} = 41.8$$

The median is the average of 39.8 and 40.2.

$$\text{Median} = \frac{39.8 + 40.2}{2} = \frac{80}{2} = 40$$

The mode is the most frequently occurring value, 39.8.

7.

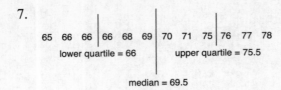

65 66 66 | 66 68 69 | 70 71 75 | 76 77 78

lower quartile = 66   upper quartile = 75.5

median = 69.5

Since there are 12 temperatures, the median is halfway between the sixth and the seventh. Since there are 6 temperatures in the lower half, the lower quartile is halfway between the third and the fourth. The lower quartile is 66. Since there are 6 temperatures in the top half, the upper quartile is halfway between 75 and 76. The upper quartile is 75.5. The interquartile range is $75.5 - 66 = 9.5$.

8. 50 percent
Since the median is the middle temperature, half or 50 percent of the temperatures are below the median.

9. 50 percent
The lower quartile is at the 25 percent level and the upper quartile is at the 75 percent level. The amount between them is 50 percent.

10. A. Cool Coffee Café had the most customers in a day. They had 100 customers.
B. Cool Coffee Café's median is 80. Jiving Java Coffee Shoppe's median is 60. Cool Coffee Café has a higher median.
C. The lower quartile of the Jiving Java Coffee Shoppe is 50.

11.

Mathville Baseball Team Batting Averages

.100 .120 .140 .160 .180 .200 .220 .240 .260 .280 .300

The median is .257. The lower quartile is .150. The upper quartile is .283. The minimum value is .123. The maximum value is .298. The interquartile range is .133.

12. **Percent of Free Throws Made by Mack Shack**

2	4
3	9
4	1 3 3 5 6 6 7
5	3 5 6 7 9
6	0 4 6
7	2 9
8	5

Key
2 | 4 represents 24%

A. Mack Shack scored at least 60 percent in 6 of his games.
B. The median is 54 percent.

13. A. The fewest number of games won by a National League team was 67.
B. The modes for the number of games won by a National League team were 67, 81, and 83.

Chapter 8 | **Probability**

EQUALLY LIKELY EVENTS

When you flip a fair coin, there are 2 possible ways that the coin can land. The side with tails is up or the side with heads is up. If the coin is flipped a lot of times, then half the time heads will be up and half the time tails will be up. In mathematics, we say that the probability that heads will be up when you flip a coin is $\frac{1}{2}$, and the probability that tails will be up is $\frac{1}{2}$.

The spinner shown below is divided into three equal parts, labeled red, blue, and yellow.

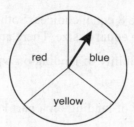

The spinner can stop on any one of the 3 colors. If the spinner is spun a lot of times, then about $\frac{1}{3}$ of the time it will stop on red, $\frac{1}{3}$ of the time it will stop on blue, and $\frac{1}{3}$ of the time it will stop on yellow. The probability that the spinner will land on red is $\frac{1}{3}$, the probability that the spinner will land on blue is $\frac{1}{3}$, and the probability that the spinner will land on yellow is $\frac{1}{3}$.

Both of the choices when flipping a coin are equally likely. Each of the 3 choices when spinning the spinner is equally likely. The spinner below does not have choices that are equally likely because it is not divided into parts that are the same size.

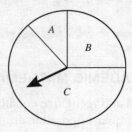

PROBABILITY

To make the spinner shown above have equal parts, divide it into 8 equal parts, as shown below.

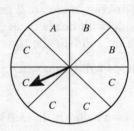

Now there are 8 equal choices. Some of the parts are labeled with the same letter, but all of the parts are equal in size. There are more parts labeled C than there are parts labeled A or B. The probability of landing on a part labeled C is $\frac{5}{8}$.

$$\text{probability of landing on a part labeled } C = \frac{\text{number of equal parts labeled } C}{\text{total number of parts}}$$

In the same way, the probability of landing on a part labeled B is $\frac{2}{8} = \frac{1}{4}$. The probability of landing on a part labeled A is $\frac{1}{8}$.

Example

A small bag contains 5 red jellybeans, 7 green jellybeans, 9 yellow jellybeans, and 10 black jellybeans. If you put your hand into the bag without looking and remove one jellybean, what is the probability that it will be yellow?

$$\text{probability of choosing a yellow jellybean} = \frac{\text{number of yellow jellybeans in the bag}}{\text{total number of jellybeans}}$$

The probability of choosing a yellow jellybean is $\frac{9}{5+7+9+10} = \frac{9}{31}$.

What is the probability of choosing a jellybean that is not red?

$$\text{probability of choosing a jellybean that is not red} = \frac{\text{number of yellow jellybeans that are not red}}{\text{total number of jellybeans}}$$

The jellybeans that are not red are the jellybeans that are green, yellow, and black, a total of $7 + 9 + 10 = 26$. The probability of choosing a jellybean that is not red is $\frac{26}{31}$.

The probability of choosing a red jellybean plus the probability of choosing a jellybean that is not red is $\frac{5}{31} + \frac{26}{31} = \frac{31}{31} = 1$. The probability of choosing a jellybean that is not red is $1 -$ probability of choosing a red jellybean.

> In general, the probability of not A is $1 -$ probability of A.

Exercises

1. A set of index cards is numbered 1 through 9. If you pick one card from the set without looking, what is the probability you pick a card with an even number written on it?

2. James has 150 pens. If he chooses a pen at random, the probability that he chooses a red pen is .7, the probability that he chooses a blue pen is .2, and the probability that he chooses a black pen is .1. What is the probability that James will choose a red or a blue pen?

Solutions to Exercises

1. probability of choosing a card with an even number $= \dfrac{\text{how many even numbers}}{\text{how many numbers}}$

The are four even numbers, 2, 4, 6, and 8. There are 9 numbers written on the set of index cards. The probability of choosing a card with an even number on it is $\frac{4}{9}$.

2. The probability of choosing a red or a blue pen is the probability of choosing a red pen plus the probability of choosing a blue pen, $.7 + .2 = .9$.

Practice Problems

1. Carol is playing a game with the spinner shown below.

What is the probability that the spinner does not land on a vowel?

2. Eric has a bag with marbles that are all the same size. The bag contains 10 green marbles, 7 blue marbles, 2 white marbles, and 4 black marbles. If Eric reaches into the bag without looking and removes 1 marble, what is the probability that the marble will be green? What is the probability that the marble removed will not be blue? What is the probability that the marble removed will be either black or white?

(Solutions are found at the end of the chapter.)

BASIC COUNTING PRINCIPLE

If you want to buy a new car, there are many different options available. Consider only two different options. Choose the kind of music options: CD player and tape player, just CD player, or neither CD player nor tape player. Choose the color choices: black, red, gray, or blue. How many different combinations of cars are available?

Choose a CD player and a tape player; then there are 4 color choices for the car that give 4 different cars. Choose only a CD player; then there are 4 color choices for the car that give 4 different cars. Choose neither a CD player nor a tape player; then there are 4 color choices for the car that give 4 different cars. Combine all of the different car choices, and there are 12 different cars to choose from, as shown below.

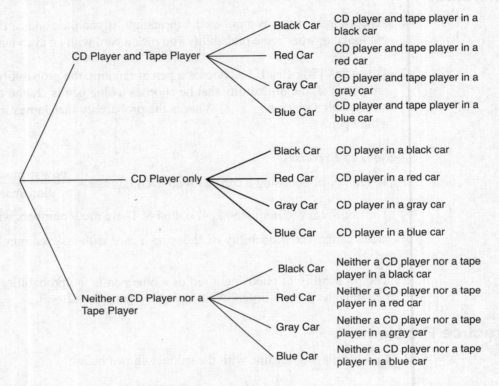

The total number of choices for a car with 3 different options for a music system and 4 different options for the color of the car is the number of choices for the first option multiplied by the number of choices for the second option.

Basic Counting Principle: If there are A ways to make the first choice, B ways to make the second choice, and C ways to make the third choice, then there are $A \cdot B \cdot C$ ways to make the first choice followed by the second choice followed by the third choice. The Basic Counting Principle works for any number of choices to be made.

Example

At a restaurant there are given 3 choices for salad, 2 choices for main dish, and 4 choices for dessert. Show all the different combinations of meals could you make.

First, choose the type of salad followed by the type of main dish followed by the type of dessert, as shown below.

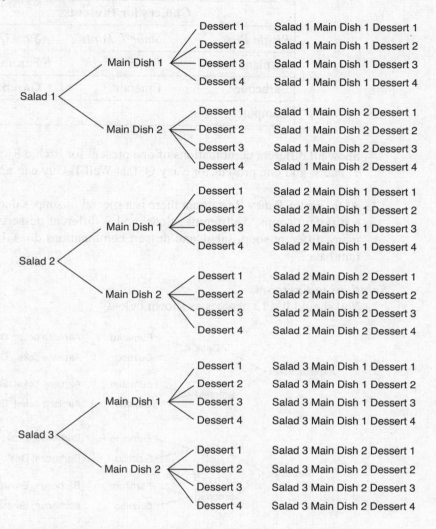

There are 24 different meal combinations that you can choose from.

Using the Basic Counting Principle with this problem, there are 3 options for the first choice of salad, 2 options for the second choice of main dish, and 4 options for the third choice of dessert. There are $4 \cdot 2 \cdot 3 = 24$ total combinations for meal choices.

Example

As you dress to go outside, you have to choose between 11 different hats, 4 different coats, and 5 different pairs of gloves. How many different hat, coat, and glove combinations are possible?

There are too many choices to list, so use the Basic Counting Principle to determine the number of combinations possible. Multiply the number of choices for hats by the number of choices for coats by the number of choices for gloves to get $11 \cdot 4 \cdot 5 = 220$ different combinations.

Exercises

1. Well T. Guy is purchasing gifts for three of his friends. His choices are shown in the following table.

Choices for Presents		
Richie Rich	*Smar T. Pants*	*Suzy Q.*
Airplane ticket	Desk	Furniture
Barbeque	Emerald	Gazebo
Computer		

Show all different combinations of one present for Richie Rich, one present for Smar T. Pants, and one present for Suzy Q. that Well T. Guy can make.

2. At the Fancy Pancy Restaurant there is a special of soup, salad, and dessert. There are 6 different soups, 5 different salads, and 9 different desserts to choose from. How many different soup, salad, and dessert combinations does the Fancy Pancy Restaurant have?

Solutions to Exercises

1. You should list 12 choices as shown below.

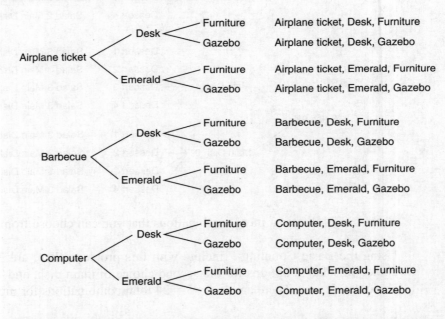

2. Using the Basic Counting Principle, there are 6 choices for the soup, 5 choices for the salad, and 9 choices for the dessert. This gives a total of $6 \cdot 5 \cdot 9 = 270$ combinations.

Suppose there are 29 different types of tea at the Gourmet Tea Shoppe. You need to make a basket of teas using 4 different teas. How many different baskets can be made?

You certainly do not want to begin by making a figure. Use the Basic Counting Principle instead. There are 29 choices for the first tea. Once that tea is chosen there will only be 28 choices for the second tea. Once those two teas have been chosen, then there will only be 27 choices for the third tea and 26 choices for the fourth tea. There will be $29 \cdot 28 \cdot 27 \cdot 26 = 570,074$ different baskets of tea.

Exercises

1. You are making a bag of treats for your English teacher. You want the bag to have 1 piece of candy, 1 pack of gum, and 1 bag of chips. At the store there are 21 different kinds of candy, 12 different types of gum, and 5 different bags of chips. How many different combinations of candy, gum, and chips can you make?

2. You now want to make a bag of treats for your Math teacher. To make it bigger than the bag for your English teacher, you want to have 2 different pieces of candy, 2 different packs of gum, and 1 bag of chips. At the store there are 21 different kinds of candy, 12 different kinds of gum, and 5 different bags of chips. How many different combinations of 2 different pieces of candy, 2 different packs of gum, and 1 bag of chips can you make?

Solutions to Exercises

1. There are 21 different types of candy, 12 different kinds of gum, and 5 different kinds of bags of chips. This would make a total of $21 \cdot 12 \cdot 5 = 1,260$ combinations of candy, gum, and chips.

2. For the first type of candy, you have 21 choices. Once you have chosen the first piece of candy, there are 20 types of candy left. For the first pack of gum, you have 12 choices. Once you have chosen the first pack of gum, you have 11 types of gum left. You only want one bag of chips, so you have 5 choices. There are $21 \cdot 20 \cdot 12 \cdot 11 \cdot 5 = 277,200$ combinations.

Practice Problems

3. At the Not So Cheap Movie Theater there are 4 choices for the size of popcorn, 3 choices for the size of beverage, and 5 choices of beverage. How many combinations of popcorn and beverage are possible?

4. Sandwich Shop offers a variety of meats, toppings, and bread for their sandwiches, as shown in the following table.

Meats	Toppings	Bread
Ham	Lettuce	Sourdough
Turkey	Mayonnaise	Wheat
Roast Beef	Pickle	White
	Mustard	Rye

How many different combinations of one meat, one topping, and one kind of bread are possible?

5. At a party each person has one present to open. You notice that there are 4 people ready to open their presents—Carol, Bob, Jules, and Pam. If the presents are opened one at a time, in how many different orders can the presents be opened?

6. The school cafeteria is making a sign advertising the large number of plate lunch choices available. The cook tells you each day there will be 5 different main dish choices, 4 salad choices, 6 fruit choices, 2 dessert choices, and 3 beverage choices. If a plate lunch contains 1 main dish, 1 salad, 1 fruit, 1 dessert, and 1 beverage, how many combinations are possible?

(Solutions are found at the end of the chapter.)

SAMPLE GQE QUESTIONS FOR CHAPTER 8

1. Suzy is playing a game with the spinners shown below.

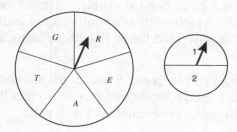

List all of the different combinations of numbers and letters that are possible with the spinners.

2. James has a bag with 240 marbles colored purple, green, white, and red. James reaches into the bag without looking and removes 1 marble. The probability of choosing a purple marble is .3, the probability of choosing a green marble is .4, and the probability of choosing a white marble is .1. What is the probability of choosing a green or a red marble?

3. Look at the spinner shown below.

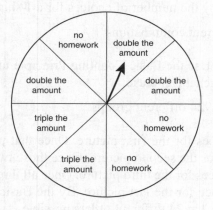

Each section of the spinner is equal in size. Your teacher asks you to spin to decide the amount of homework for tonight. What is the probability that the spinner will land on double the amount of homework or triple the amount of homework?

4. You are the lucky winner of a prize package from Disc World. The prize package includes 1 music CD, 1 movie DVD, and 1 computer CD. You have to choose from 35 music CDs, 38 movie DVDs, and 25 computer CDs. How many different combinations could you make?

5. James says he has 7 different combinations of jeans and a T-shirt from the table of clothes shown below.

Jeans	T-shirts
Black	Math is Cool T-shirt
Light Blue	I Love Math T-shirt
Dark Blue	Math Rules T-shirt
	We All Do Math Everyday T-shirt

Sam disagrees with James by stating there are 12 different combinations of jeans and a T-shirt. Who is correct? Explain how you determined the number of different combinations.

6. Rachel is decorating her folder with stickers. She has collected 29 unusual stickers, but only wants to put 3 different stickers on the folder. How many distinctive groups of 3 stickers could Rachel make?

7. Sally is designing a presentation on her computer. She wants to show 4 different pictures, one at a time. She can show the pictures in any order. How many different orders are possible?

Answers Explained

1. G1, G2, R1, R2, E1, E2, A1, A2, T1, T2. With each letter, you can spin either a 1 or a 2. Since there are 5 letters and 2 numbers, there should be 10 possible choices.

2. .6. To find the probability of choosing a green or a red marble, you need the probability of choosing a green marble and the probability of choosing a red marble. The probability of choosing a green marble is .4. Find the probability of choosing a red marble.

   Remember the probability of choosing a purple, green, white, or red marble adds up to 1. You know all of the probabilities except for the red marble, so add them up, $.3 + .4 + .1 = .8$. The probability of choosing a red marble is $1 - .8 = .2$.

   The probability of choosing a green or a red marble is $.4 + .2 = .6$.

3. $\frac{5}{8}$. The probability of landing on double the amount of homework or triple the amount of homework =

   $$\frac{\text{number of spaces labeled double} + \text{number of spaces labeled triple}}{\text{total number of spaces}} = \frac{3+2}{8} = \frac{5}{8}.$$

4. 33,250 different prize packages. Use the Basic Counting Principle to find the number of prize packages.

   $$35 \cdot 38 \cdot 25 = 33{,}250$$

5. Sam is correct. To determine the number of combinations of jeans and a T-shirt, multiply the number of choices for jeans by the number of choices for a T-shirt.

   $$3 \cdot 4 = 12 \text{ different combinations}$$

6. 21,924 different groups of 3 stickers. Use the Basic Counting Principle to find how many different groups of 3 stickers Rachel can make.

   $$29 \cdot 28 \cdot 27 = 21{,}924 \text{ different groups}$$

7. 24 different orders. There are 4 choices for the first picture. Once that picture has been shown there are only 3 choices for the second picture. Once those two pictures have been shown there are only 2 choices for the third picture. Once all three of those have been shown, there is only 1 choice for the last picture. By the Basic Counting Principle, there are a total of $4 \cdot 3 \cdot 2 \cdot 1 = 24$ different orders possible.

Solutions to Practice Problems

1. probability the spinner does not land on a vowel = $\frac{5}{8}$. There are five spaces without vowels: M, T, H, C, and L. There are a total of eight spaces.

   $$\text{probability of not landing on a vowel} = \frac{\text{number of spaces without vowels}}{\text{total number of spaces}} = \frac{5}{8}$$

2. $\frac{10}{23}$. To find the probability that Eric will remove a green marble, use the number of green marbles and the total number of marbles. The probability that Eric will remove a green marble =

$$\frac{\text{number of green marbles}}{\text{total number of marbles}} = \frac{10}{23}.$$

$\frac{16}{23}$. There are two ways to find the probability of removing a marble that is not blue.

probability of removing a marble that is not blue =

$$\frac{\text{number of marbles that are not blue}}{\text{total number of marbles}} = \frac{10 + 2 + 4}{23} = \frac{16}{23}$$

probability of removing a marble that is not blue = 1 – probability of removing a marble that is blue = $1 - \frac{7}{23} = \frac{16}{23}$

$\frac{6}{23}$. To find the probability of removing a marble that is either black or white, use the total number of black marbles and white marbles, and the total number of marbles.

probability of removing either a black or white marble =

$$\frac{\text{number of black and white marbles}}{\text{total number of marbles}} = \frac{2 + 4}{23} = \frac{6}{23}$$

3. 60 combinations. Use the Basic Counting Principle. There would be $4 \cdot 3 \cdot 5 = 60$ combinations of popcorn and beverage.

4. 48 combinations. Using the Basic Counting Principle, there would be $3 \cdot 4 \cdot 4 = 48$ combinations.

5. 24 different orders. There are 4 choices for the first person to open a present. Once the first person has opened the present, then there are only 3 people left. Once the first 2 people have opened their presents, there are only 2 people left. Finally, there is only 1 person left to open a present. So, there are $4 \cdot 3 \cdot 2 \cdot 1 = 24$ different orders for all 4 people to open their presents.

6. 720 possible combinations. Multiply the number of choices for the main dish times the number of choices for salad times the number of choices for fruit times the number of choices for dessert times the number of choices for beverage.

$$5 \cdot 4 \cdot 6 \cdot 2 \cdot 3 = 720 \text{ possibilities}$$

Chapter 9 | Algebra 1: Relations and Functions

> **INDIANA'S ACADEMIC MATHEMATICS STANDARDS ADDRESSED:**
>
> A1.3.1 Sketch a reasonable graph for a given relationship.
> A1.3.2 Interpret a graph representing a given situation.
> A1.3.3 Understand the concept of a function, decide if a given relation is a function, and link equations to functions.
> A1.3.4 Find the domain and range of a relation.

INTERPRETING GRAPHS

Graphs can visually describe many different relationships. The graph below shows the relationship between the temperature outside and the time of day.

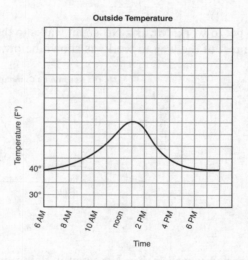

You can tell that the temperature started about 40°, went up to about 60° by 1 P.M., and dropped down to about 40° by 6 P.M.

The scales are left off of the graph below showing the temperature outside of the school.

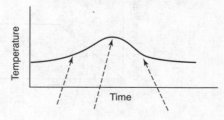

You cannot tell exact temperatures or times because there are no numbers. You do know that the temperature was rising at the first dotted arrow, hit its highest temperature at the second dotted arrow, and was dropping at the third dotted arrow. All of that information is available on a graph without numbers.

The graph below shows what has happened to the length of Harry Guy's hair over a few months.

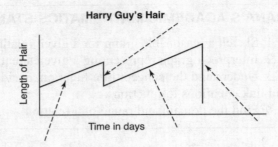

What is happening at each dotted arrow? At the first arrow, Harry Guy's hair is growing at a regular rate. At the second dotted arrow, Harry Guy gets a haircut. His hair gets immediately shorter. At the third arrow, Harry Guy's hair is growing at a steady rate. This time Harry Guy lets his hair grow longer than the first time he had it cut. At the fourth dotted arrow, Harry Guy gets his hair cut much shorter than he did the first time he had it cut.

Exercises

1. To get to work, Mr. Biz Nessman walks to the bus stop, rides the bus, and then walks the rest of the way to work, as shown below.

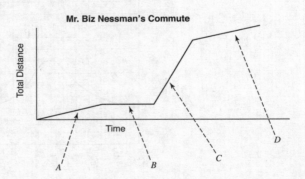

Describe what Mr. Biz Nessman is doing during each section on the graph. Why is section *C* steeper than section *A*?

2. A researcher keeps track of the height of a pelican as he searches and finds a fish. A graph of the pelican's height above the water level is shown below.

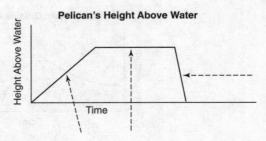

Describe what the pelican is doing during each of the three sections of the graph.

Solutions to Exercises

1. In section *A*, Mr. Biz Nessman is walking to the bus stop. In section *B*, Mr. Biz Nessman is waiting for the bus to arrive. In section *C*, Mr. Biz Nessman is riding the bus. In section *D*, Mr. Biz Nessman is walking the rest of the way to work. Notice that the line segments in section *A* and section *D* are slanted the same way because he is walking at the same speed. Section *C* is steeper because the bus is covering distance more quickly.

2. In the first section, the pelican is taking off from the water. He is flying higher and higher. In the second section, the pelican is staying at the same height. In the third section the pelican is diving toward the water.

Practice Problems

1. Racy Rabbit and Tardy Turtle are running a 60-yard dash. The graph below shows the distance each of them traveled over time.

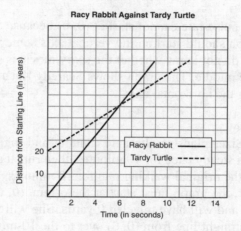

Answer each of the questions using the information from the graph.

A. Who finished the 60-yard race first?
B. How many seconds sooner did the first animal arrive at the finish line?
C. Who was ahead after 4 seconds?
D. Who was ahead after 6 seconds?

2. Athel Eat kept track of his heart rate during his 30-minute exercise program. Describe what is happening during each section of the graph below.

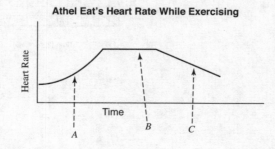

(Solutions are found at the end of the chapter.)

SKETCHING GRAPHS

Sometimes you will be asked to sketch a graph from information given. Your graph may not be exactly the same as the one shown, but a few essential parts must be in your graph. The graph below shows Scaredy Kat's pulse rate as she watches a scary movie.

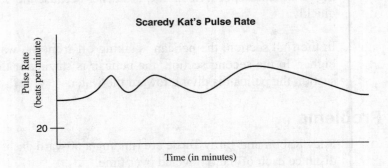

The pulse rate cannot start at (0, 0) because Scaredy Kat's pulse rate cannot be zero. The graph should start on the vertical axis somewhere above the horizontal axis.

You do not know whether the scary movie is scary all the time or just scary at the end or in the middle. The graph shows Scaredy Kat's pulse rate going up and down but never touching the horizontal axis.

Example

Eckser Sizer plans to alternate running and jogging for her exercise program. She will walk at a steady pace for 10 minutes, then run at a faster pace for 10 minutes, and, finally, walk at the original walking rate for 10 minutes.

First, the graph must start at the origin, (0, 0). Eckser Sizer will begin walking at 0 minutes and will have traveled 0 yards. She will walk at a steady pace for 10 minutes, so draw a straight line from (0, 0) over to the 10-minute line, as shown below.

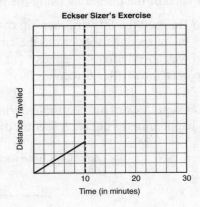

During the second section, Eckser Sizer is running at a faster pace, so the line should be steeper. Draw a straight line from the point where she began running at the 10-minute line to the 20-minute line, as shown below.

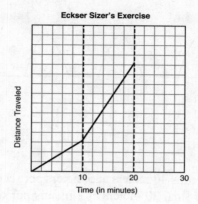

For the third section, Eckser Sizer returns to walking at the same rate that she did for the first 10 minutes. Draw a straight line from where she began walking at the 20-minute line to the 30-minute line. The line in the third section should be slanted exactly the same way as the line in the first section. The graph is shown below.

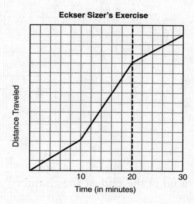

Exercise

Mr. Ner D. Mathteacher is out on the playground watching his students jumping rope. Mr. Mathteacher is graphing the distance from Jum Peen's feet to the ground as she jumps rope.

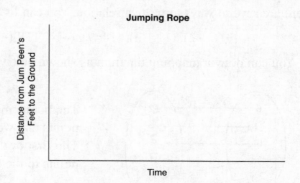

Use the graph above to sketch a graph that shows the distance from Jum Peen's feet to the ground as she jumps rope.

Solution to Exercise

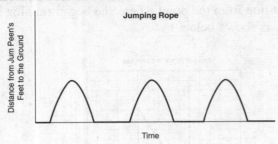

Your graph may look different than the graph shown above, but there are essential parts that it must have. First, the graph must have repeated curves up into the air. If your line segments are straight instead of curved, that is acceptable in this case. Second, there must be periodic straight lines drawn on the horizontal axis.

Practice Problems

3. Speedy runs 40 meters at 5 meters per second. Slowman gets a 20-meter head start and runs the final 20 meters at 2 meters per second. Use the grid below to sketch a graph representing this situation. Be sure to label Speedy's graph and Slowman's graph.

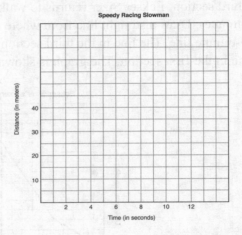

(Solution is found at the end of the chapter.)

RELATIONS AND FUNCTIONS

Definition: A relation is a set of ordered pairs.

There are several ways to write a relation. You can list a set of ordered pairs.

$$\{(6, -2), (11, -3), (14, 9), (-14, 11), (-14, 20), (-21, -31)\}$$

You can draw a mapping diagram as shown below.

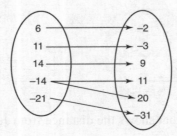

The points in the mapping diagram are the same points that were listed in the set of ordered pairs. The first oval contains the *x*-value and the arrow points to the appropriate *y*-value.

You can also write a relation in a table.

x	y
6	−2
11	−3
14	9
−14	11
−14	20
−21	−31

Finally, you can display a relation on a graph, as shown below.

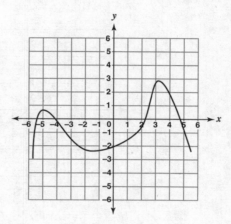

You can write a relation as an equation, $y = x^2$.
A function is a special kind of relation.

Definition: A function is a relation where for every x there is exactly one value for y.

A function cannot have two points with the same x-value and different y-values. A function can have two points with the same y-value and different x-values.

To classify relations as functions when you are given a list, a mapping diagram, or a table, check the x-values. If there are two x-values that are the same but have different y-values, then the relation is not a function. If all of the x-values are different, then the relation is a function.

To classify relations as functions when you are given a graph, look for two points directly above one another. Two points that have the same x-value but have different y-values will be on the same vertical line. If a vertical line intersects more than one point on the graph, then the relation is not a function. If a vertical line intersects the graph in only one point, then the relation is a function.

To classify a relation as a function when you are given an equation, then graph the equation and look at the graph.

Exercises

1. Is the relation {(4, 3), (5, 2), (7, 4), (8, 3)} a function? Why or why not?

2. Is the relation in the mapping diagram below a function? Why or why not?

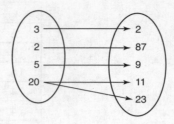

3. Is the relation in the graph below a function? Why or why not?

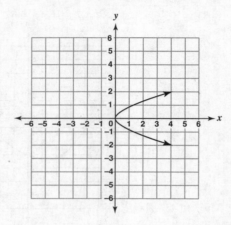

Solutions to Exercises

1. The relation is a function because all of the *x*-values are different. It does not matter that two of the *y*-values are the same.

2. The relation is not a function. There are two points that have the same *x*-value but have different *y*-values, (20, 11) and (20, 23).

3. The relation is not a function. There are points that have the same *x*-value but have different *y*-values. For example, points (4, 2) and (4, –2) make the relation not a function.

Practice Problems

4. Change one number in the relation {(5, 7), (5, 9), (8, 9)} so that the relation will be a function.

5. Is the relation in the table below a function? Why or why not?

x	*y*
6	5
11	5
14	5
–14	11

(Solutions are found at the end of the chapter.)

DOMAIN AND RANGE OF RELATIONS

Definition: The domain of a relation is the set of all of the *x*-values in the relation. The range is the set of all *y*-values in the relation.

When relations are given in a list, a mapping diagram, or a table, finding the domain and the range is quite simple.

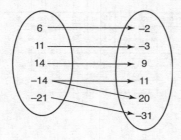

The domain in the diagram on the left is {6, 11, 14, –14, –21}. The range in the diagram is {–2, –3, 9, 11, 20, –31}.

When a graph is drawn or a relation is described by an equation or words, then finding the domain and range of a relation is a little more difficult. Consider the graph below.

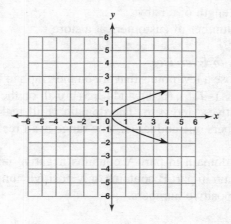

List points on the graph that are in different quadrants. In this case, points (0, 0), (4, 2), and (4, –2) are all on the graph.

Consider first the values for *x*. Notice there are no negative numbers listed for the *x*-value. This graph only has *x*-values that are zero or positive numbers. For the graph shown above, the domain is zero and all positive numbers. A shorter way to say that is all nonnegative numbers.

Consider the values for *y*. Notice that you have listed positive numbers, negative numbers, and zero. The range for the relation in the graph above is all real numbers.

If a relation is described with an equation, then it is easiest to graph the equation to find the domain and the range.

If a relation is described in words, then you must think about the situation. Suppose you want to graph the speed a person drives on a 3-hour trip. The domain will be the number of hours, which will be all real numbers from 0 to 3. The range will be the speed the person travels, all real numbers from 0 to 65.

Exercises

1. What are the domain and range of the relation of the graph below?

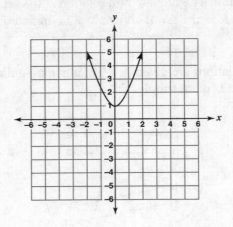

2. With each situation, determine the combination of negative numbers, positive numbers, or zero that are in the domain.

   A. Temperatures outside of an Indianapolis hotel in December
   B. Length of a baby
   C. Number of customers at a store

Solutions to Exercises

1. Choose a few points that are on the graph to help you determine the domain and range. Points (–2, 5), (0, 1), and (2, 5) are all on the graph. The values for x include positives, negatives, and zero. The domain is all real numbers. The values for y only include numbers one and larger. The range is all real numbers greater than or equal to one.

2. The domain in part A contains negative numbers, positive numbers, and zero. The domain in part B contains only positive numbers. The domain in part C contains zero and positive numbers.

Practice Problems

6. What are the domain and range of the relation graphed below?

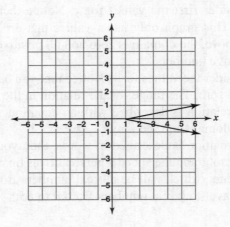

(Solution is found at the end of the chapter.)

SAMPLE GQE QUESTIONS FOR CHAPTER 9

1. Mr. Clean takes a bath between 7:30 and 8:00. The graph below shows the depth of the water as Mr. Clean is taking a bath.

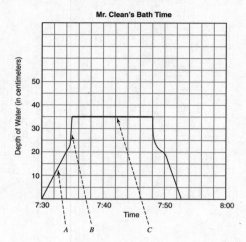

Describe what is happening during section *A*, section *B*, and section *C*.

2. Kreigh Zee is in the middle of graphing the speed at which his wife is driving. For 15 minutes she continues to drive at 30 miles per hour. Then she gets to the interstate and takes 5 minutes to increase her speed to 65 miles per hour. After traveling for 45 more minutes, she stops at a rest stop. Use the graph below to sketch Mrs. Kreigh Zee's speed on this part of the trip.

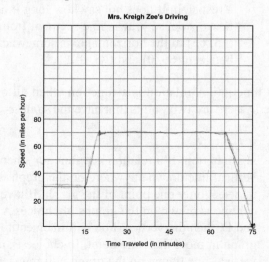

3. Choose the relation that is not a function. Explain your reasoning.

   A. {(5, 6), (4, 9), (7, 2)}
   B. {(7, 0), (5, 0), (9, 0)}
   C. {(3, 5), (9, 7), (4, 12)}
   D. {(0, 8), (5, 3), (0, 9)}

4. Choose the relation that is a function. Explain your reasoning.

   A.

   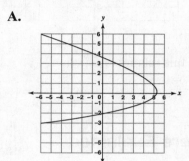

   B.

   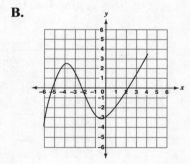

   C.

   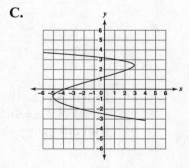

   D.

   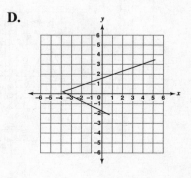

5. What are the domain and the range of the graphed relation?

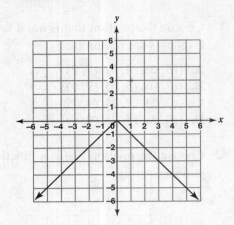

Is this relation a function?

Answers Explained

1. In section *A*, Mr. Clean is filling the bathtub with water. In section *B*, Mr. Clean is getting into the bathtub. In section *C*, Mr. Clean is sitting in the bathtub.

2.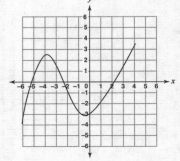

   Mrs. Kreigh Zee drives at 30 miles per hour for the first 15 minutes. She is not starting from a standstill, so draw a line segment from (0, 30) to (15, 30). She accelerates for 5 minutes from 30 miles per hour to 65 miles per hour. Draw a line segment from (15, 30) to (20, 65). She drives at that rate for 45 more minutes. Draw a line segment from (20, 65) to (65, 65). She then stops at a rest stop. It does not say how long it takes her to stop, so draw a line segment from (65, 65) to the horizontal axis somewhere between (66, 0) and (75, 0).

3. D. {(0, 8), (5, 3), (0, 9)} is not a function. A relation is a function when all of the values of *x* are different. In choice D there are two points that have the same *x*-value with different *y*-values.

4. B.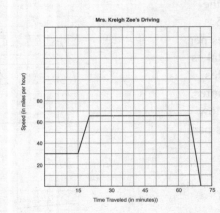

   is a function. If a relation is a function, then a vertical line drawn anywhere on the graph will intersect only one point of the graph. If a vertical line is drawn on the graphs of choices A, C, and D, then the vertical line will intersect the graph in more than one point. In choice B, any vertical line drawn on the graph will intersect the graph in one point.

5. The domain is all real numbers. The range is zero and all negative real numbers. To help find the domain and range of a graph, list a few of the points on the graph. Points (–6, –6), (0, 0), and (6, –6) are all on the graph. Notice that the x-values are negative, positive, and zero. Notice that the y-values are only negative numbers and zero.
The relation is a function. If you drew a vertical line anywhere on the graph, the relation would intersect the vertical line in only one point.

Solutions to Practice Problems

1. A. Racy Rabbit crossed the finish line in 9 seconds. Tardy Turtle crossed the finish line in 12 seconds. Racy Rabbit crossed the finish line first.
B. Racy Rabbit finished 3 seconds before Tardy Turtle.
C. Racy Rabbit was about 27 yards from the starting line after 4 seconds. Tardy Turtle was 33 yards from the starting line after 4 seconds. Tardy Turtle was ahead.
D. After 6 seconds, the two animals were at the same distance from the starting line.

2. In section A, Athel Eat begins his exercise program and his heart rate increases. During section B, Athel Eat keeps exercising at a steady rate so his heart rate stays the same. In section C, Athel Eat is in his cool-down time so his heart rate is slowing back down to its normal rate.

3.

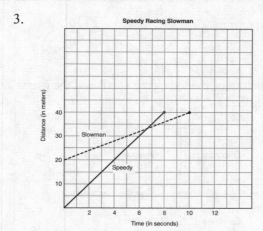

Speedy begins at (0, 0) and travels at 5 meters every second. Draw a straight line from (0, 0) to (8, 40). Slowman has a 20-meter head start so he begins at (0, 20). Slowman travels at 2 meters every second, so draw a straight line from (0, 20) to (10, 40).

4. Change the x-value in either (5, 7) or (5, 9) to any number other than 5 or 8: all the x-values will be different. It does not matter that the y-values are the same.

5. Yes. All of the x-values are different, so the relation is a function.

6. To find the domain and range of a graph, try listing points in different quadrants. Points (1, 0), (6, 1), and (6, –1) are all on the graph. The domain is all real numbers greater than or equal to one. The range is all real numbers.

Chapter 10

Algebra 1: Graphing Linear Equations and Inequalities

INDIANA'S ACADEMIC MATHEMATICS STANDARDS ADDRESSED:

A1.4.1 Graph a linear equation.

A1.4.2 Find the slope, *x*-intercept, and *y*-intercept of a line given its graph, its equation, or two points on the line.

A1.4.3 Write the equation of a line in slope-intercept form. Understand how the slope and *y*-intercept of the graph are related to the equation.

A1.4.4 Write the equation of a line given appropriate information.

A1.4.5 Write the equation of a line that models a data set and use the equation (or the graph of the equation) to make predictions. Describe the slope of the line in terms of the data, recognizing that the slope is the rate of change.

A1.4.6 Graph a linear inequality in two variables.

8.6.5 Represent two-variable data with a scatterplot on the coordinate plane and describe how the data points are distributed. If the pattern appears to be linear, draw a line that appears to best fit the data and write the equation of that line.

GRAPHING POINTS

The coordinate plane, shown below, is divided by the *x*-axis and the *y*-axis into four quadrants.

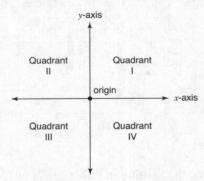

The *x*- and *y*-axes are numbered as shown on the grid below.

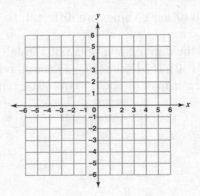

The point where the x-axis and y-axis meet is the origin, (0, 0). Points are graphed using ordered pairs, (x, y). The x-value is the distance to the left or right from the origin. The y-value is the distance up or down. To graph the point (5, 2), begin at the origin, move 5 spaces to the right, and 2 spaces up. The point (5, 2) is shown below.

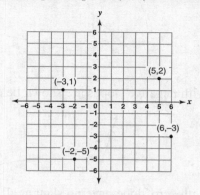

Points (−3, 1), (−2, −5), and (6, −3) are also shown.

Exercise

Use the coordinate plane below to graph and label the points B(−3, −1), R(−1, 4), I(2, 4), G(5, 0), H(1, −4), and T(0, −3).

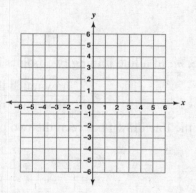

Solution to Exercise

1.

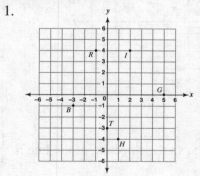

Be sure that your points are labeled as shown above. Be careful when graphing points like G(5, 0) and T(0, −3). With point G, move 5 to the right and do not move up or down. With the point T, do not move to the left or right and move down 3.

SLOPE OF A LINE

The slope, *m*, of a line is the way in which the line is slanted. Lines with positive slopes are shown below.

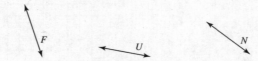

Lines with negative slopes are shown below.

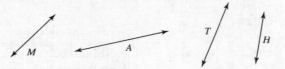

Lines with zero slope have no slant at all, as shown below.

Lines with an undefined slope are vertical, straight up and down, as shown below.

The slope, *m*, of a line joining two points is defined by

$$m = \frac{\text{how much the } y \text{ changes}}{\text{how much the } x \text{ changes}}.$$

Consider the line drawn between the two points (–2, 1) and (3, 2) on the graph below.

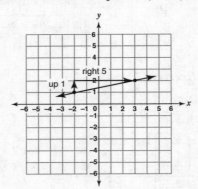

The *y* changes 1 as the *x* changes 5, so the slope, *m*, of the line is $\frac{1}{5}$.

When using a graph to find the slope, always begin at the lowest point on the graph and count up to the level of the second point. Then count to the right (a positive direction) or to the left (a negative direction).

Example

The slope of the line joining the two points on the graph below is found by starting at the point (–1, –4).

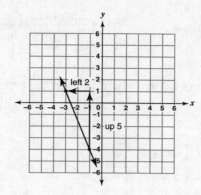

Count up 5 spaces to be level with the point (–3, 1) and left 2 spaces to the point (–3, 1). The slope of the line is $m = \dfrac{5}{-2} = -\dfrac{5}{2}$. When you count 2 spaces to the left, you are moving negative 2.

Exercises

1. Find the slope of the line joining the points graphed below.

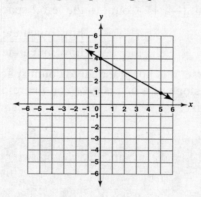

2. Graph the points (–2, –3) and (–4, 1). Find the slope of the line containing those two points.

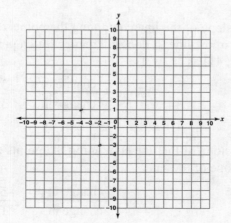

Solutions to Exercises

1. Begin at the lowest point on the line (5, 1). Count up 3 spaces to the level of the point (0, 4). Count 5 spaces left to get to the point (0, 4).

$$m = \frac{\text{how much the } y \text{ changes}}{\text{how much the } x \text{ changes}} = \frac{3}{-5}$$

It does not matter where you put the negative sign in your answer. $\dfrac{3}{-5} = \dfrac{-3}{5} = -\dfrac{3}{5}$

2.

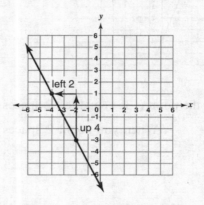

Begin at the lower of the two points (–2, –3). Count up 4 spaces to the same level as the point (–4, 1). Count to the left 2 spaces, –2, to the point (–4, 1).

$$m = \frac{\text{how much the } y \text{ changes}}{\text{how much the } x \text{ changes}} = \frac{4}{-2}$$

Of course, it is not practical to count spaces to find the slope of all lines. The formula for the slope of a line is given on the reference sheet, so it is not necessary for you to memorize.

> **Definition:** Let (x_1, y_1) and (x_2, y_2) be two points in the plane.
>
> $$\text{slope} = \frac{\text{change in } y}{\text{change in } x} = \frac{y_2 - y_1}{x_2 - x_1}, \text{ where } x_1 \neq x_2$$

Example

Find the slope of the line containing the points (23, 41) and (31, 33). Use the formula for the slope of the line. It does not matter which point you designate as (x_1, y_1). Consider the point (23, 41) as (x_1, y_1).

$$\text{slope} = \frac{\text{change in } y}{\text{change in } x} = \frac{33 - 41}{31 - 23} = \frac{-8}{8} = -1$$

Example

Find the slope of the line containing the points (37, –58) and (12, –58). Use the formula for the slope of the line.

$$\text{slope} = \frac{\text{change in } y}{\text{change in } x} = \frac{-58 - -58}{12 - 37} = \frac{0}{-25} = 0$$

Be Careful! Many mistakes are made when reducing fractions with zero in the numerator or the denominator.

$$\frac{0}{\text{any number other than } 0} = 0$$

$$\frac{\text{any number other than } 0}{0} = \text{undefined}$$

Division by zero is not possible.

The line joining $(37, -58)$ and $(12, -58)$ is a horizontal line. A horizontal line has 0 slope.

Exercises

1. Find the slope of the line joining the points $(0, -3)$ and $(15, -13)$.

2. What is the slope of the line segment that goes through the points $(7, 100)$ and $(7, 43)$?

Solutions to Exercises

1. $\text{slope} = \dfrac{\text{change in } y}{\text{change in } x} = \dfrac{-13 - (-3)}{15 - 0} = \dfrac{-10}{15} = -\dfrac{2}{3}$

2. $\text{slope} = \dfrac{\text{change in } y}{\text{change in } x} = \dfrac{43 - 100}{7 - 7} = \dfrac{-57}{0} = \text{undefined}$

The line segment joining $(7, 100)$ and $(7, 43)$ is a vertical line. The slope of a vertical line is undefined. It should make sense that the slope is undefined because division by zero is undefined.

Practice Problems

1. Find the slope of the line drawn below.

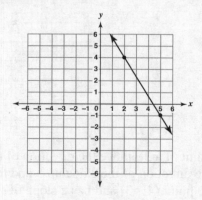

2. Find the slope of the line connecting $(-7, -2)$ and $(-5, -12)$.

(Solutions are found at the end of the chapter.)

GRAPHING LINES USING A POINT AND THE SLOPE OF THE LINE

When you are asked to graph a line, you will be given different types of information.

Example

Draw the line that has a slope of $\frac{3}{4}$ and passes through the point (–5, –2). On the grid below, graph the point (–5, –2).

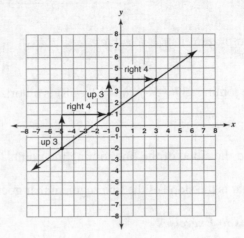

From that point, (–5, –2), move the direction of the slope up 3 and right 4. From that new point (–1, 1), you can go up 3 and right 4. The line drawn through all three points has a slope of $\frac{3}{4}$ and passes through the point (–5, –2).

Example

Draw the line that has a slope of –5 and contains the point (4, –3). Remember that a slope of –5 can be written $-5 = \frac{-5}{1} = \frac{5}{-1}$. Begin by graphing the point (4, –3) as shown below.

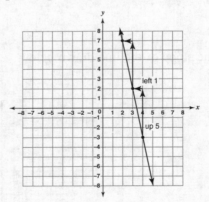

From the point (4, –3), move the direction of the slope up 5 and left 1. From that point (3, 2), move the direction of the slope up 5 and to the left 1, as shown above. The line drawn contains the point (4, –3) and has a slope of –5.

Exercises

1. On the coordinate plane below, graph the line that has a slope of $-\dfrac{1}{5}$ and contains the point (6, 1).

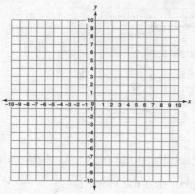

2. On the coordinate plane below, graph the line that has 0 slope and contains the point (2, 4).

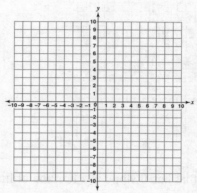

3. On the coordinate plane below, draw a line through the point (−2, 3) that has an undefined slope.

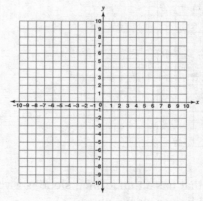

Solutions to Exercises

1.

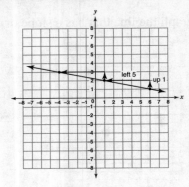

First graph the point (6, 1), as shown above. Then use the slope to find other points. Remember that the slope is $-\dfrac{1}{5} = \dfrac{-1}{5} = \dfrac{1}{-5}$, which is up 1 and left 5.

2.

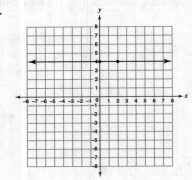

A line that has 0 slope is a horizontal line. Graph the point (2, 4). Draw a horizontal line as shown.

3.

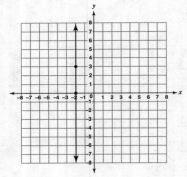

A line that has an undefined slope is a vertical line. Graph the point (–2, 3). Draw a vertical line as shown above.

Practice Problem

3. On the coordinate plane below, graph the line that has a slope of -3 and contains the point $(4, -1)$

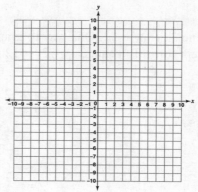

(The solution is found at the end of the chapter.)

GRAPHING LINES USING AN EQUATION

The reference sheet gives slope-intercept form as $y = mx + b$, where m is the slope of the line and b is the y-intercept.

> **Definition:** The y-intercept is the point where the line crosses the y-axis. The ordered pair is $(0, b)$.

To graph a line using an equation in slope-intercept form, use the same process as when you are given a point and the slope.

Example
Graph the line $y = 4 - \dfrac{1}{2}x$. Rewrite $y = 4 - \dfrac{1}{2}x$ in slope-intercept form.

$$y = -\frac{1}{2}x + 4$$

The slope is $-\dfrac{1}{2} = \dfrac{-1}{2} = \dfrac{1}{-2}$ and the y-intercept is 4. Graph the y-intercept $(0, 4)$. Use the

slope to find another point by counting up 1 and left 2. The line $y = -\dfrac{1}{2}x + 4$ is shown below.

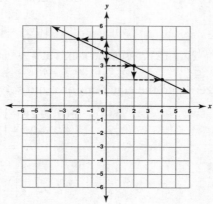

You could have moved the other direction with the slope. Think of the slope as $\frac{-1}{2}$. Move down 1 space and to the right 2 spaces, as shown above with the dotted gray arrows.

Example

Graph the line $3x - 2y = 12$. One way to graph this line is by changing $3x - 2y = 12$ into slope-intercept form. To change $3x - 2y = 12$ into slope-intercept form, solve the equation for y.

$$3x - 2y = 12$$
$$-2y = -3x + 12$$
$$y = \frac{-3}{-2}x + \frac{12}{-2}$$
$$y = \frac{3}{2}x - 6$$

The slope of $y = \frac{3}{2}x - 6$ is $\frac{3}{2}$ and the y-intercept is -6. Graph the y-intercept $(0, -6)$ and then use the slope to find the other points. The line $3x - 2y = 12$ is graphed below.

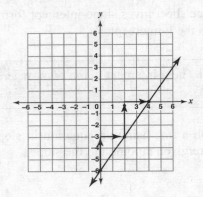

A second way to graph $3x - 2y = 12$ is to use the x- and y-intercepts.

Definition: The x-intercept is the point where the line crosses the x-axis. The ordered pair is $(a, 0)$. The y-intercept is the point where the line crosses the y-axis. The ordered pair is $(0, b)$.

Substitute zero into the equation for y and solve for x.

$$3x - 2(0) = 12$$
$$3x = 12$$
$$x = 4$$

The line $3x - 2y = 12$ crosses the x-axis at 4, which means that it goes through the point $(4, 0)$.

Substitute zero into the equation for x and solve for y.

$$3(0) - 2y = 12$$
$$-2y = 12$$
$$y = -6$$

The line $3x - 2y = 12$ crosses the y-axis at -6, which means that it goes through the point $(0, -6)$.

Graph the x-intercept and the y-intercept. Draw the line connecting the two intercepts. The line $3x - 2y = 12$ is drawn through the x-intercept $(4, 0)$ and the y-intercept $(0, -6)$, as shown below.

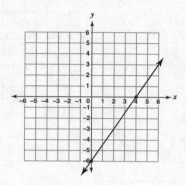

When given an equation in a form other than slope-intercept form, either change the equation to slope-intercept form and graph the line using the new equation or find the x-intercept and the y-intercept and use those to graph the line.

There are two special cases when graphing the equation of a line.

Graph the line $x = 4$. This equation cannot be changed into slope-intercept form because there is not a y in the equation. The line $x = 4$ means that all of the x-values are 4. The y-values can be any number. Ordered pairs on the line include $(4, 1)$, $(4, 2)$, $(4, 0)$, $(4, -1)$ and $(4, 4)$. The graph of the line $x = 4$ is shown below.

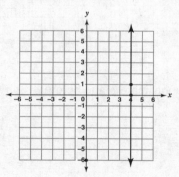

The slope of the line $x = 4$ is undefined because it is a vertical line. The equation cannot be changed into slope-intercept form because the slope is undefined.

Graph the line $y = 2$. This equation can be changed into slope-intercept form.

$$y = 2$$
$$y = 0x + 2$$

The slope of the line $y = 2$ is 0, and the y-intercept is 2. Graph the y-intercept $(0, 2)$. A line with 0 slope is a horizontal line. The graph of $y = 2$ is a horizontal line through the point $(0, 2)$, as shown below.

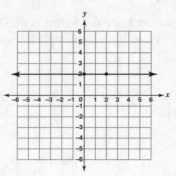

Exercises

1. On the coordinate plane below, graph the equation $y = -\dfrac{1}{2}x - 4$. What is its slope?

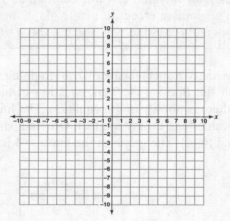

2. On the coordinate plane below, graph the equation $-x + 2y = -6$. What is its slope?

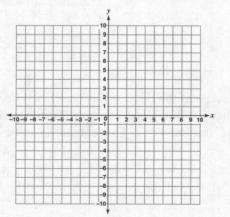

Solutions to Exercises

1.

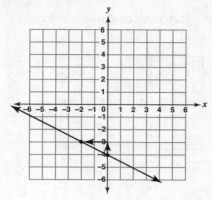

The equation $y = -\dfrac{1}{2}x - 4$ is in slope-intercept form. The slope is $-\dfrac{1}{2} = \dfrac{-1}{2} = \dfrac{1}{-2}$.

The y-intercept is –4, which is the point (0, –4). Graph the y-intercept first. Use the slope to find another point on the line, as shown.

2. Changing $-x + 2y = -6$ into slope-intercept form gives the equation

$$y = \dfrac{1}{2}x - 3$$

$$-x + 2y = -6$$
$$2y = x + -6$$

$$y = \dfrac{1}{2}x - 3$$

The slope of $y = \dfrac{1}{2}x - 3$ is $\dfrac{1}{2}$ and the y-intercept is –3. Graph the y-intercept (0, –3). Use the slope to find another point, as shown below.

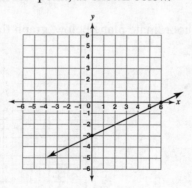

Find the x- and y-intercepts, (6, 0) and (0, –3).

$$-x + 2(0) = -6$$
$$-x = -6$$
$$x = 6$$

$$-(0) + 2y = -6$$
$$-2y = -6$$
$$y = -3$$

Graph each intercept and draw the line containing both intercepts, as shown.

Practice Problems

4. On the coordinate plane below, graph the equation $y = 4x - 1$.

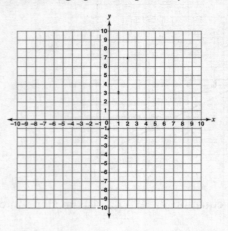

5. On the coordinate plane below, graph the equation $2y - x = 4$.

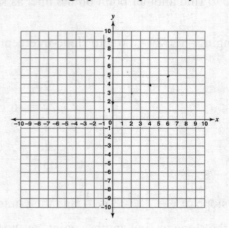

6. On the coordinate plane below, graph the equation $x = 1\frac{1}{2}$.

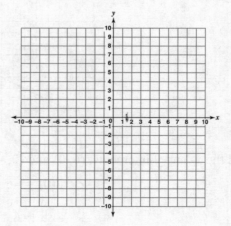

(Solutions are found at the end of the chapter.)

WRITING EQUATIONS OF LINES

The reference sheet gives point-slope form of the equation of a line.

> **Definition:** $y - y_1 = m(x - x_1)$ where m is the slope and $(x_1 - y_1)$ is a point on the line.

The form is called point-slope form because you use the point and the slope. To write the equation of a line, you need at least two pieces of information, the slope and a point on the line. The point on the line can be the y-intercept or any other point. Once you have the information, substitute it in the places shown in point-slope form, as shown below.

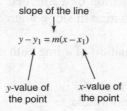

slope of the line

$$y - y_1 = m(x - x_1)$$

y-value of x-value of
the point   the point

Example

Write the equation of the line that passes through the point $(4, 2)$ and has a slope of 5. Substitute $x_1 = 4$, $y_1 = 2$, and the slope into the point-slope form of the equation of a line.

$$y - y_1 = m(x - x_1)$$
$$y - 2 = 5(x - 4)$$

This $y - 2 = 5(x - 4)$ is an answer to the problem. Sometimes you will be asked to give your answer in slope-intercept form. Use the answer you found in point-slope form and change it into slope-intercept form.

$$y - 2 = 5(x - 4)$$
$$y - 2 = 5x - 20$$
$$y = 5x - 18$$

The equation $y - 2 = 5(x - 4)$ in point-slope form is equivalent to $y = 5x - 18$ in slope-intercept form.

Example

Write the equation of the line that passes through the points $(1, 7)$ and $(-3, -1)$. To write the equation of a line, you must have the slope. Find the slope of the line using the slope formula.

$$\text{slope} = \frac{\text{change in } y}{\text{change in } x} = \frac{y_2 - y_1}{x_2 - x_1} = \frac{-1 - 7}{-3 - 1} = \frac{-8}{-4} = 2$$

Use the slope and one of the points in the point-slope form of the equation of the line. It does not matter whether you use $(1, 7)$ or $(-3, -1)$. The two answers are equivalent equations.

$$y - y_1 = m(x - x_1)$$ $$y - y_1 = m(x - x_1)$$
$$y - 7 = 2(x - 1)$$ $$y - (-1) = 2(x - (-3))$$
$$y + 1 = 2(x + 3)$$

Either $y - 7 = 2(x - 1)$ or $y + 1 = 2(x + 3)$ is an acceptable answer if the question does not ask you to give your answer in slope-intercept form. Sometimes you may be asked to give your answer in slope-intercept form.

$$y - 7 = 2(x - 1) \qquad\qquad y + 1 = 2(x + 3)$$
$$y - 7 = 2x - 2 \qquad\qquad y + 1 = 2x + 6$$
$$y = 2x + 5 \qquad\qquad\quad y = 2x + 5$$

Exercises

1. Write the equation of the line passing through the points $(-2, 5)$ and $(1, -1)$. Give your answer in slope-intercept form.

2. Write the equation of a line that contains the point $(3, -9)$ and has a slope of $-\frac{2}{3}$. Write the equation in slope-intercept form.

3. Write the equation of a line with a slope of -3 and a y-intercept of $\frac{2}{3}$.

Solutions to Exercises

1. Use the slope formula.

$$\text{slope} = \frac{\text{change in } y}{\text{change in } x} = \frac{y_2 - y_1}{x_2 - x_1} = \frac{-1 - 5}{1 - (-2)} = \frac{-6}{3} = -2$$

Use one of the points, the slope, and the point-slope form of the equation of a line to find the equation of the line.

$$y - 5 = -2(x - (-2))$$
$$y - 5 = -2(x + 2)$$
$$y - 5 = -2x - 4$$
$$y = -2x + 1$$

2. Since you are given a point and the slope, use the point-slope form of the equation of a line.

$$y - (-9) = -\frac{2}{3}(x - 3)$$

$$y + 9 = -\frac{2}{3}(x + -3)$$

$$y + 9 = -\frac{2}{3}x + 2$$

$$y = -\frac{2}{3}x - 7$$

3. Since you are given the slope and the y-intercept, use the slope-intercept form of the equation of a line.

$$y = mx + b$$

$$y = -3x + \frac{2}{3}$$

Practice Problems

7. Write the equation of a line that has a slope of $\frac{1}{2}$ and passes through the point (6, 4).

8. Write the equation of a line that passes through (6, 1) and (8, –4). Write the equation in slope-intercept form.

9. Write the equation of a line with a slope of 7 and a y-intercept of 2.

10. Write the equation of a line with a slope of 0 and a y-intercept of 3.

11. Write the equation of the line drawn below. You may leave the equation in point-slope form.

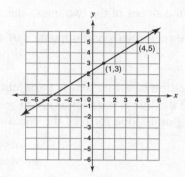

(Solutions are found at the end of the chapter.)

WRITING EQUATIONS OF PARALLEL LINES AND PERPENDICULAR LINES

Parallel lines are lines in the same plane that never meet. The lines shown below are parallel.

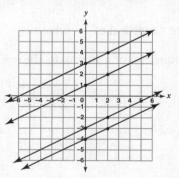

> Parallel lines have the same slope.

Parallel lines are slanted the same way. Compare the equations of the lines shown above. The equations of the lines are $y = \frac{1}{2}x + 3$, $y = \frac{1}{2}x + 1$, $y = \frac{1}{2}x - 3$, and $y = \frac{1}{2}x - 4$. The slopes of the lines are $\frac{1}{2}$.

Perpendicular lines are lines that intersect at a 90° angle. The lines shown below are perpendicular.

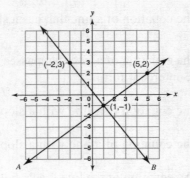

Perpendicular lines have slopes that are opposite reciprocals.

Compare the slopes of the two lines. The slope of line A is $\frac{3}{4}$. The slope of line B is $-\frac{4}{3}$. The slopes are reciprocals and opposites of each other.

Example

Write the equation of a line that is parallel to the line $y = 5x + 1$ and passes through the point (2, 4). If the line is parallel to $y = 5x + 1$, then it has the same slope. The slope of the line is 5. Use point-slope form of the equation of a line to write the equation.

$$y - 4 = 5(x - 2)$$
$$y - 4 = 5x - 10$$
$$y = 5x - 6$$

Example

Write the equation of a line that is perpendicular to the line $y = \frac{1}{2}x - 4$ and contains the point (7, –1). The slope of $y = \frac{1}{2}x - 4$ is $\frac{1}{2}$. The slope of a line perpendicular to $y = \frac{1}{2}x - 4$ is the opposite reciprocal of $\frac{1}{2}$, which is –2. Use point-slope form of the equation of a line to write the equation of a line with slope –2 that passes through (7, –1).

$$y - (-1) = -2(x - 7)$$
$$y + 1 = -2x + 14$$
$$y = -2x + 13$$

Exercises

1. Write the equation of a line that is parallel to $y = \frac{3}{4}x - 21$ and contains the point (8, 12). Write your answer in slope-intercept form.

2. Write the equation of a line perpendicular to $y = \frac{3}{4}x - 21$ that contains the point (9, 12).

Solutions to Exercises

1. The slope of a line parallel to $y = \frac{3}{4}x - 21$ is $\frac{3}{4}$. Use point-slope form of the equation

of a line to write the equation of a line with slope $\frac{3}{4}$ that passes through (8, 12).

$$y - 12 = \frac{3}{4}(x - 8)$$

$$y - 12 = \frac{3}{4}x - 6$$

$$y = \frac{3}{4}x + 6$$

2. The slope of the line $y = \frac{3}{4}x - 21$ is $\frac{3}{4}$. The slope of a perpendicular line is $-\frac{4}{3}$. Use point-slope form of the equation of a line to write the equation.

$$y - 12 = -\frac{4}{3}(x - 9)$$

$$y - 12 = -\frac{4}{3}x + 12$$

$$y = -\frac{4}{3}x + 24$$

Practice Problems

12. Write the equation of a line that contains the point (12, 15) and is parallel to the line $y = \frac{2}{3}x - 37$. Write your answer in slope-intercept form.

13. Write the equation of a line that is perpendicular to $y = \frac{4}{5}x - 10$ and passes through the point (4, 16). Write the answer in slope-intercept form.

(Solutions are found at the end of the chapter.)

MODELING DATA WITH LINEAR EQUATIONS

Many times linear data are given in the form of points on a graph. Linear means that there is a straight line that fits the data. You can use the information in a graph to write the equation of a line that goes through most of the points.

Example

Dogs age differently than humans. Some people think that a dog ages 1 year for every 7 human years. The graph below shows the actual relationship between dog years and human years.

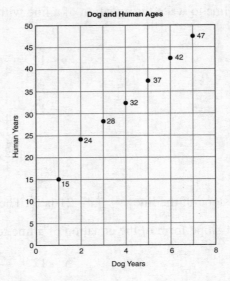

A 3-year-old dog is approximately 28 human years old. The relationship looks linear. Draw a line through the points such that the line has as many points above it as below it. The graph below shows a line that goes through some of the points and has the same number of points above and below the line.

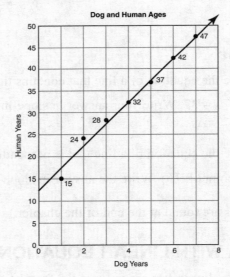

Choose two of the points that look like they are directly on the line. In this case, (4, 32) and (6, 42) are directly on the line. Find the equation of the line containing those two points. Find the slope.

$$\text{slope} = \frac{\text{change in } y}{\text{change in } x} = \frac{42-32}{6-4} = \frac{10}{2} = 5$$

Use point-slope form of the equation of a line.

$$y - 32 = 5(x - 4)$$
$$y - 32 = 5x - 20$$
$$y = 5x + 12$$

In the equation $y = 5x + 12$, y represents human years and x represents dog years. So if a dog is 2 years old, then he would be about $y = 5(2) + 12 = 10 + 12 = 22$ human years. You see that the equation gave us a number close to the actual point on the graph, (2 dog years, 24 human years).

You could use the equation $y = 5x + 12$ to find the human age of a dog's age that is not on the graph. For example, what would be the human age of a dog that is 10 years old? The dog would be $y = 5(10) + 12 = 50 + 12 = 62$ human years old.

You can interpret the slope to make sense for the situation of dog and human ages. If you put labels with the slope, then it has meaning.

$$\text{slope} = \frac{\text{change in } y \text{ human years}}{\text{change in } x \text{ dog years}} = \frac{10 \text{ human years}}{2 \text{ dog years}} = 5 \text{ human years per dog year}$$

This would mean that a dog ages 5 human years for every dog year.

Exercise

As people grow older, the size of their pupils tends to get smaller. The average diameter (in millimeters) of a person's pupils is given in the graph shown below.

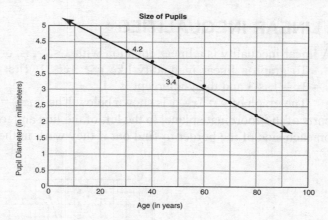

Write the equation of the line drawn through the points above. Write the equation in slope-intercept form.

Solution to Exercise

The line drawn goes through the points (30, 4.2) and (50, 3.4). Find the equation of the line containing those two points. Find the slope.

$$\text{slope} = \frac{\text{change in } y}{\text{change in } x} = \frac{3.4 - 4.2}{50 - 30} = \frac{-.8}{20} = -.04$$

Use point-slope form of the equation of a line to write the equation.

$$y - 4.2 = -.04(x - 30)$$
$$y - 4.2 = -.04x + 1.2$$
$$y = -.04x + 5.4$$

In the equation $y = -.04x + 5.4$, y represents pupil diameter and x represents age in years.

Practice Problem

14. The graph below shows the relationship between the brain weight of selected small animals and their maximum life expectancy.

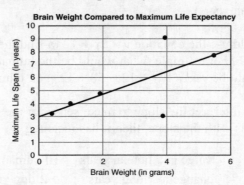

Brain Weight Compared to Maximum Life Expectancy

Use the points (1, 3.9) and (5.5, 7.5) to write the equation of the line drawn above. Interpret the meaning of the slope in this situation. Use your equation to predict the maximum life span of a cat with a brain weight of 25 grams.

(The solution is found at the end of the chapter)

GRAPHING LINEAR INEQUALITIES

A linear inequality is a linear equation with a $<$, $>$, $\leq$, or $\geq$ sign instead of an equal sign.

To graph a linear inequality like $y \geq 3x - 1$, first graph $y = 3x - 1$. The slope of $y = 3x - 1$ is 3 and the y-intercept is -1.

The graph of $y = 3x - 1$ is shown below. The line $y = 3x - 1$ is the border between two parts of the coordinate plane, to the left of the line and to the right of the line. Pick a point on each side of the border to find out if they work in the inequality $y \geq 3x - 1$.

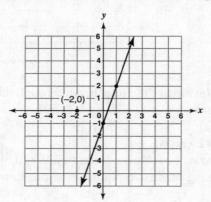

On the left side of the line, choose the point $(-2, 0)$, as shown on the graph above. Find out if it works in the inequality $y \geq 3x - 1$.

$$0 \overset{?}{\geq} 3(-2) - 1$$
$$0 \overset{?}{\geq} -6 - 1$$
$$0 \geq -7$$

The point $(-2, 0)$ makes the inequality $y \geq 3x - 1$ true.

On the right side of the line, choose the point (2, 0), as shown below, and find out if it works in the inequality $y \geq 3x - 1$.

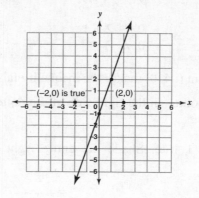

$$0 \overset{?}{\geq} 3(2) - 1$$
$$0 \overset{?}{\geq} 6 - 1$$
$$0 \not\geq 5$$

The point (2, 0) makes the inequality $y \geq 3x - 1$ false.

All of the points on the left side of the line $y = 3x - 1$ make the inequality $y \geq 3x - 1$ true, while all of the points on the right side of the line $y = 3x - 1$ make the inequality $y \geq 3x - 1$ false. Shade the left side of the line. The graph of $y \geq 3x - 1$ is shown below.

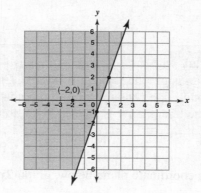

The inequality $y < \dfrac{1}{2}x + 1$ means $y = \dfrac{1}{2}x + 1$ is not included in the graph. To show

that you do not want to include the points that are on the line $y = \dfrac{1}{2}x + 1$, you graph the

line $y = \dfrac{1}{2}x + 1$ with a dotted line.

The dotted line divides the graph into two parts, points that will make the inequality

$y < \dfrac{1}{2}x + 1$ true and points that will make the inequality $y < \dfrac{1}{2}x + 1$ false. To determine

the side with the true points, pick a point on the left side, say (–4, 0) and pick a point on

the right side of the line, say (0, 0). Test each point in the inequality $y < \dfrac{1}{2}x + 1$ to find
out if the inequality will be true.

$$0 \overset{?}{<} \frac{1}{2}(-4) + 1 \qquad\qquad 0 \overset{?}{<} \frac{1}{2}(0) + 1$$

$$0 \overset{?}{<} -2 + 1 \qquad\qquad 0 \overset{?}{<} 0 + 1$$

$$0 \not< -1 \qquad\qquad 0 < 1$$

The point (–4, 0) makes the inequality false. The point (0, 0) makes the inequality true. All of the points on the right side of the dotted line $y = \frac{1}{2}x + 1$ will make the inequality

$y < \frac{1}{2}x + 1$ true. Shade the entire right side of the dotted line, as shown below.

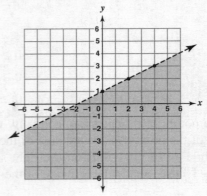

An inequality with a < or > sign is graphed using a dotted line. An inequality with a ≤ or ≥ sign is graphed with a solid line. Choose points to decide which side of the solid or dotted line contains all of the points that make the inequality true. If (0, 0) is not on the line, then it is an easy point to use.

Exercise

1. On the coordinate plane below, graph $2x + y > 6$.

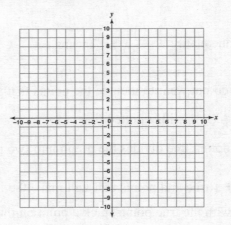

Solution to Exercise

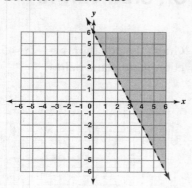

Graph the dotted line $2x + y = 6$. Choose a point on each side of the line and test it in the inequality $2x + y > 6$. The point $(0, 0)$ does not make the inequality $2x + y > 6$ true. The point $(5, 1)$ does make the inequality $2x + y > 6$ true. Shade the side of the dotted line that contains the point $(5, 1)$, as shown above.

Practice Problems

15. On the coordinate plane below, graph $y > -2$.

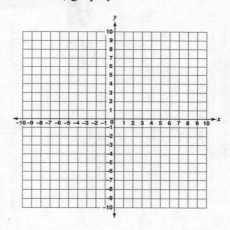

16. On the coordinate plane below, graph $y \le -\dfrac{2}{3}x - 4$.

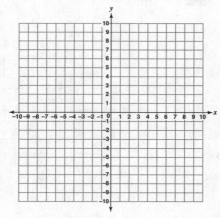

(Solutions are found at the end of the chapter.)

SAMPLE GQE QUESTIONS FOR CHAPTER 10

1. What is the slope of the line passing through the points (–8, 2) and (3, 5)?

A. $\frac{3}{11}$

B. $\frac{-3}{11}$

C. $\frac{11}{3}$

D. $\frac{-11}{3}$

2. What is the slope of the line in the graph below?

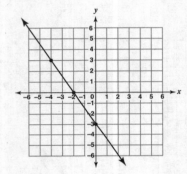

A. $\frac{3}{2}$

B. $\frac{-3}{2}$

C. $\frac{2}{3}$

D. $\frac{-2}{3}$

3. Graph the equation $y = -2x - 5$ on the coordinate plane below.

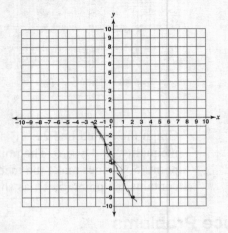

4. Graph the equation $2x - 3y = 12$ on the coordinate plane below.

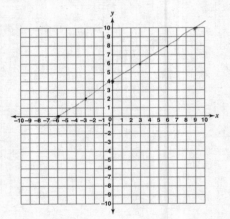

5. What is the equation of the line graphed below?

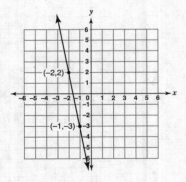

6. What is the equation of the straight line that has a slope of 5 and passes through the point (2, –3)?

   A. $y = -5x - 13$
   B. $y = -5x + 13$
   C. $y = 5x + 13$
   D. $y = 5x \ominus 13$

   mistake

7. What is the equation of the line that passes through the points (5, –4) and (–5, 10)?

8. On the grid below draw the line that has a slope of 2 and passes through the point (–2, –3).

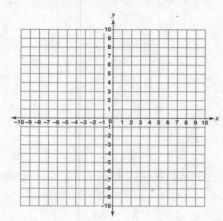

9. What is the equation of a line that is parallel to the line $y = 6x - 5$ and contains the point (4, –5)?

   A. $y = -6x + 19$
   B. $y = 6x + 34$
   C. $y = 6x - 29$
   D. $y = -6x - 26$

10. What is the equation of a line that is perpendicular to the line $y = \frac{1}{2}x + 9$ and contains the point (8, 6)?

   A. $y = -\frac{1}{2}x + 2$

   B. $y = -\frac{1}{2}x + 10$

   C. $y = -2x + 22$
   D. $y = -2x + 20$

11. Which of the following is a graph of $2x + 3y > 6$?

A.

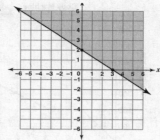

B.

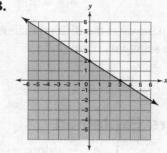

C.

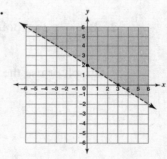

D.

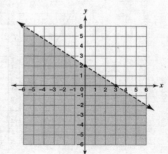

12. The graph below shows the number of copies a photocopy machine made over time.

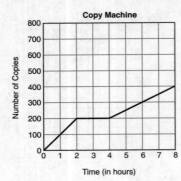

Copy Machine

Explain what the different slopes of the three line segments represent.

Answers Explained

1. **A** $\dfrac{3}{11}$. slope $= \dfrac{\text{change in } y}{\text{change in } x} = \dfrac{5-2}{3-(-8)} = \dfrac{3}{11}$

2. **B** $\dfrac{-3}{2}$. Two points on the graph are $(-2, 0)$ and $(0, -3)$. Count the spaces on the graph or use the slope formula. The slope is $\dfrac{3}{-2} = -\dfrac{3}{2} = \dfrac{-3}{2}$.

$$\text{slope} = \dfrac{\text{change in } y}{\text{change in } x} = \dfrac{-3-0}{0-(-2)} = \dfrac{-3}{2}$$

3.
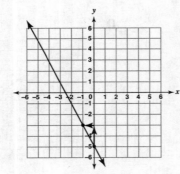

The slope of the line is -2 and the y-intercept is -5. Graph the y-intercept, $(0, -5)$. The slope is $-2 = -\dfrac{2}{1} = \dfrac{-2}{1} = \dfrac{2}{-1}$. Count up 2 spaces and left 1 space, as shown.

4.
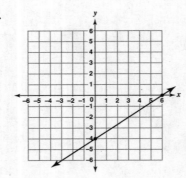

Find the x-intercept and the y-intercept.

$$2x - 3(0) = 12 \qquad 2(0) - 3y = 12$$
$$2x = 12 \qquad\qquad -3y = 12$$
$$x = 6 \qquad\qquad y = -4$$

The line crosses the x-axis at $(6, 0)$. The line crosses the y-axis at $(0, -4)$. Draw the line between the two intercepts, as shown on the left.

If you did not remember how to graph using the intercepts, then change the equation into slope-intercept form.

$$2x - 3y = 12$$
$$-3y = -2x + 12$$
$$y = \frac{2}{3}x + -4$$

Then use the slope $\frac{2}{3}$ and the y-intercept -4 to graph the line.

5. $y = -5\,x - 8$. Two points are given, $(-2, 2)$ and $(-1, -3)$.

$$\text{slope} = \frac{\text{change in } y}{\text{change in } x} = \frac{-3 - 2}{-1 - (-2)} = \frac{-5}{1} = -5$$

$$y - 2 = -5(x - (-2))$$
$$y - 2 = -5(x + 2)$$
$$y - 2 = -5x + -10$$
$$y = -5x - 8$$

6. **D** $y = 5x - 13$

$$y - (-3) = 5(x - 2)$$
$$y + 3 = 5x - 10$$
$$y = 5x - 13$$

7. $y = -\frac{7}{5}x + 3$

$$\text{slope} = \frac{\text{change in } y}{\text{change in } x} = \frac{10 - (-4)}{-5 - 5} = \frac{14}{-10} = -\frac{7}{5}$$

$$y - (-4) = -\frac{7}{5}(x - 5)$$

$$y + 4 = -\frac{7}{5}x + 7$$

$$y = -\frac{7}{5}x + 3$$

8.

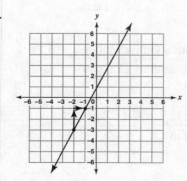

Graph the point $(-2, -3)$. The slope is $2 = \frac{2}{1}$.

From the point $(-2, -3)$ count up 2 spaces and right 1 space. Draw the line joining the two points, as shown.

9. **C** $y = 6x - 29$. Parallel lines have the same slope. The slope of $y = 6x - 5$ is 6. Use the point-slope form of the equation of a line.

$$y - (-5) = 6(x - 4)$$
$$y + 5 = 6x + -24$$
$$y = 6x - 29$$

10. **C** $y = -2x + 22$. Perpendicular lines have slopes that are opposite reciprocals.

The slope of $y = \frac{1}{2}x + 9$ is $\frac{1}{2}$. The slope of a perpendicular line is -2.

Use point-slope form of the equation of a line.

$$y - 6 = -2(x - 8)$$
$$y - 6 = -2x + 16$$
$$y = -2x + 22$$

11. **C** When you graph an inequality with a <, use a dotted line. The point (0, 0) does not make the inequality true. The point (4, 2) does make the inequality true. Shade the side of the dotted line that contains the point (4, 2).

12. For the first 2 hours, the copy machine worked at the rate of $\frac{100 \text{ copies}}{1 \text{ hour}}$. For the next 2 hours, the copy machine worked at the rate of $\frac{0 \text{ copies}}{1 \text{ hour}}$. For the last 4 hours, the copy machine worked at the rate of $\frac{50 \text{ copies}}{1 \text{ hour}}$.

Solutions to Practice Problems

1. $m = -\frac{5}{3}$. Two points on the line are (2, 4) and (5, -1). Finding the slope by counting is shown below.

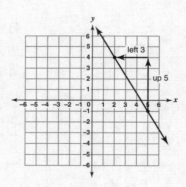

$$m = \frac{\text{how much the } y \text{ changes}}{\text{how much the } x \text{ changes}} = \frac{5}{-3}$$

Or use the slope formula.

$$\text{slope} = \frac{\text{change in } y}{\text{change in } x} = \frac{-1-4}{5-2} = \frac{-5}{3} = -\frac{5}{3}$$

2. slope = $\dfrac{\text{change in } y}{\text{change in } x} = \dfrac{-12-(-2)}{-5-(-7)} = \dfrac{-10}{2} = -5$

3.

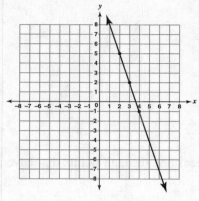

First, graph the point $(4, -1)$ as shown above. Then use the slope to find other points.

4.

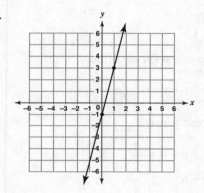

The slope of the line is 4 and the y-intercept is -1. Graph the y-intercept $(0, -1)$. Then use the slope to find another point on the line, as shown.

5.

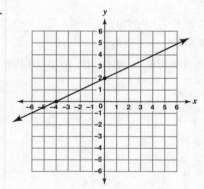

Change the equation $2y - x = 4$ into slope-intercept form. In slope-intercept form, $2y - x = 4$ is $y = \dfrac{1}{2}x + 2 = 4$. The slope is $\dfrac{1}{2}$ and the y-intercept is 2. Or find the x- and y-intercepts of the line $2y - x = 4$. The x-intercept is -4 and the y-intercept is 2.

6.

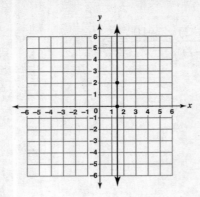

The graph of $x = 1\frac{1}{2}$ is a vertical line through points that have an x-value of $1\frac{1}{2}$.

Points included on the line are $\left(1\frac{1}{2}, 2\right)$, $\left(1\frac{1}{2}, 0\right)$, and $\left(1\frac{1}{2}, -2\right)$.

7. $y - 4 = \frac{1}{2}(x - 6)$ or $y = \frac{1}{2}x + 1$

Use the point-slope form of the equation of a line.

$$y - 4 = \frac{1}{2}(x - 6)$$

You can change to slope-intercept form.

$$y - 4 = \frac{1}{2}x - 3$$

$$y = \frac{1}{2}x + 1$$

8. Find the slope using the formula.

$$\text{slope} = \frac{\text{change in } y}{\text{change in } x} = \frac{y_2 - y_1}{x_2 - x_1} = \frac{-4 - 1}{8 - 6} = \frac{-5}{2}$$

Use one of the points, the slope, and the point-slope form of the equation of a line to write the equation.

$$y - y_1 = m(x - x_1)$$

$$y - 1 = -\frac{5}{2}(x - 6)$$

$$y - 1 = -\frac{5}{2}x + 15$$

$$y = -\frac{5}{2}x + 16$$

9. $y = 7x + 2$. Substitute the slope of 7 and y-intercept of 2 into slope-intercept form of the equation of a line.

10. $y = 3$. Substitute the slope of 0 and y-intercept of 3 into slope-intercept form.

11. $y - 3 = \frac{2}{3}(x - 1)$ or $y - 5 = \frac{2}{3}(x - 4)$. You are given two points on the line, (1, 3) and (4, 5). Use the slope formula, or count spaces on the graph to find that the slope is $\frac{2}{3}$. Use one of the points, the slope, and the point-slope form of the equation of a line to write the equation.

$$y - 3 = \frac{2}{3}(x - 1) \text{ or } y - 5 = \frac{2}{3}(x - 4)$$

12. $y = \frac{2}{3}x + 7$

The slope of the line $y = \frac{2}{3}x - 37$ is $\frac{2}{3}$, so the slope of a parallel line is $\frac{2}{3}$.

$$y - 15 = \frac{2}{3}(x - 12)$$

$$y - 15 = \frac{2}{3}x - 8$$

$$y = \frac{2}{3}x + 7$$

13. $y = -\frac{5}{4}x + 21$

The slope of $y = \frac{4}{5}x - 10$ is $\frac{4}{5}$, so the slope of a perpendicular line is $-\frac{5}{4}$.

$$y - 16 = -\frac{5}{4}(x - 4)$$

$$y - 16 = -\frac{5}{4}x + 5$$

$$y = -\frac{5}{4}x + 21$$

14. $y = .8x + 3.1$. Find the equation of the line containing the two points. Find the slope.

$$\text{slope} = \frac{\text{change in } y}{\text{change in } x} = \frac{7.5 - 3.9}{5.5 - 1} = \frac{3.6}{4.5} = .8$$

Use point-slope form of the equation of a line.

$$y - 3.9 = .8(x - 1)$$
$$y - 3.9 = .8x - .8$$
$$y = .8x + 3.1$$

In the equation $y = .8x + 3.1$, y represents maximum life span of the animal and x represents brain weight of the animal.

Put labels with the slope.

$$\text{slope} = \frac{\text{change in } y \text{ maximum life span}}{\text{change in } x \text{ brain weight}} = \frac{3.6 \text{ years}}{4.5 \text{ grams}} = .8 \text{ years per gram}$$

This would mean that an animal gains .8 years of life for every gram increase in brain weight.

Use the equation $y = .8x + 3.1$ to find the maximum life span of a cat with a brain weight of 25 grams.

$$y = .8x + 3.1$$
$$y = .8(25) + 3.1$$
$$y = 20 + 3.1$$
$$y = 23.1$$

Using the formula, the maximum life span of a cat is 23.1 years.

15.

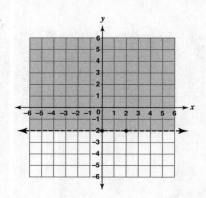

Graph the dotted line $y = -2$. The point $(0, 0)$ does make the inequality true. The point $(0, -5)$ does not make the inequality true. Shade the side of the dotted line that contains the point $(0, 0)$.

16.

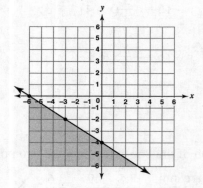

Graph the solid line $y = -\frac{2}{3}x - 4$. The point $(0, 0)$ does not make the inequality true. The point $(-3, -4)$ does make the inequality true. Shade the side of the solid line that contains the point $(-3, -4)$.

Chapter 11 | Algebra 1: Pairs of Linear Equations and Inequalities

INDIANA'S ACADEMIC MATHEMATICS STANDARDS ADDRESSED:

A1.5.1 Use a graph to estimate the solution of a pair of linear equations in two variables.

A1.5.2 Use a graph to find the solution set of a pair of linear inequalities in two variables.

A1.5.3 Understand and use the substitution method to solve a pair of linear equations in two variables.

A1.5.4 Understand and use the addition or subtraction method to solve a pair of linear equations in two variables.

A1.5.5 Understand and use multiplication with the addition or subtraction method to solve a pair of linear equations in two variables.

A1.5.6 Use pairs of linear equations to solve word problems.

ESTIMATING THE SOLUTION TO A PAIR OF LINEAR EQUATIONS

The solution to a pair of linear equations is the point or points that satisfy both equations. Using a graph to estimate the solution means to look at the graphs of two linear equations and find the point or points where they intersect. You may want to review graphing of lines in Chapter 10.

When two linear equations are graphed on the same set of axes, two lines are drawn. Two lines in a plane can intersect in one point, in no points, or in an infinite number of points.

If the two lines intersect in one point, then the solution to the pair of equations is the point where the two lines intersect, as shown below.

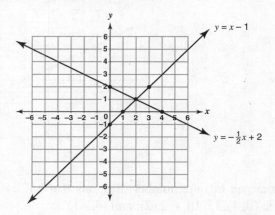

The lines graphed are $y = x - 1$ and $y = -\frac{1}{2}x + 2$. They intersect in one point, (2, 1).

This means that the point (2, 1) is graphed on both lines and works in both equations. You can check to see if (2, 1) works in both equations by substituting 2 for x and 1 for y.

$$y = x - 1 \qquad\qquad y = -\frac{1}{2}x + 2$$

$$1 \overset{?}{=} 2 - 1 \qquad\qquad 1 \overset{?}{=} -\frac{1}{2} \cdot 2 + 2$$

$$1 = 1 \qquad\qquad 1 \overset{?}{=} -1 + 2$$

$$\qquad\qquad\qquad\qquad 1 = 1$$

If two lines in a plane do not intersect, then that means that they are parallel, as shown below.

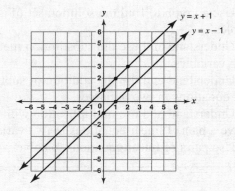

The lines graphed are $y = x + 1$ and $y = x - 1$. The lines are parallel and will never intersect. Remember parallel lines have the same slope. Refer to Chapter 10 to review equations of parallel lines. The solution to this pair of equations is the empty set, $\varnothing$, because there are no points that satisfy both equations.

If two lines are the same line when they are graphed, then there are an infinite number of points that satisfy both equations. The two equations may not look the same, but when graphed, you see that they are the same line. The pair of lines, $y = -\frac{1}{2}x + 1$ and $2x + 4y = 4$ are graphed below.

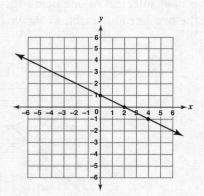

There are an infinite number of points that will satisfy both equations. Points that work include $(0, 1)$, $(2, 0)$, $(-2, 2)$, and $(4, -1)$.

Exercise

Graph both lines $x + 4y = 2$ and $y = \frac{1}{2}x + 2$ on the coordinate plane below.

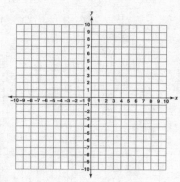

Use your graph to find the solution to the system of equations.

Solution to Exercise
Both lines are graphed below.

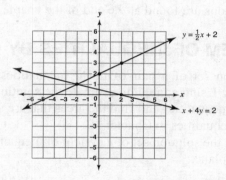

The solution is the point where the two lines meet, $(-2, 1)$.

Practice Problems

1. Look at the system of equations below.

$$x = y + 1$$

$$y = \frac{1}{2}x - 2$$

Graph the system of equations on the coordinate plane below.

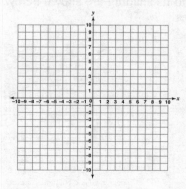

What is the solution to the system of equations?

2. On the coordinate plane provided, graph the system of equations.

$$-x + \frac{1}{2}y = -2$$

$$2x - y = 6$$

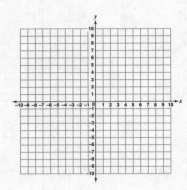

Find the solution to the system of equations.

(Solutions are found at the end of the chapter.)

SOLVING A SYSTEM OF INEQUALITIES BY GRAPHING

The solution set of a pair of linear inequalities is (are) the point(s) that satisfy both inequalities. Using a graph to estimate the solution set means to look at the graphs of the two linear inequalities and find the areas that are shaded in both graphs. Review graphing of linear inequalities in Chapter 10.

To find the solution set of a system of inequalities, graph each inequality on the same coordinate plane.

$$x + 2y > 4$$
$$y \geq x - 1$$

To graph $x + 2y > 4$, first graph $x + 2y = 4$ with a dotted line.

> **Be Careful!** An inequality with a $<$ or $>$ sign is graphed using a dotted line. An inequality with a $\leq$ or $\geq$ sign is graphed with a solid line.

Then, shade the side of the line that has the points that makes $x + 2y > 4$ true. The dotted line and its shading are shown below.

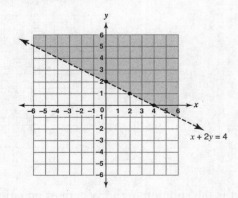

$x + 2y = 4$

To graph $y \geq x - 1$, first graph $y = x - 1$ with a solid line. Then, shade the side of the line that has the points that make $y \geq x - 1$ true. The solid line and its shading are shown below.

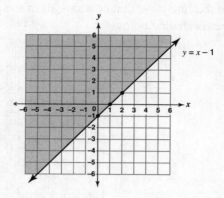

The final solution is the area shaded in both figures above. The final answer is shown below.

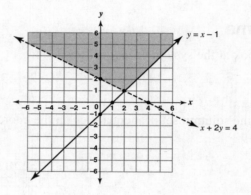

Exercise

Look at the system of inequalities below.

$$x - y \geq 0$$
$$y < 2$$

Graph the solution set of the system of inequalities on the coordinate plane below.

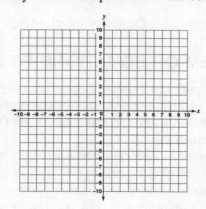

Solution to Exercise

First, graph $x - y = 0$ with a solid line. Then, shade the side of the line that contains all of the points that satisfy the inequality $x - y \geq 0$. Draw a dotted line on $y = 2$. Then, shade the side of the line $y = 2$ that contains all of the points that satisfy the inequality $y < 2$. The solution set is shaded below.

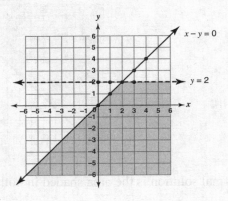

Practice Problems

3. Look at the system of inequalities below.

$$y \geq 2x - 3$$
$$y < 7 - 3x$$

Graph the solution set of inequalities on the coordinate plane below.

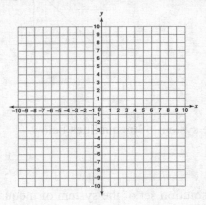

(Solution is found at the end of the chapter.)

SOLVING A SYSTEM OF EQUATIONS BY SUBSTITUTION

Although graphing a system of equations does provide a solution, the answer will only be as accurate as the graph drawn. If the solution is not an ordered pair composed of integers, then the solution will be an estimate. Algebra can be used to find the exact answer to any system of equations.

Substitution is a process familiar to anyone who has seen a sporting event. The coach decides that one player needs to take the place of another player, so a substitution is made. As one player exits the game, another player enters in that position. Substitution works the same way in solving systems of equations. As one variable exits from an equation, another variable expression enters in its place.

Example

Solve $y = 1 - 2x$ and $7x + 2y = 17$ using substitution.

The first equation states that y is the same as $1 - 2x$. That means that y can be taken out of the equation as the expression is substituted in its place, as shown below.

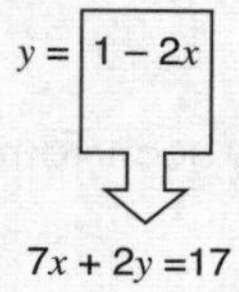

$$y = \boxed{1 - 2x}$$
$$\downarrow$$
$$7x + 2y = 17$$

The new equation will become $7x + 2(1 - 2x) = 17$. Now the equation only has one variable, so it is easy to solve for x.

$$7x + 2(1 - 2x) = 17$$
$$7x + 2 - 4x = 17$$
$$3x + 2 = 17$$
$$3x = 15$$
$$x = 5$$

The solution to a system of equations is the point that satisfies both equations. We must find the value of y when x is 5. By substituting 5 for x in one of the equations, the value of y can be found. It does not matter which equation is used because the point must make both equations true. The work to find y using both equations is shown below.

$$y = 1 - 2x \qquad\qquad 7x + 2y = 17$$
$$y = 1 - 2(5) \qquad\qquad 7(5) + 2y = 17$$
$$y = 1 - 10 \qquad\qquad 35 + 2y = 17$$
$$y = -9 \qquad\qquad\qquad 2y = -18$$
$$\qquad\qquad\qquad\qquad y = -9$$

The solution to the system of equations is (5, –9).

On the GQE, not all of the systems of equations are ready to use the substitution process. Sometimes one of the equations will need to be solved for a variable so that substitution can be used.

Example

Solve $-\dfrac{1}{2}x + y = 6$ and $2y - 4x = -27$ using the substitution process.

Neither equation is arranged so that substitution can be used immediately. One of the equations must be solved for either x or y. The best choice would be to solve the first equation for y because it would require adding $\dfrac{1}{2}x$ to both sides of the equal sign. Solving for any of the other variables in either equation would require more than one step.

$$-\frac{1}{2}x + y = 6$$

$$y = \frac{1}{2}x + 6$$

Now use the substitution process.

$$2(\frac{1}{2}x + 6) - 4x = -27$$
$$x + 12 - 4x = -27$$
$$-3x + 12 = -27$$
$$-3x = -39$$
$$x = 13$$

Since the value of x has been found, find the value of y when x is 13.

$$2y - 4(13) = -27$$
$$2y - 52 = -27$$
$$2y = 25$$
$$y = \frac{25}{2} \text{ or } 12\frac{1}{2}$$

The solution to the system of equations is $(13, 12\frac{1}{2})$.

Exercises

1. Solve $x = 2 - 5y$ and $7y - 2x = 30$ using substitution.

2. Find the solution to the system of equations using substitution.

$$x - 3y = 2$$
$$-3x + 4y = -41$$

Solutions to Exercises

1. The first equation states that x is the same as $2 - 5y$. Substitute $2 - 5y$ in place of x in $7y - 2x = 30$.

$$7y - 2(2 - 5y) = 30$$

Be Careful! $7y - 2(2 - 5y)$ is the same as $7y + -2(2 + -5y)$. Use the distributive property to multiply each term in the parentheses by -2.

$$7y - 4 + 10y = 30$$
$$17y - 4 = 30$$
$$17y = 34$$
$$y = 2$$

Find the value of x when y is 2.

$$x = 2 - 5(2)$$
$$x = 2 - 10$$
$$x = -8$$

The solution to the system of equations is $(-8, 2)$.

2. Neither equation is ready to use the substitution process, so one of the equations must be solved for a variable. The best choice is to solve the first equation for x.

$$x - 3y = 2$$
$$x = 3x + 2$$

Now substitute $3y + 2$ for the x in the equation $-3x + 4y = -41$.

$$-3(3y + 2) + 4y = -41$$
$$-9y - 6 + 4y = -41$$
$$-5y - 6 = -41$$
$$-5y = -35$$
$$y = 7$$

Find the value of x when $y = 7$.

$$x - 3(7) = 2$$
$$x - 21 = 2$$
$$x = 23$$

The solution to the system of equations is $(23, 7)$.

Practice Problems

4. Look at the system of equations below.

$$x = \frac{1}{2}y + 3$$
$$4x - 5y = -15$$

What is the solution to the system of equations?

5. Solve the system of equations $7x - 4y = 21$ and $y - 2x = -1$.

(Solutions are found at the end of the chapter.)

SOLVING A SYSTEM OF EQUATIONS BY ADDITION

Sometimes the substitution process becomes messy. Consider the system of equations shown below.

$$5x - 3y = 22$$
$$4x + 3y = -4$$

If either equation is solved for a variable, then lots of fractions will appear. Since fractions cause foolish mistakes, another method can be used with this problem. This process is called the addition or subtraction method.

To solve the system of equations shown below by using the addition method, just add the two equations together.

$$5x - 3y = 22$$
$$\underline{4x + 3y = -4}$$
$$9x \quad\quad = 18$$

Since $-3y + 3y = 0y = 0$, the equation will have one variable, x. The equation $9y = 18$ is simple to solve. We find that $x = 2$. Find the value of y when x is 2.

$$5(2) - 3y = 22$$
$$10 - 3y = 22$$
$$-3y = 12$$
$$y = -4$$

The solution to the system is $(2, -4)$.

Example

Look at the system of equations shown below.

$$7x + 2y = -11$$
$$4y - 7x = 41$$

Mistakes abound when the equations are added together as they are now. Rearranging the terms in the second equation will put the like terms under each other.

$$4y - 7x = 41$$
$$4y + -7x = 41$$
$$-7x + 4y = 41$$

Rewrite the original problem.

$$7x + 2y = -11$$
$$\underline{-7x + 4y = \quad 41}$$
$$6y = \quad 30$$
$$y = \quad 5$$

Find the value of x when y is 5.

$$7x + 2(5) = -11$$
$$7x + 10 = -11$$
$$7x = -21$$
$$x = -3$$

The solution to the system of linear equations is $(-3, 5)$.

Exercises

1. Solve the system of equations $x + 4y = -1$ and $x - 4y = 7$ using the addition method.

2. Use the addition method to solve the system of equations $7x - 5y = -11$ and $5y = 3x - 1$.

Solutions to the Exercises

1.

> **Be Careful!** Remember that $x + x$ means $1x + 1x = 2x$.

$$x + 4y = -1$$
$$\underline{x - 4y = 7}$$
$$2x = 6$$
$$x = 3$$

To find the value of y, substitute 3 for the x in one of the equations.

$$3 + 4y = -1$$
$$4y = -4$$
$$y = -1$$

The solution to the system of equations is $(3, -1)$.

2. Observe that the equations do not have the variables lined up in the same order. The equation $5y = 3x - 1$ should be changed so that the term containing the x is first and the term containing the y is second.

$$5y = 3x - 1$$
$$-3x + 5y = -1$$

Now the system of equations can easily be solved using the addition method.

$$7x - 5y = -11$$
$$\underline{-3x + 5y = -1}$$
$$4x = -12$$
$$x = -3$$

To find the value of y, substitute -3 for the x in one of the equations.

$$7(-3) - 5y = -11$$
$$-21 - 5y = -11$$
$$-5y = 10$$
$$y = -2$$

The solution to the system of equations is $(-3, -2)$.

Practice Problems

6. Find the solution to the system of equations $7y + 3x = -16$ and $10y = 3x + 50$.

(Solutions are found at the end of the chapter.)

SOLVING A SYSTEM OF EQUATIONS BY MULTIPLICATION WITH ADDITION

Most pairs of equations on the GQE are not so simple. The new process involves one additional step, the use of multiplication before the addition process is applied.

Example
Find the solution to the system of equations $4x + 3y = -10$ and $5x - 6y = 46$.

Before you jump into solving the pair of equations, look carefully at the problem. Notice that neither equation could be easily solved for x or for y, so the substitution process is not the best choice. If you add the two equations together as they are now, neither one of the variables is eliminated.

To solve the system of equations using multiplication combined with the addition method, look at the coefficients of the variables. The coefficients of the x's are 4 and 5, while the coefficients of the y's are 3 and -6. The coefficients of the x's or of the y's need to be opposites of each other. By doing that, a variable will be eliminated when the equations are added.

Look at the coefficients of the y's. What number do both 3 and -6 go into? They both go into 6.

The coefficients of y, 3 and -6, could be changed into opposites of each other if you multiplied 3 by 2.

Multiply each term of the first equation by 2.

$$2 \cdot 4x + 2 \cdot 3y = -10$$
$$8x + 6y = -20$$

The system of equations has changed. Add the two equations.

$$8x + 6y = -20$$
$$\underline{5x - 6y = 46}$$
$$13x = 26$$
$$x = 2$$

To find the value of y, substitute 2 for x in one of the original equations.

$$4(2) + 3y = -10$$
$$8 + 3y = -10$$
$$3y = -18$$
$$y = -6$$

The solution to the system of equations is $(2, -6)$.

Example

Some cases are more complicated. Look at the system of equations.

$$7x + 3y = 6$$
$$2x + 5y = -19$$

Look at the variable terms. The coefficients of x are 7 and 2. The coefficients of y are 3 and 5. Choose the variable that you would like to eliminate and work to make the coefficients opposite of each other.

The coefficients of x are 7 and 2. Think of a number that both 7 and 2 go into. Both 7 and 2 go into 14. The coefficients need to be opposite of each other, so make one of the terms $14x$ and one of them $-14x$.

Multiply each term of the first equation by 2 and each term of the second equation by -7.

$$2 \cdot 7x + 2 \cdot 3y = 2 \cdot 6$$
$$-7 \cdot 2x + -7 \cdot 5y = -7 \cdot -19$$

Use the addition process to solve the system.

$$14x + 6y = 12$$
$$\underline{-14x - 35y = 133}$$
$$-29y = 145$$
$$y = -5$$

The value of x can be found by substituting -5 for the y in one of the original equations.

$$7x + 3(-5) = 6$$
$$7x - 15 = 6$$
$$7x = 21$$
$$x = 3$$

The solution to the system is $(3, -5)$.

Exercise

Look at the system of equations.

$$9x + 10y = -40$$
$$2x + 3y = -5$$

What is the solution to the system?

Solution to Exercise

The coefficients of the terms with the x are 9 and 2. The coefficients of the terms with the y are 10 and 3. With this problem you can eliminate either the x-terms or the y-terms.

Eliminate the terms with the x.

$$2 \cdot 9x + 2 \cdot 10y = 2 \cdot -40$$
$$-9 \cdot 2x + -9 \cdot 3y = -9 \cdot -5$$
$$18x + 20y = -80$$
$$\underline{18x - 27y = 45}$$
$$-7y = -35$$
$$y = 5$$

> Both 9 and 2 go into 18. Multiply the top equation by 2 and the bottom equation by –9.

Find the value of x by substituting 5 for the y in one of the original equations.

$$9x + 10(5) = -40$$
$$9x + 50 = -40$$
$$9x = -90$$
$$x = -10$$

The solution to the system of equations is (–10, 5).

Practice Problems

7. Find the solution to the system of equations shown below.

$$4x + 3y = 3$$
$$5x - 2y = -25$$

(Solution is found at the end of the chapter.)

SOLVING WORD PROBLEMS USING A SYSTEM OF EQUATIONS

The GQE has word problems that require you to write a system of equations, and then, solve the equations to find the answer to the word problem. Points are given for writing a system of equations that describe the word problem, and points are given for solving your equations.

   The hardest part of solving word problems is writing the equations. Refer to Chapter 3 for a review of writing equations for word problems.

Example

Ticket prices for two different amusement parks are shown in the table below.

	Adult Admission	Child (under 16) Admission
Lotsa Fun Amusement Park	$10.50	$5.50
Cra-Z Times Roller Coasters	$12.50	$5.00

Mr. Cool Teacher wants to take his ninth grade class, with chaperones, on a field trip to one of two places. At Lotsa Fun Amusement Park, the bill would be $130.50. At Cra-Z Times Roller Coasters, the bill would be $127.50. Write a system of equations that could be used to find the number of adults (A) going on the trip and the number of children (C) going on the trip. Do not solve your system of equations.

Read the problem several times. Decide what you are looking for in the problem. The phrase near the end of the problem asks you to find the number of adults (A) going on the trip and the number of children (C) going on the trip. The problem gives you the variables to use. A will represent the number of adults on the field trip, and C will represent the number of children on the field trip.

To find the total bill for Lotsa Fun Amusement Park, multiply the admission fee for an adult (10.50) times the number of adults (A), and add that to the admission fee for a child (5.50) times the number of children (C).

$$10.50 \cdot A + 5.50 \cdot C = 130.50$$

To find the total bill for Cra-Z Times Roller Coasters, multiply the admission fee for an adult (12.50) times the number of adults (A), and add that to the admission fee for a child (5.00) times the number of children (C).

$$12.50 \cdot A + 5.00 \cdot C = 127.50$$

The system of equations that could be used to solve the word problem is

$$10.50A + 5.50C = 130.50$$
$$12.50A + 5.00C = 127.50$$

Example
Mr. Gath R. Coins has collected 2,036 nickels and dimes over the past year. He received $141.25 when he cashed in all of his coins. Write a pair of equations that could be used to find the number of nickels, n, and the number of dimes, d, he had. Use your equations to find the number of nickels Gath R. Coins had.

There were a total of 2,036 coins, so the number of nickels and the number of dimes added together would be 2,036.

$$n + d = 2,036$$

Each nickel is worth five cents and that can be written 5¢ or $.05. To find out how much all of the nickels are worth, you would multiply 5¢ or $.05 by the number of nickels. Remember we chose to use n to represent the number of nickels Gath R. Coins had. His nickels would be worth $5n$ cents or $.05n$ dollars.

Each dime is worth ten cents and that can be written as 10¢ or $.10. To find out how much all of the dimes are worth, you would multiply 10¢ or $.10 by the number of dimes. Remember we chose to use d to represent the number of dimes Gath R. Coins had. His dimes would be worth $10d$ cents or $.10d$ dollars.

The total amount of money Gath R. Coins had was $141.25, which is the amount of money the nickels are worth plus the amount of money the dimes are worth.
The second equation would be $.05n + .10d = 141.25$.

Solve the system of equations.

$$n + d = 2{,}036$$
$$.05n + .10d = 141.25$$

The substitution method is shown. Solve the first equation for n.

$$n + d = 2{,}036$$
$$n = 2{,}036 - d$$

Substitute $2{,}036 - d$ into the second equation for n.

$$.05(2{,}036 - d) + .10d = 141.25$$
$$101.8 - .05d + .10d = 141.25$$
$$.05d + 101.8 = 141.25$$
$$.05d = 39.45$$
$$d = 789$$

Substitute 789 for the d in the first equation to find the number of nickels.

$$n + 789 = 2{,}036$$
$$n = 1{,}247$$

Gath R. Coins collected a total of 2,036 coins, 1,247 nickels and 789 dimes.

Exercises

1. Mr. Wrought N. Teeth visited his dentist to have work done. Mr. Teeth could have 5 fillings and 1 root canal for a cost of $1,472. Or, Mr. Teeth could have 2 fillings and 3 root canals for a cost of $2,758.50. Write a system of equations that could be used to find the cost of 1 filling, f, and the cost of 1 root canal, r. Do not solve the system of equations.

2. Phil Thee Rich has a wad of bills in his money clip. He has 49 more hundred-dollar bills than he does fifty-dollar bills. Phil Thee Rich has $10,150 in his money clip. Write a system of equations that could be used to find the number of hundred-dollar bills (h) and the number of fifty-dollar bills (f) that Phil Thee Rich has.

Solutions to Exercises

1. The problem wants you to find the cost of a filling and the cost of a root canal. The phrase *5 fillings and 1 root canal for a cost of $1,472* translates to $5f + 1r = 1{,}472$. The phrase *2 fillings and 3 root canals for a cost of $2,758.50* translates to $2f + 3r = 2{,}758.50$.
   You have written the system of equations.

$$5f + 1r = 1{,}472$$
$$2f + 3r = 2{,}758.50$$

2. h is the number of hundred-dollar bills that Phil Thee Rich has and f is the number of fifty-dollar bills he has.
There are 49 more hundred-dollar bills than there are fifty-dollar bills.

> **Be Careful!** There are fewer fifty-dollar bills than hundred-dollar bills. That means you need to add 49 to the number of fifty-dollar bills to make it the same as the number of hundred-dollar bills.

The equation for this idea is $h = f + 49$.

The money is worth a total of \$10,150.
Fifty-dollar bills are worth \$50. If you have f fifty-dollar bills, then they would be worth $50f$. Similarly, hundred-dollar bills are worth \$100. If you have h hundred-dollar bills, then they would be worth $100h$.
The equation for this idea is $50f + 100h = 10,150$.

Practice Problems

8. Ms. T. Cher wrote a 100-point algebra test with 26 questions. Some of the questions are worth 3 points and some of the questions are worth 5 points. Write a system of equations that could be used to find the number of three-point questions T and the number of five-point questions F on the test. Solve your system of equations to find the number of each type of question on the test.

9. Not 2 Healthy Candy Store is making a mixture of gummy bears and gummy worms to sell at the Fun Fair. Gummy bears cost \$1.89 a pound and gummy worms cost \$2.08 per pound. The 19-pound mixture needs to cost a total of \$38. Write a system of equations that could be used to find the number of pounds of gummy bears (B) and the number of pounds of gummy worms (W) that should be used. Solve your system of equations to find the number of pounds of gummy bears that would be needed.

(Solutions are found at the end of the chapter.)

SAMPLE GQE QUESTIONS FOR CHAPTER 11

1. Look at the graph below.

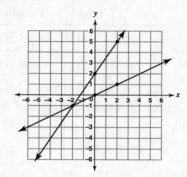

What is the solution to the system of linear equations?

2. Look at the system of equations below.

$$y = x + 2$$

$$y = -\frac{1}{3}x - 2$$

Graph the system of equations on the coordinate plane below.

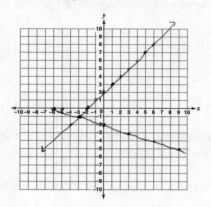

Use your graph to find the solution to the two equations. Describe one way to check the accuracy of your solution.

3. Look at the system of inequalities below.

$$x + 2y \leq 6$$

$$x - 4y > 0$$

Graph the solution set of the system of inequalities on the coordinate plane below.

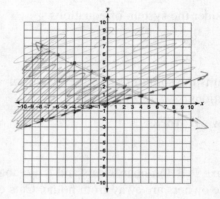

4. Look at the system of equations below.

$$x = 3 - y$$

$$4x + 3y = -3$$

What is the solution to the system of equations?

Show All Work

5. What is the solution to the system of equations shown below?

$$2x + 3y = -8$$

$$4x - 3y = -106$$

Show All Work

6. Look at the system of equations below.

$$4x + 15y = -10$$

$$7y - 4x = 54$$

What is the solution to the system of equations?

Show All Work

7. Look at the system of equations below.

$$3x - 2y = 13$$
$$7x - 6y = 25$$

What is the solution set to the system of equations?

Show All Work

8. Consider the system of equations.

$$8x + 9y = -5$$
$$5x + 6y = -2$$

What is the solution set to the system of equations?

Show All Work

9. Deluxe Pets Inn boards cats and large dogs while owners are away from home. Cats cost $25 per night and large dogs cost $42 a night. Tuesday night there are 14 animals staying for a total charge of $452. Write a system of equations that could be used to find the number of cats staying Tuesday night at the Deluxe Pets Inn. Use your equations to find the number of cats staying Tuesday night at Deluxe Pets Inn.

10. The costs to board a large dog at two different kennels is shown in the table below.

	Cost per Night	Cost per Playtime
Deluxe Pets Inn	$45.00	$3.50
Pet Hostel	$28.00	$4.50

The equation below represents some of this information.

$$c = 45 + 3.5p$$

Explain what each variable, coefficient, and constant in the equation represent.

11. The Smith family reunion had 9 adults and 6 children. At the B-Full Buffet, the total bill was $106.50. The Johnson family reunion had 7 adults and 12 children. Their bill at B-Full Buffet was $108.50. Write a system of equations that could be used to find the cost of an adult, a, at B-Full Buffet and the cost of a child, c, at B-Full Buffet. You do not need to solve your system of equations.

Answers Explained

1. $(-2, -1)$. On a graph of a system of equations, the solution is found where the two lines intersect. The point of intersection of the lines graphed is $(-2, -1)$.

2. $(-3, -1)$. The line $y = x + 2$ has a slope of 1 and a y-intercept of 2, as shown below.

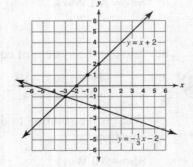

Also shown is the line $y = -\frac{1}{3}x - 2$ with a slope of $-\frac{1}{3}$

and a y-intercept of -2. The point where the two lines intersect is $(-3, -1)$, so the solution to the system of equations is $(-3, -1)$.

Check the accuracy of your solution by substituting the values for x and for y into the system of equations. Both equations must be true. If your answer did not work in one of the two equations, then you know that it is wrong.

3.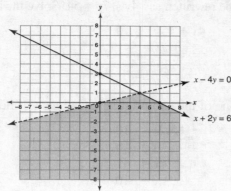

First, graph the inequality $x + 2y \leq 6$. Remember to use a solid line and shade the side that has the points that satisfy the inequality $x + 2y \leq 6$.

Graph the line $x - 4y > 0$ with a dotted line. Shade the side that has all the points that make the inequality $x - 4y > 0$ true.

The solution set of the system of inequalities is the portion of the graph that is shaded.

4. (−12, 15). Use the substitution process. Replace x in the second equation with $3 - y$.

$$4(3 - y) + 3y = -3$$
$$12 - 4y + 3y = -3$$
$$-y + 12 = -3$$
$$-y = -15$$
$$y = 15$$

Either equation in the system of equations can be used to find the value of x.

$$x = 3 - y$$
$$x = 3 - 15$$
$$x = -12$$

5. (−19, 10). Look at the two equations before you begin solving. Notice that the coefficients of the y's are opposites. Use the addition method.

$$2x + 3y = -8$$
$$\underline{4x + 3y = -106}$$
$$6x = -114$$
$$x = -19$$

Substitute −19 for the x in the first equation to find the value of y.

$$2(-19) + 3y = -8$$
$$-38 + 3y = -8$$
$$3y = 30$$
$$y = 10$$

6. (−10, 2). Always look at the system of equations to make a plan for solving. Make a plan.

> **Be Careful!** The variables in the two equations are not in the same order. Rearrange the terms of the second equation so that the term with the x is first.

The second equation can be rewritten as $-4x + 7y = 54$. Solve the system of equations.

$$4x + 15y = -10$$
$$\underline{-4x + 7y = 54}$$
$$22y = 44$$
$$y = 2$$

Replace the y with 2 in one of the two original equations to find the value of x.

$$4x + 15(2) = -10$$
$$4x + 30 = -10$$
$$4x = -40$$
$$x = -10$$

7. (7, 4). Use the multiplication with addition method to solve. Notice that the y's could be eliminated if you multiplied the first equation by -3 before you added the equations.

$$(-3)3x - (-3)2y = (-3)13$$
$$-9x + 6y = -39$$

Solve the system of equations.

$$-9x + 6y = -39$$
$$\underline{7x - 6y = 25}$$
$$-2x = -14$$
$$x = 7$$

You can now substitute 7 for the x in either of the original equations to find the value of y.

$$3(7) - 2y = 13$$
$$21 - 2y = 13$$
$$-2y = -8$$
$$y = 4$$

8. (–4, 3). You can choose which variable to eliminate because they both involve the same amount of work. Multiply the first equation by 5 and the second equation by -8. The coefficient of the x in the first equation will be 40 and the coefficient in the second equation will be -40.

$$5(8x) + 5(9y) = 5(-5)$$
$$-8(5x) + -8(6y) = -8(-2)$$

Simplify each equation and solve the system of equations.

$$40x + 45y = -25$$
$$\underline{-40x - 48y = 16}$$
$$-3y = -9$$
$$y = 3$$

Replace the y with 3 in one of the original equations to find the value of x.

$$8x + 9(3) = -5$$
$$8x + 27 = -5$$
$$8x = -32$$
$$x = -4$$

9. 8 cats and 6 dogs. Since the problem asks you for a system of equations, you will be given points for writing two equations. c represents the number of cats at the Deluxe Pets Inn on Tuesday night and d represents the number of dogs. The total number of cats and dogs at the Inn is 14, so one equation would be $c + d = 14$. The total charge for Tuesday night is $452. Cats cost $25 a night and dogs cost $42 a night. The second equation would be $25c + 42d = 452$.

Solve the system of equations.

$$c + d = 14$$
$$25c + 42d = 452$$

$c + d = 14$ can be changed to $d = 14 - c$ and then substituted into the second equation.

$$25c + 42(14 - c) = 452$$
$$25c + 588 - 42c = 452$$
$$-17c + 588 = 452$$
$$-17c = -136$$
$$c = 8$$

There are 8 cats staying at the Deluxe Pets Inn on Tuesday evening.

10. c represents the total bill for a large dog to stay one night at a kennel. 45 is the cost (in dollars) for boarding a large dog one night at Deluxe Pets Inn. 3.5 is the cost (in dollars) for each playtime opportunity given a large dog at Deluxe Pets Inn. p represents the number of playtimes the owner requests for the large dog.

11. $9a + 6c = 106.50$ and $7a + 12c = 108.50$. The Smith family reunion spent a total of $106.50 for 9 adults and 6 children. Each adult cost a dollars, so the Smith family adults cost $9a$. Each child costs c dollars, so the Smith family children cost $6c$. The total Smith family bill is $106.50.

$$9a + 6c = 106.50$$

Similarly, the total Johnson family bill is $7a + 12c = 108.50$.

Solutions to Practice Problems

1.

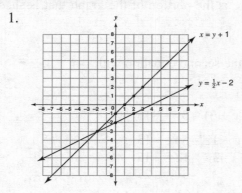

The solution to the system of equations is the point of intersection, $(-2, -3)$, as shown.

2. The solution is the empty set because the two lines are parallel as shown below.

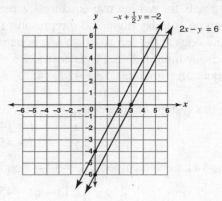

3.

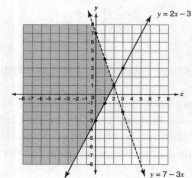

Graph $y = 2x - 3$ with a solid line. Shade the side of the line that contains the points that make $y \geq 2x - 3$ true.

Graph $y = 7 - 3x$ with a dotted line.

Hint: Rewrite $y = 7 - 3x$ as $y = -3x + 7$ so it is easier to find the slope and the y-intercept.

Shade the side of the line that contains the points that make $y < 7 - 3x$ true.

The solution graphed above is the portion of the graph that is shaded in both areas previously graphed.

4. $(7\frac{1}{2}, 9)$. Replace the x in the second equation with $\frac{1}{2}y + 3$.

$$4(\frac{1}{2}y + 3) - 5y = -15$$
$$2y + 12 - 5y = -15$$
$$-3y + 12 = -15$$
$$-3y = -27$$
$$y = 9$$

Find the value of x when y is 9.

$$4x - 5(9) = -15$$
$$4x - 45 = -15$$
$$4x = 30$$
$$x = 7\frac{1}{2}$$

5. $(-17, -35)$. Solve the second equation for y.

$$y - 2x = -1$$
$$y = 2x - 1$$

Substitute $2x - 1$ for the y in the first equation.

$$7x - 4(2x - 1) = 21$$
$$7x - 8x + 4 = 21$$
$$-x + 4 = 21$$
$$-x = 17$$
$$x = -17$$

Find the value of y when x is -17.

$$7(-17) - 4y = 21$$
$$-119 - 4y = 21$$
$$-4y = 140$$
$$y = -35$$

6. $(-10, 2)$. Observe the order of the variables in each equation. In the first equation the y-variable is first, and the x-variable is second. Use the commutative property to rearrange the first equation so that the x is first and the y is second.

$$+3x + 7y = -16$$

In the second equation, the x's and the y's are not on the same side of the equal sign. Move the $3x$ to the left side.

$$-3x + 10y = 50$$

Solve the system of equations.

$$3x + 7y = -16$$
$$\underline{-3x + 10y = 50}$$
$$17y = 34$$
$$y = 2$$

Find the value of x when y is 2.

$$10(2) = 3x + 50$$
$$20 = 3x + 50$$
$$-30 = 3x$$
$$-10 = x$$

7. $(-3, 5)$. It is easy to get the coefficients of the y-terms to be opposites. Multiply each term of the first equation by 2.

$$4x + 3y = 3$$
$$2 \cdot 4x + 2 \cdot 3y = 2 \cdot 3$$
$$8x + 6y = 6$$

Multiply each term of the second equation by 3.

$$5x - 2y = -25$$
$$3 \cdot 5x - 3 \cdot 2y = 3 \cdot -25$$
$$15x - 6y = -75$$

Add the two equations.

$$8x + 6y = 6$$
$$\underline{15x - 6y = -75}$$
$$23x = -69$$
$$x = -3$$

Find the value of y when x is -3.

$$4(-3) + 3y = 3$$
$$-12 + 3y = 3$$
$$3y = 15$$
$$y = 5$$

8. There would be 15 three-point questions and 11 five-point questions.
T represents the number of three-point questions and F represents the number of five-point questions. There are a total of 26 questions.

$$T + F = 26$$

Every three-point question is worth 3 points. If there are T three-point questions, then they will be worth $3T$ points. Every five-point question is worth 5 points. If there are F five-point questions, then they will be worth $5F$ points. The total points on the test is 100.

$$3T + 5F = 100$$

This system can easily be solved by the multiplication with addition method.

$$T + F = 26$$
$$3T + 5F = 100$$

Multiply the first equation by -3.

$$T + F = 26$$
$$-3 \cdot T + -3 \cdot F = -3 \cdot 26$$
$$-3T + -3F = -78$$

Solve the system of equations.

$$-3T + -3F = -78$$
$$\underline{3T + 5F = 100}$$
$$2F = 22$$
$$F = 11$$

There are 11 five-point questions. Since there is a total of 26 questions, there are 15 three-point questions.

9. You are trying to find the number of pounds of gummy bears and gummy worms that you need to make a 19-pound mixture worth $38. The problem says to use B for the number of pounds of gummy bears and to use W for the number of pounds of gummy worms.
The weight of gummy bears and gummy worms is 19 pounds.

$$B + W = 19$$

Gummy bears cost $1.89 for every pound and you are going to buy B pounds. The cost of B pounds of gummy bears is $1.89B$. Gummy worms cost $2.08 per pound and you are going to buy W pounds. The cost of W pounds of gummy worms is $2.08W$. The total cost for the mixture is $38.

$$1.89B + 2.08W = 38$$
$$B + W = 19$$

Chapter 12 | Algebra 1: Polynomials

POLYNOMIALS

A polynomial is an expression with whole number powers of the variable. Examples of polynomials include $5x^3$, $4y - 6w$, $7c^5 - 2c^8 + 6c^{10}$, and 7. Notice that the powers of the variables are whole numbers.

Examples of expressions that are not polynomials include $7w^{-5}$, $\dfrac{4}{x^{10}}$, $\sqrt{y}$, and $\dfrac{x^3}{w^5}$.

Some of these expressions look like they have whole number powers of the variable, but they do not. Remember that $\dfrac{4}{x^{10}} = 4x^{-10}$. The power of the variable is not a whole number.

Definition: A term is a product of constants and variables.

A term is numbers and letters multiplied together. In the polynomial $7c^5 - 2c^8 + 6c^{10}$, there are three terms, $7c^5$, $-2c^8$, and $6c^{10}$. In the polynomial $4y - 6w$, there are two terms, $4y$ and $-6w$. Each part of the polynomial separated by an addition or subtraction sign is called a term.

Like terms have the same variables with the same powers. For example, $5x^3$ and $-7x^3$ are like terms, but $5x^3$ and $5x^2$ are not like terms. $4x^3y^2$ and $7x^3y^2$ are like terms, but $4x^3y^2$ and $7x^2y^3$ are not like terms.

Definition: The coefficient of a term is the numerical part of the term.

The coefficient of $5x^3$ is 5 and the coefficient of $-7x^3$ is -7. The coefficient of $4x^3y^2$ is 4. Before finding the coefficient of a term in a polynomial, change subtraction problems to addition problems. Then the coefficient of a term will be the number with the variable. For example, in the polynomial $7c^5 - 2c^8$, the coefficient of c^5 is 7, and the coefficient of c^8 is -2.

ADDING POLYNOMIALS

To add polynomials together, look for the like terms. Add the coefficients of the like terms and keep the variable and its power the same.

Example

To add $(5x^3 + 2x^2 - 4x - 5) + (-7x^3 - 8x^2 - 7x + 2)$, change subtraction problems to addition problems.

$$(5x^3 + 2x^2 + -4x + -5) + (-7x^3 + -8x^2 + -7x + 2)$$

Find the like terms. $5x^3$ and $-7x^3$ are like terms. $2x^2$ and $-8x^2$ are like terms. $-4x$ and $-7x$ are like terms.

$$(\underline{5x^3} + \underline{2x^2} + \underline{-4x} + \underline{-5}) + (\underline{-7x^3} + \underline{-8x^2} + \underline{-7x} + \underline{2})$$

Underlining each pair of like terms in the same way helps organize the work. Add together the coefficients of the like terms, keeping the variables and their powers the same.

$$-2x^3 - 6x^2 - 11x - 3 \text{ or } -2x^3 + -6x^2 + -11x + -3$$

Answers will be given as $-2x^3 - 6x^2 - 11x - 3$, but it is also correct to have it written as $-2x^3 + -6x^2 + -11x + -3$ or even a different order for the terms.

Exercise

Simplify: $(8c^5 - 4c^3 - c) + (12c + c^5 - 5c^3)$

Solution to Exercise

Rewrite the problem using only addition signs. Remember that $-c = -1c$ and $c^5 = 1c^5$. Underline like terms in preparation for adding like terms.

$$(\underline{8c^5} + \underline{-4c^3} + (\underline{-1c}) + (\underline{12c} + \underline{1c^5} + \underline{-5c^3})$$

Add the coefficients of the like terms, keeping the variables and their powers the same.

$$9c^5 - 9c^3 + 11c$$

Practice Problems

1. Simplify: $(7y - 8x + 2y^2 + 8x^2) + (5x^2 - 12x) + (5y^2 - 12y)$

(Solutions are found at the end of the chapter.)

SUBTRACTING POLYNOMIALS

Subtraction of polynomials uses the same process of adding like terms as addition of polynomials. But first, change the subtraction problem into an addition problem.

To simplify $(5x - 3) - (4x + 7)$, rewrite the problem as $(5x - 3) - 1(4x + 7)$.

Remember: $(4x + 7) = 1(4x + 7)$. One multiplied by any number gives you the number you started with.

Rewrite $(5x - 3) - 1(4x + 7)$ as $(5x + -3) + -1(4x + 7)$. Use the distributive property to simplify the problem.

$$(5x + -3) + -1(4x + 7)$$
$$(5x + -3) + -1 \cdot 4x + -1 \cdot 7$$
$$(\underline{5x} + \underline{-3}) + \underline{-4x} + \underline{-7}$$
$$x - 10$$

Example

$$(-7x^3 + 8x + -7) - (10x^3 - 7x^2 + 4x - 12)$$

Rewrite the problem by inserting a one between the subtraction sign and the beginning of the second parenthesis.

$$(-7x^3 + 8x - 7) - 1(10x^3 - 7x^2 + 4x - 12)$$

Although this looks like a small step, it is important to remind you to use the distributive property. Now change all subtraction problems to addition problems.

$$(-7x^3 + 8x + -7) + -1(10x^3 + -7x^2 + 4x + -12)$$

Use the distributive law to simplify.

$$(-7x^3 + 8x + -7) + -1 \cdot 10x^3 + -1 \cdot -7x^2 + -1 \cdot 4x + -1 \cdot -12)$$
$$(\underline{-7x^3} + \underline{\underline{8x}} + \underline{-7}) + \underline{-10x^3} + 7x^2 + \underline{\underline{-4x}} + \underline{\underline{12}}$$
$$-17x^3 + 7x^2 + 4x + 5$$

Exercise

Subtract: $(8w^2 - 7w) - (4w - 12)$

Solution to Exercise
Place 1 in front of the second parenthesis because $-(4w - 12) = -1(4w - 12)$.

$$(8w^2 - 7w) - (4w - 12) = (8w^2 - 7w) - 1(4w - 12)$$

Rewrite subtraction problems as addition problems.

$$(8w^2 + -7w) + -1(4w + -12)$$

Use the distributive property to simplify.

$$(8w^2 + -7w) + -1 \cdot 4w + -1 \cdot -12$$
$$(8w^2 + \underline{-7w}) + \underline{-4w} + 12$$

Add like terms.

$$8w^2 - 11w + 12$$

Practice Problems

2. Simplify: $(7y^4 - 3y^2 - 2) + (4y^2 - 12) - (9y^4 - y^2 - 8)$

(Solution is found at the end of the chapter.)

MULTIPLYING MONOMIALS

A monomial is the same as a term. To multiply monomials, use the law of exponents.

> **Remember:** $a^b \cdot a^c = a^{b+c}$
>
> **Example:** $7^{10} \cdot 7^4 = 7^{10+4} = 7^{14}$ or $x^{10} \cdot x^4 = x^{10+4} = x^{14}$

Review the work on exponents in Chapter 2.

$5x^3 \cdot 7x^4$ means $(5 \cdot x^3) \cdot (7 \cdot x^4)$. Since all of the operations are multiplication, use the commutative and associative properties of multiplication to regroup the numbers and the variables.

$$5x^3 \cdot 7x^4 = (5 \cdot x^3) \cdot (7 \cdot x^4) = (5 \cdot 7) \cdot (x^3 \cdot x^4)$$

Now simplify the problems inside the parentheses.

$$5x^3 \cdot 7x^4 = (5 \cdot x^3) \cdot (7 \cdot x^4) = (5 \cdot 7) \cdot (x^3 \cdot x^4) = 35x^7$$

Most students do not show all of the work and that is fine. When you multiply $5x^3 \cdot 7x^4$, think of multiplying $5 \cdot 7$ and $x^3 \cdot x^4$. The reasons and steps are shown for this problem, but they will not be shown again.

Example

Multiply: $9x^8 \cdot 2x^3$

Think of multiplying $9 \cdot 2$ and $x^8 \cdot x^3$. When you multiply variables with powers, you add the powers.

$$9x^8 \cdot 2x^3 = 18x^{11}$$

Exercises

1. Simplify: $3w^5 \cdot 2w^3 \cdot 5w$ 2. Simplify: $5x^4y^3 \cdot 3x^6y$

Solution to Exercises

1. $30w^9$

   Think of multiplying $3 \cdot 2 \cdot 5$ and $w^5 \cdot w^3 \cdot w$. Remember $w = w^1$.

2. $15x^{10}y^4$

$$5x^4y^3 \cdot 3x^6y$$
$$(5 \cdot 3) \cdot (x^4 \cdot x^6) \cdot (y^3 \cdot y)$$
$$15x^{10}y^4$$

DIVIDING MONOMIALS

To divide monomials, remember the law of exponents.

> **Remember:** $\dfrac{a^b}{a^c} = a^{b-c}$
>
> **Example:** $\dfrac{5^{10}}{5^2} = 5^{10-2} = 5^8$ or $\dfrac{w^{10}}{w^2} = w^{10-2} = w^8$

Review the exponent work in Chapter 2.

Example

Divide: $\dfrac{18x^{10}}{6x^2}$

Think of this problem as $\dfrac{18x^{10}}{6x^2} = \dfrac{18}{6} \cdot \dfrac{x^{10}}{x^2}$. Now divide and simplify.

$$\frac{18x^{10}}{6x^2} = \frac{18}{6} \cdot \frac{x^{10}}{x^2} = 3x^8$$

A typical mistake made by students is dividing the exponents when dividing $\dfrac{x^{10}}{x^2}$.

Exercises

1. Divide: $\dfrac{72y^8}{9y^2}$ 2. Simplify: $(x^4w^3) \div (xw^2)$

Solutions to Exercises

1. Regroup and simplify.

$$\frac{72y^8}{9y^2} = \frac{72}{9} \cdot \frac{y^8}{y^2} = 8y^6$$

2. $(x^4w^3) \div (xw^2)$ can be written as the fraction $\dfrac{x^4w^3}{xw^2}$. Regroup and simplify.

$$(x^4w^3) \div (xw^2) = \frac{x^4w^3}{xw^2} = \frac{x^4}{x} \cdot \frac{w^3}{w^2} = x^3w$$

Practice Problems

3. Multiply: $(10w^8) \cdot (5w^2)$ 5. Simplify: $\dfrac{x^{20}w^{30}c^{40}}{x^{10}w^{30}c^{10}}$

4. Simplify: $x^5w^3c^{10} \cdot x^{20}w^{10}c$

(Solutions are found at the end of the chapter.)

TAKING POWERS OF MONOMIALS

To take the power of a monomial, remember the law of exponents.

Remember: $(a^b)^c = a^{bc}$

Example: $(w^{11})^3 = w^{11 \cdot 3} = w^{33}$ or $(3^{11})^3 = 3^{11 \cdot 3} = 3^{33}$

$$(ab)^c = a^c b^c$$

Example: $(5x)^4 = 5^4 x^4$

Example

Simplify: $(3x^5)^4$

Think of the problem as $(3)^4 \cdot (x^5)^4$. Work each part separately.

$$3^4 = 3 \cdot 3 \cdot 3 \cdot 3 = 81$$
$$(x^5)^4 = x^5 \cdot x^5 \cdot x^5 \cdot x^5 = x^{20}$$
$$(3x^5)^4 = (3)^4 \cdot (x^5)^4 = 81x^{20}$$

Example

Simplify: $(x^{10}w^5c^2)^7$

Think of the problem as $(x^{10})^7 \cdot (w^5)^7 \cdot (c^2)^7$. Simplify each part separately.

$$(x^{10}w^5c^2)^7 = (x^{10})^7 \cdot (w^5)^7 \cdot (c^2)^7 = x^{70}w^{35}c^{14}$$

Exercises

1. Simplify: $(2w^{20})^3$

2. Simplify: $(c^5a^{10}r^{15})^{10}$

Solutions to Exercises

1. Rewrite $(2w^{20})^3$ as $(2)^3 \cdot (w^{20})^3$. Simplify each parenthesis.

$$(2w^{20})^3 = (2)^3 \cdot (w^{20})^3 = 8w^{60}$$

2. $(c^5a^{10}r^{15})^{10} = (c^5)^{10} \cdot (a^{10})^{10} \cdot (r^{15})^{10} = c^{50}a^{100}r^{150}$

TAKING ROOTS OF MONOMIALS

To find the root of a monomial, remember how to write a root using exponents.

Remember: $\sqrt[y]{b^x} = b^{\frac{x}{y}}$ for any positive number b.

Simplify: $\sqrt{x^{40}}$

Remember that $\sqrt{x^{40}} = \sqrt[2]{x^{40}}$. Using exponents, $\sqrt{x^{40}} = \sqrt[2]{x^{40}} = x^{\frac{40}{2}}$. Then simplify to x^{20}.

Example

Calculate: $\sqrt{16x^{16}}$

Since there is a number and a variable under the square root sign, rewrite the problem as two separate square roots.

$$\sqrt{16x^{16}} = \sqrt{16} \cdot \sqrt{x^{16}}$$

Work each part separately.

$$\sqrt{16} = 4$$

$$\sqrt{x^{16}} = \sqrt[2]{x^{16}} = x^{\frac{16}{2}} = x^8$$

$$\sqrt{16x^{16}} = \sqrt{16} \cdot \sqrt[2]{x^{16}} = 4x^8$$

Typical wrong answers given by many students are $\sqrt{16x^{16}} \neq 8x^8$ and $\sqrt{16x^{16}} \neq 4x^4$.

Exercises

1. Simplify: $\sqrt[3]{c^{21}x^{15}}$

2. Find the cube root of $c^9 w^{81}$.

3. Simplify: $\sqrt[4]{16w^{20}}$

Solutions to Exercises

1. $\sqrt[3]{c^{21}x^{15}} = c^{\frac{21}{3}} x^{\frac{15}{3}} = c^7 x^5$

2. The cube root of a number means what multiplied by itself three times will give the number. Written in symbols the cube root is $\sqrt[3]{}$.

$$\sqrt[3]{c^9 w^{81}} = c^{\frac{9}{3}} w^{\frac{81}{3}} = c^3 w^{27}$$

3. $\sqrt[4]{16w^{20}} = \sqrt[4]{16} \cdot \sqrt[4]{w^{20}} = 2 \cdot w^{\frac{20}{4}} = 2w^5$

To find $\sqrt[4]{16}$, think of what number multiplied by itself four times will give you 16.

Practice Problems

6. Simplify: $(w^3 c^4)^5$

7. Simplify: $\sqrt[10]{w^{50} c^{40} x^{30}}$

8. Change $(5x^4)^3$ to simplest form.

9. Change $\sqrt[3]{64c^{36}}$ to simplest form.

(Solutions are found at the end of the chapter.)

MULTIPLYING POLYNOMIALS

The distributive property and multiplication of monomials are the basis for multiplication of polynomials.

Distributive Property: $a \cdot (b + c) = a \cdot b + a \cdot c$ or $a(b + c) = ab + ac$

If you need review with the distributive property, go to Chapter 4. If you need review with multiplication of monomials, go to the first part of this chapter.

Example
Multiply: $5x^2(4x^2 + 7x - 3)$
Change the subtraction problem to an addition problem. Use the distributive property as shown below.

$$5x^2(4x^2 + 7x + -3) = 5x^2 \cdot 4x^2 + 5x^2 \cdot 7x + 5x^2 \cdot -3$$

$$= 20x^4 + 35x^3 - 15x^2.$$

Remember, $5x^2 \cdot 4x^2 = (5 \cdot 4) \cdot (x^2 \cdot x^2) = 20x^4$ and $5x^2 \cdot 7x = (5 \cdot 7) \cdot (x^2 \cdot x) = 35x^3.$

Example

To multiply $(2x + 3)(5x + 4)$, use the distributive property. Think of $(2x + 3)$ as what you are multiplying by, as shown below.

$$(2x + 3)(5x + 4) = (2x + 3) \cdot 5x + (2x + 3) \cdot 4$$

Use the distributive property for each underlined part on the left side to get the respective underlined part on the right side, as shown below.

$$\underline{(2x + 3) \cdot 5x} + \underline{\underline{(2x + 3) \cdot 4}} = \underline{5x \cdot 2x + 5x \cdot 3} + \underline{\underline{4 \cdot 2x + 4 \cdot 3}}$$

Then simplify each product.

$$\underline{5x \cdot 2x} + \underline{5x \cdot 3} + \underline{\underline{4 \cdot 2x}} + \underline{\underline{4 \cdot 3}} = \underline{10x^2} + \underline{15x} + \underline{\underline{8x}} + \underline{\underline{12}}$$

Add like terms.

$$(2x + 3)(5x + 4) = 10x^2 + 23x + 12$$

Example

Find the product $(x + 3)(5x + 6)$.

Use the distributive property. Think of multiplying by $(x + 3)$, as shown below.

$$(x + 3)(5x + 6) = (x + 3) \cdot 5x + (x + 3) \cdot 6$$

Use the distributive property on the right side of the equation.

$$(x + 3) \cdot 5x + (x + 3) \cdot 6 = 5x \cdot x + 5x \cdot 3 + 6 \cdot x + 6 \cdot 3$$
$$\underline{5x \cdot x} + \underline{5x \cdot 3} + \underline{\underline{6 \cdot x}} + \underline{\underline{6 \cdot 3}} = \underline{5x^2} + \underline{15x} + \underline{\underline{6x}} + \underline{\underline{18}}$$
$$5x^2 + \underline{15x + 6x} + 18 = 5x^2 + 21x + 18$$

Exercises

1. Find the product $x^3(4x^2 - 6x - 9)$.   3. Multiply: $(5x - 4)(3x + 2)$

2. Multiply: $(3w + 5)(w + 2)$

Solutions to Exercises

1. Use the distributive property.

$$x^3(4x^2 + - 6x + -9) = x^3 \cdot 4x^2 + x^3 \cdot -6x + x^3 \cdot -9$$

Simplify each product.

$$\underline{x^3 \cdot 4x^2} + \underline{\underline{x^3 \cdot -6x}} + \underline{\underline{x^3 \cdot -9}} = \underline{4x^5} + \underline{\underline{-6x^4}} + \underline{\underline{-9x^3}}$$

2. Use the distributive property.

$$(3w + 5)(w + 2) = (3w + 5) \cdot w + (3w + 5) \cdot 2$$

Use the distributive property again.

$$(3w + 5) \cdot w + (3w + 5) \cdot 2 = w \cdot 3w + w \cdot 5 + 2 \cdot 3w + 2 \cdot 5$$

Simplify by multiplying.

$$w \cdot 3w + w \cdot 5 + 2 \cdot 3w + 2 \cdot 5 = 3w^2 + 5w + 6w + 10$$

Add like terms.

$$3w^2 + 5w + 6w + 10 = 3w^2 + 11w + 10$$

3.

$$(5x - 4)(3x + 2) = (5x + -4) \cdot 3x + (5x + -4) \cdot 2$$
$$3x \cdot 5x + 3x \cdot -4 + 2 \cdot 5x + 2 \cdot -4$$
$$15x^2 + -12x + 10x + -8$$
$$15x^2 - 2x - 8$$

Practice Problems

10. Multiply: $c^3 y^5 (c^5 - y^2 + 2)$ 12. Multiply: $(4x + 3)^2$

11. Multiply: $(5 - 3x)(4x + 7)$ 13. Multiply: $(7x - 3y)(2x + 4y)$

(Solutions are found at the end of the chapter.)

DIVIDING POLYNOMIALS BY MONOMIALS

Adding fractions that have common denominators is very simple.

$$\frac{7}{15} + \frac{4}{15} = \frac{7+4}{15} = \frac{11}{15}$$

Add the numerators while keeping the denominators the same. Dividing a polynomial by a monomial uses the same process except backwards. You will take a problem like $\frac{7+4}{15}$ and break it into $\frac{7}{15} + \frac{4}{15}$.

To divide $(25x^4 + 15x^3 + 20x^2)$ by $5x$, you write the problem like a fraction.

$$\frac{25x^4 + 15x^3 + 20x^2}{5x}$$

Then break apart the fraction into three separate fractions. The numerator of each separate fraction is a term of the polynomial in the numerator. The denominator is the same in all three fractions.

$$\frac{25x^4}{5x} + \frac{15x^3}{5x} + \frac{20x^2}{5x}$$

Simplify each fraction using the rules for dividing monomials.

$$5x^3 + 3x^2 + 4x$$

Example

Simplify: $\dfrac{8w^5 y^6 + 6w^4 y^7 - 12w^3 y^8}{2w^2 y^6}$

$$\frac{8w^5y^6 + 6w^4y^7 - 12w^3y^8}{2w^2y^6} = \frac{8w^5y^6 + 6w^4y^7 + -12w^3y^8}{2w^2y^6}$$

$$\frac{8w^5y^6}{2w^2y^6} + \frac{6w^4y^7}{2w^2y^6} + \frac{-12w^3y^8}{2w^2y^6}$$

Simplify your answer by dividing the monomials in each fraction.

$$4w^3 + 3w^2y - 6wy^2$$

Exercise

Divide $(8c^{28} - 12c^{21} + 16c^{14} + 20c^7)$ by $4c^7$.

Solutions to Exercises
Rewrite the problem as a fraction.

$$\frac{8c^{28} - 12c^{21} + 16c^{14} + 20c^7}{4c^7}$$

Separate the fraction.

$$\frac{8c^{28} - 12c^{21} + 16c^{14} + 20c^7}{4c^7} = \frac{8c^{28}}{4c^7} + \frac{-12c^{21}}{4c^7} + \frac{16c^{14}}{4c^7} + \frac{20c^7}{4c^7}$$

Simplify each fraction.

$$2c^{21} - 3c^{14} + 4c^7 + 5$$

Practice Problems

14. Simplify: $\dfrac{36x^{36} + 18x^{18} + 27x^{27} + 9x^9}{9x^9}$

(Solution is found at the end of the chapter.)

MONOMIAL FACTORS OF A POLYNOMIAL

When we reduced fractions like $\dfrac{12}{18}$, we looked for the largest number that was a factor of both 12 and 18. The factors of 12 are 1, 2, 3, 4, 6, and 12. The factors of 18 are 1, 2, 3, 6, 9, and 18. The largest factor in both lists is 6. Six is called the greatest common factor because it is the largest number that goes into both 12 and 18.

Monomial factors of a polynomial involve finding the greatest common factor of the terms of a polynomial. We want to find the largest term (both number and variable) that goes into all of the terms of the polynomial.

To find the monomial factor of $12w + 18$, find the largest factor of $12w$ and 18. The factors of $12w$ are 1, 2, 3, 4, 6, 12, and w. The factors of 18 are 1, 2, 3, 6, 9, and 18. The largest term (both number and variable) that is in both lists is 6.

Now factor 6 out of the polynomial $12w + 18$. Think of factoring like doing the distributive property in reverse order.

$$\text{Factor: } ab + ac = a(b + c)$$

We want to factor 6 out of $12w + 18$.

$$12w + 18 = 6 \cdot (\underline{} + \underline{})$$

To find the number that goes in the first blank, think what term multiplied by 6 gives $12w$. You know that $6 \cdot 2w = 12w$.

$$12w + 18 = 6 \cdot (2w + \underline{})$$

To find the number that goes into the second blank, think what term multiplied by 6 gives you 18. You know that $6 \cdot 3 = 18$.

$$12w + 18 = 6 \cdot (2w + \underline{3})$$

Now you have factored $12w + 18$ into $6(2w + 3)$.

You can always check your work by using the distributive property on your answer. You should get the original problem.

$$6(2w + 3) = 6 \cdot 2w + 6 \cdot 3 = 12w + 18$$

Example

Factor: $12x^3 + 8x^2$

The factors of $12x^3$ are 1, 2, 3, 4, 6, 12, x, x^2, and x^3. The factors of $8x^2$ are 1, 2, 4, 8, x, and x^2. The largest term (both number and variable) in both lists is $4x^2$. Use $4x^2$ to factor $12x^3 + 8x^2$.

$$12x^3 + 8x^2 = 4x^2(\underline{} + \underline{})$$

Think what term multiplied by $4x^2$ would give you $12x^3$. You know that $4x^2 \cdot 3x = 12x^3$.

$$12x^3 + 8x^2 = 4x^2(\underline{3x} + \underline{})$$

Think what term multiplied by $4x^2$ is $8x^2$. You know that $4x^2 \cdot 2 = 8x^2$.

$$12x^3 + 8x^2 = 4x^2(\underline{3x} + \underline{2})$$

Exercises

1. Factor: $15x^4y^2 + 25x^3y^4 - 20x^2y^6$

2. What is the greatest common factor of the terms $21w^8c^4$ and $15w^6c^8$?

Solutions to Exercises

1. The factors of $15x^4y^2$ are 1, 3, 5, 15, x, x^2, x^3, x^4, y, and y^2.
    The factors of $25x^3y^4$ are 1, 5, 25, x, x^2, x^3, y, y^2, y^3, and y^4.
    The factors of $20x^2y^6$ are 1, 2, 4, 5, 10, 20, x, x^2, y, y^2, y^3, y^4, y^5 and y^6.
    The largest term (both number and variable) in all three lists is $5x^2y^2$.

    Use $5x^2y^2$ to factor $15x^4y^2 + 25x^3y^4 - 20x^2y^6$.

    $$5x^2y^2(\underline{3x^2} + \underline{5xy^2} - \underline{4y^4})$$

    $3x^2$ multiplied by $5x^2y^2$ gives $15x^4y^2$.
    $5xy^2$ multiplied by $5x^2y^2$ gives $25x^3y^4$.
    $4y^4$ multiplied by $5x^2y^2$ gives $20x^2y^6$.

2. The factors of $21w^8c^4$ include 1, 3, 7, 21, all the integer powers of w from w^1 to w^8, and all the integers powers of c from c^1 to c^4.

The factors of $15w^6c^8$ are 1, 3, 5, 15, all the integers powers of w from w^1 to w^6, and all the integers powers of c from c^1 to c^8.

The greatest common factor (largest number and variable) in both lists is $3w^6c^4$.

Practice Problem

15. Factor: $42x^3 + 49x^2 - 35x$

(Solution is found at the end of the chapter.)

ANALYZING THE PRODUCT OF BINOMIALS AND FOIL

Earlier in this chapter we multiplied $(x + 7)(2x + 1)$ to get $2x^2 + 15x + 7$. The polynomials $(x + 7)$ and $(2x + 1)$ are called binomials because they are composed of two terms. The answer $2x^2 + 15x + 7$ is called a quadratic because the highest power of the variable is two.

We need to take a close look at the relationship between the binomials we are multiplying together and the quadratic we get as an answer.

$$(3x + 5)(2x + 1)$$
$$(3x + 5) \cdot 2x + (3x + 5) \cdot 1$$
$$\underline{2x \cdot 3x} + \underline{2x \cdot 5 + 1 \cdot 3x} + 1 \cdot \underline{\underline{5}}$$
$$\underline{6x^2} + \underline{10x + 3x} + \underline{\underline{5}}$$
$$\underline{6x^2} + \underline{13x} + \underline{\underline{5}}$$

Notice that the first term of the answer, $6x^2$, came from multiplying $3x \cdot 2x$, the first terms of each binomial. Also notice that the last term of the answer, 5, came from multiplying $5 \cdot 1$, the last terms of each binomial. The difficult part is noticing how we got the middle term, $13x$. The middle term is a combination of $10x + 3x$, which came from $2x \cdot 5 + 1 \cdot 3x$, as shown below.

$$(3x + 5)(2x + 1)$$

In the problem $(\underline{7x} + \underline{3})(\underline{2x} + \underline{5}) = \underline{14x^2} + 41x + \underline{15}$, notice that multiplying together the first terms of each binomial gives the first term of the answer. Also, notice that multiplying together the last terms of each binomial gives you the last term of the answer. The middle term of the answer is a combination of $6x + 35x$, which came from $2x \cdot 3 + 5 \cdot 7x$.

When multiplying binomials $(ax + b)(cx + d)$, the first term of the answer will be $ax \cdot cx$ and the last term of the answer will be $b \cdot d$. The middle term will be $cx \cdot b + d \cdot ax$.

There is a shortcut way to remember the order in which to multiply binomials. The acronym FOIL represents the steps used to multiply the binomials $(ax + b)(cx + d)$.

F tells you to multiply together the first terms of each binomial, $ax \cdot cx$.
O tells you to multiply together the outside terms, $ax \cdot d$.
I tells you to multiply together the inside terms, $b \cdot cx$.
L tells you to multiply together the last terms of each binomial, $b \cdot d$.

The figure below shows FOIL.

$$(ax + b)(cx + d)$$

To use FOIL to multiply $(7x - 2)(8x + 4)$, change the subtraction problem to an addition problem $(7x + -2)(8x + 4)$.

F tells you to multiply together the first terms of each binomial, $7x \cdot 8x = 56x^2$.
O tells you to multiply together the outside terms, $7x \cdot 4 = 28x$.
I tells you to multiply together the inside terms, $-2 \cdot 8x = -16x$.
L tells you to multiply together the last terms of each binomial, $-2 \cdot 4 = -8$.

The answer is the sum of all of the parts, $56x^2 + 28x - 16x - 8 = 56x^2 + 12x - 8$.

Hopefully, multiplying binomials is easier using FOIL. Practice using FOIL until you can just write down the sum of the parts and then the final answer.

Exercises

1. Use FOIL to multiply $(6x - 5)(6x + 5)$.

2. Find the product $(10x + 8)(4x + 3)$.

Solutions to Exercises

1. Rewrite $(6x - 5)(6x + 5)$ as $(6x + -5)(6x + 5)$.
   F tells you to multiply $6x \cdot 6x = 36x^2$.
   O tells you to multiply $6x \cdot 5 = 30x$.
   I tells you to multiply $-5 \cdot 6x = -30x$.
   L tells you to multiply $-5 \cdot 5 = -25$.
   The answer is $36x^2 + 30x - 30x - 25 = 36x^2 - 25$. Notice that the middle term dropped out because $30x + -30x = 0x = 0$.

2. F tells you to multiply $10x \cdot 4x = 40x^2$.
   O tells you to multiply $10x \cdot 3 = 30x$.
   I tells you to multiply $8 \cdot 4x = 32x$.
   L tells you to multiply $8 \cdot 3 = 24$.
   The answer is $40x^2 + 30x + 32x + 24 = 40x^2 + 62x + 24$.

Practice Problems

16. Multiply: $(7x + 3)(5x - 9)$ 17. Simplify: $(6x - 7)^2$

(Solutions are found at the end of the chapter.)

FACTORING QUADRATICS

Factoring a quadratic means to break the quadratic into two binomials multiplied together.
Factoring and multiplying are reverse operations of each other.
Multiply $(x + 7)(2x + 1)$ to get the quadratic $2x^2 + 15x + 7$.
Factor the quadratic $2x^2 + 15x + 7$ to get $(x + 7)(2x + 1)$.
Some quadratics can be factored. Not all quadratics can be factored.

Factor $5x^2 + 11x + 2$ means find two binomials that multiply together to give you $5x^2 + 11x + 2$.

$$5x^2 + 11x + 2 = (\_\_\_ + \_\_\_)(\_\_\_ + \_\_\_)$$

You know that $5x^2$ came from multiplying together the first term of each binomial, as shown below.

$$5x^2 + 11x + 2 = (\_\_ + \_\_)(\_\_ + \_\_)$$

The choices for these numbers are $5x$ and $1x$, $5x \cdot 1x = 5x^2$. Put these as the first term of each binomial.

$$5x^2 + 11x + 2 = (5x + \_\_\_)(1x + \_\_\_)$$

You know that the 2 came from multiplying together the last term of each binomial, as shown below.

$$5x^2 + 11x + 2 = (5x + \_\_)(1x + \_\_)$$

The only choices for these numbers are 2 and 1. You cannot just put these two numbers in the open places because the placement of the 2 and the 1 makes a difference. You can either put the 2 in the first binomial and the 1 in the second binomial or vice versa. Try putting the 2 in the first binomial and the 1 in the second binomial. Use FOIL to multiply together the binomials to check if their product equals the original problem.

$$5x^2 + 11x + 2 \overset{?}{=} (5x + \underline{2})(1x + \underline{1})$$
$$5x^2 + 11x + 2 \overset{?}{=} 5x^2 + 5x + 2x + 2$$
$$5x^2 + 11x + 2 \neq 5x^2 + 7x + 2$$

Notice that the first term on each side is the same and the last term on each side is the same. The middle term is not the same on both sides. The 2 and the 1 were put in the wrong places, so switch them and try again.

$$5x^2 + 11x + 2 \overset{?}{=} (5x + \underline{1})(1x + \underline{2})$$
$$5x^2 + 11x + 2 \overset{?}{=} 5x^2 + 10x + 1x + 2$$
$$5x^2 + 11x + 2 = 5x^2 + 11x + 2$$

Since the two sides are the same, $5x^2 + 11x + 2$ factors into $(5x + 1)(1x + 2)$.

Example

Factor: $7x^2 - 2x - 5$

$$7x^2 + -2x + -5 = (\_\_\_ + \_\_\_)(\_\_\_ + \_\_\_)$$

You know that $7x^2$ came from multiplying the first term of each binomial. The only factors that could be used are $1x$ and $7x$ because $1x \cdot 7x = 7x^2$.

$$7x^2 + -2x + -5 = (1x + \_\_\_)(7x + \_\_\_)$$

There are two choices for -5, $1 \cdot -5 = -5$ or $-1 \cdot 5 = -5$. Remember that where you place the numbers also makes a difference. At this point, try placing one of the pairs into the binomial and use the FOIL process to check if it is correct. Try placing 1 in the first binomial and -5 in the second binomial.

$$7x^2 + -2x + -5 \overset{?}{=} (1x + \underline{1})(7x + \underline{-5})$$
$$7x^2 + -2x + -5 \overset{?}{=} 7x^2 + -5x + 7x + -5$$
$$7x^2 + -2x + -5 \neq 7x^2 + 2x + -5$$

Again, the first term on each side is the same and the last term on each side is the same. The middle terms are not the same. But the middle terms are similar; one is $-2x$ and one is $2x$. When the middle terms are opposites of each other, then the numbers are in the correct places and the signs are in the wrong places. Keep the numbers in the same places and switch the signs of the numbers.

$$7x^2 + -2x + -5 \overset{?}{=} (1x + \underline{-1})(7x + \underline{5})$$
$$7x^2 + -2x + -5 \overset{?}{=} 7x^2 + 5x + -7x + -5$$
$$7x^2 + -2x + -5 = 7x^2 + -2x + -5$$

$7x^2 - 2x - 5$ factors into $(1x + -1)(7x + 5)$ or $(1x - 1)(7x + 5)$.

Example
Factor: $8x^2 + 10x + 3$

$$8x^2 + 10x + 3 = (\underline{\quad} + \underline{\quad})(\underline{\quad} + \underline{\quad})$$

You know that the first term, $8x^2$, can be factored into $4x \cdot 2x = 8x^2$ or $1x \cdot 8x = 8x^2$. Before just guessing which pair to try first, look at the last term of the quadratic, 3. You know that 3 factors into only $1 \cdot 3 = 3$. This time place the 1 and the 3 as the second term of each binomial.

$$8x^2 + 10x + 3 = (\underline{\quad} + 1)(\underline{\quad} + 3)$$

Now guess which pair of factors of $8x^2$ would work. Try $1x$ and $8x$.

$$8x^2 + 10x + 3 \overset{?}{=} (\underline{1x} + 1)(\underline{8x} + 3)$$
$$8x^2 + 10x + 3 \overset{?}{=} 8x^2 + 3x + 8x + 3$$
$$8x^2 + 10x + 3 \neq 8x^2 + 11x + 3$$

The middle term is not the same on the left side and the right side. So try switching the placement of the $1x$ and the $8x$. Keep the second term of each binomial the same. Keeping organized work is essential when factoring quadratics.

$$8x^2 + 10x + 3 \overset{?}{=} (\underline{8x} + 1)(\underline{1x} + 3)$$
$$8x^2 + 10x + 3 \overset{?}{=} 8x^2 + 24x + 1x + 3$$
$$8x^2 + 10x + 3 \neq 8x^2 + 25x + 3$$

Once again, the middle terms are not the same. Now you know that $8x^2$ will not use the factors $1x$ and $8x$. Try using the factors $4x$ and $2x$.

$$8x^2 + 10x + 3 \overset{?}{=} (\underline{2x} + 1)(\underline{4x} + 3)$$
$$8x^2 + 10x + 3 \overset{?}{=} 8x^2 + 6x + 4x + 3$$
$$8x^2 + 10x + 3 = 8x^2 + 10x + 3$$

We have finally factored $8x^2 + 10x + 3$ into $(2x + 1)(4x + 3)$.

Example
Factor: $49x^2 - 1$

Rewrite the problem as $49x^2 + -1$. This problem looks very different because there are only two terms. The problem is missing the middle term that has an x. That means that the middle term is $0x$. Put the middle term into the problem, $49x^2 + 0x + -1$. Try factoring now. You know that $49x^2$ has two choices for factors, $1x \cdot 49x = 49x^2$ or $7x \cdot 7x = 49x^2$. Check the last term to see how many factors it has. The last term, -1, only has factors $1 \cdot -1 = -1$. It does not matter where you put the negative sign because the two numbers are the same. Place 1 and -1 in for the second terms of the binomials.

$$49x^2 + 0x + -1 = (\underline{\quad} + 1)(\underline{\quad} + -1)$$

Choose one pair of factors of $49x^2$, either $1x \cdot 49x = 49x^2$ or $7x \cdot 7x = 49x^2$, to use as the first term of each binomial.

$$49x^2 + 0x + -1 \stackrel{?}{=} (\underline{1x} + 1)(\underline{49x} + -1)$$
$$49x^2 + 0x + -1 \stackrel{?}{=} 49x^2 + -1x + 49x + -1$$
$$49x^2 + 0x + -1 \neq 49x^2 + 48x + -1$$

The middle terms are not correct. You do not need to switch the places of $1x$ and $49x$ because the second term of the binomials are opposites of each other. Try using the other pair of factors for $49x^2$, $7x \cdot 7x = 49x^2$.

$$49x^2 + 0x + -1 \stackrel{?}{=} (\underline{7x} + 1)(\underline{7x} + -1)$$
$$49x^2 + 0x + -1 \stackrel{?}{=} 49x^2 + -7x + 7x + -1$$
$$49x^2 + 0x + -1 = 49x^2 + 0x + -1$$

We have factored $49x^2 - 1$ into $(7x + 1)(7x - 1)$.

$49x^2 - 1$ is a special quadratic called the difference of two squares. The word "difference" means subtraction. The words "two squares" mean that each term is a perfect square. So the difference of two squares means you are subtracting two perfect squares. Other examples of quadratics that are the difference of two squares include $25x^2 - 49$, $100x^2 - 9$, and $x^2 - 64$.

> The difference of two squares, $a^2 - b^2$, factors into $(a + b)(a - b)$.

Example
Factor: $36x^2 - 25$
Think of $36x^2$ as a perfect square, $(6x)^2$. Think of 25 as a perfect square, 5^2. Now write $36x^2 - 25$ as the difference of two squares.

$$36x^2 - 25 = (6x)^2 - 5^2$$

Use the fact that $a^2 - b^2 = (a + b)(a - b)$ to factor the problem.

$$36x^2 - 25 = (6x + 5)(6x - 5)$$

Example
Factor: $4x^2 - 81$
Think of $4x^2$ as a perfect square, $(2x)^2$. Think of 81 as a perfect square, 9^2. Rewrite $4x^2 - 81$ as the difference of two squares, $(2x)^2 - 9^2$. Factor the difference of two squares.

$$4x^2 - 81 = (2x)^2 - 9^2$$
$$= (2x + 9)(2x - 9)$$

Exercises

1. Factor: $6x^2 + 17x + 7$

2. Factor: $5x^2 + 14x - 3$

3. Factor: $2x^2 - 13x + 11$

4. Factor: $121x^2 - 1$

Solutions to Exercises

1.

$$6x^2 + 17x + 7 = (\ \_\_\_ + \_\_\_)(\ \_\_\_ + \_\_\_)$$
$$6x^2 + 17x + 7 = (\ \_\_\_ + 7)(\ \_\_\_ + 1)$$
$$6x^2 + 17x + 7 = (\underline{2x} + 1)(\underline{3x} + 7)$$

2.

$$5x^2 + 14x - 3 = (\ \_\_\_ + \_\_\_)(\ \_\_\_ + \_\_\_)$$
$$5x^2 + 14x - 3 = (5x + \_\_\_)(1x + \_\_\_)$$
$$5x^2 + 14x - 3 = (5x + \underline{-1})(1x + \underline{3})$$

3.

$$2x^2 - 13x + 11 = (\ \_\_\_ + \_\_\_)(\ \_\_\_ + \_\_\_)$$
$$2x^2 - 13x + 11 = (2x + \_\_\_)(1x + \_\_\_)$$
$$2x^2 - 13x + 11 = (2x + \underline{11})(1x + \underline{1})$$

Check:

$$2x^2 - 13x + 11 \overset{?}{=} 2x^2 + 2x + 11x + 11$$
$$2x^2 - 13x + 11 \neq 2x^2 + 13x + 11$$

Since the middle terms are opposites of each other, change the signs of the second term in each binomial. That means to change both numbers to their opposites.

$$2x^2 - 13x + 11 = (2x + \underline{-11})(1x + \underline{-1})$$

4.

$$121x^2 - 1 = 121x^2 + 0x + -1$$
$$121x^2 - 1 = (11x + \_\_\_)(11x + \_\_\_)$$
$$121x^2 - 1 = (11x + \underline{1})(11x + \underline{-1})$$

Practice Problems

18. Factor: $5x^2 + 2x - 3$

20. Factor: $144x^2 - 25$

19. Factor: $35x^2 - 12x + 1$

(Solutions are found at the end of the chapter.)

SAMPLE GQE QUESTIONS FOR CHAPTER 12

1. Simplify: $(5x^4)^2$

   A. $10x^6$
   B. $25x^6$
   C. $10x^8$
   D. $25x^8$

2. Find the sum: $(5x^2 + 2x + 3) + (4x^2 - 6x - 9)$.

   A. $9x^4 - 4x^2 - 6$
   B. $9x^2 - 4x - 6$
   C. $20x^4 - 12x^2 - 27$
   D. $20x^2 - 12x - 27$

3. Find the difference: $7x - (2x^2 - 4x)$.

   A. $2x^2 + 3x$
   B. $2x^2 + 11x$
   C. $-2x^2 + 11x$
   D. $-2x^2 + 3x$

4. Find the cube root of $27x^{27}$.

   A. $3x^3$
   B. $9x^3$
   C. $3x^9$
   D. $9x^9$

5. Simplify: $\dfrac{18x^6w^2 + 12x^9w^8 - 24x^3w^6}{6x^3w^2}$

   A. $12x^3 + 6x^6w^6 - 18w^4$
   B. $12x^2 + 6x^3w^4 - 18w^3$
   C. $3x^3 + 2x^6w^6 - 4w^4$
   D. $3x^2 + 2x^3w^4 - 4w^3$

6. Find the product of $x^5w^4c^5 \cdot (x^2w)^4$.

   A. $x^{13}w^8c^5$
   B. $x^{40}w^{16}c^5$
   C. $x^{11}w^8c^5$
   D. $x^{30}w^{16}c^5$

7. Factor: $8x^2w^5 - 20x^3w^4 + 16x^4w^3$

8. Factor: $3x^2 + 14x - 5$

9. Factor: $121x^2 - 100w^2$

10. Multiply: $(7x - 3)(8x + 4)$

    A. $56x^2 - 12$
    B. $15x^2 + 1$
    C. $56x^2 + 4x - 12$
    D. $15x^2 + 6x + 1$

11. Find the product of $(3x + 9)^2$.

Answers Explained

1. **D** $25x^8$
$$(5x^4)^2 = (5)^2 \cdot (x^4)^2$$
$$25 \cdot x^{4 \cdot 2}$$
$$25x^8$$

2. **B** $9x^2 - 4x - 6$
$$(\underline{5x^2} + \underline{2x} + 3) + (\underline{4x^2} + \underline{-6x} + -9) = 9x^2 + -4x + -6$$

3. **C** $-2x^2 + 11x$
$$7x - (2x^2 - 4x)$$
$$7x + -1(2x^2 + -4x)$$
$$7x + -1 \cdot 2x^2 + -1 \cdot -4x$$
$$7x + -2x^2 + 4x$$
$$-2x^2 + 11x$$

4. **C** $3x^9$. The cube root of $27x^{27}$ is written in symbols as $\sqrt[3]{27x^{27}}$.

$$\sqrt[3]{27x^{27}}$$
$$\sqrt[3]{27} \cdot \sqrt[3]{x^{27}}$$
$$3 \cdot x^{\frac{27}{3}}$$
$$3x^9$$

Remember that $\sqrt[3]{27}$ means the number that multiplied by itself three times gives 27. $\sqrt[3]{27} = 3$ because $3 \cdot 3 \cdot 3 = 27$.

5. **C** $3x^3 + 2x^6w^6 - 4w^4$

$$\frac{18x^6w^2 + 12x^9w^8 - 24x^3w^6}{6x^3w^2}$$

$$\frac{18x^6w^2}{6x^3w^2} + \frac{12x^9w^8}{6x^3w^2} + \frac{-24x^3w^6}{6x^3w^2}$$

$$\frac{18}{6} \cdot \frac{x^6}{x^3} \cdot \frac{w^2}{w^2} + \frac{12}{6} \cdot \frac{x^9}{x^3} \cdot \frac{w^8}{w^2} + \frac{-24}{6} \cdot \frac{x^3}{x^3} \cdot \frac{w^6}{w^2}$$

$$3x^3 + 2x^6w^6 - 4w^4$$

6. **A** $x^{13}w^8c^5$

$$x^5w^4c^5 \cdot (x^2w)^4$$
$$x^5w^4c^5 \cdot (x^2)^4 \cdot (w)^4$$
$$x^5w^4c^5 \cdot x^8 \cdot w^4$$
$$x^{13}w^8c^5$$

7. $4x^2w^3 (2w^2 - 5xw + 4x^2)$

The factors of $8x^2w^5$ are 1, 2, 4, 8, x, x^2, w, w^2, w^3, w^4, and w^5.
The factors of $20x^3w^4$ are 1, 2, 4, 5, 10, 20, x, x^2, x^3, w, w^2, w^3, and w^4.
The factors of $16x^4w^3$ are 1, 2, 4, 8, 16, x, x^2, x^3, x^4, w, w^2, and w^3.
The largest term (both number and variable) in all three lists is $4x^2w^3$.

$$4x^2w^3(\underline{\quad} - \underline{\underline{\quad}} + \underline{\underline{\underline{\quad}}})$$

$$4x^2w^3 \cdot \underline{2w^2} = 8x^2w^5$$

$$4x^2w^3 \cdot \underline{5xw} = 20x^3w^4$$

$$4x^2w^3 \cdot 4x^2 = 16x^4w^3$$

8. $(3x - 1)(x + 5)$

$$3x^2 + 14x - 5 = (\underline{\quad} + \underline{\quad})(\underline{\quad} + \underline{\quad})$$
$$3x^2 + 14x - 5 = (3x + \underline{\quad})(1x + \underline{\quad})$$
$$3x^2 + 14x - 5 = (3x + \underline{-1})(1x + \underline{5})$$

9. $(11x + 10w)(11x - 10w)$. Think of $121x^2 - 100w^2$ as the difference of two squares, $(11x)^2 - (10w)^2$. Then use $a^2 - b^2 = (a + b)(a - b)$.

$$121x^2 - 100w^2 = (11x + 10w)(11x - 10w)$$

10. **C** $56x^2 + 4x - 12$

$$(7x + -3)(8x + 4)$$
$$7x \cdot 8x \cdot 7x \cdot 4 + -3 \cdot 8x + -3 \cdot 4$$
$$56x^2 + 28x + -24x + -12$$
$$56x^2 + 4x - 12$$

11. $9x^2 + 54x + 81$. Remember $(3x + 9)^2$ means $(3x + 9) \cdot (3x + 9)$. Use FOIL to multiply the binomials.

$$(3x + 9)(3x + 9)$$
$$3x \cdot 3x + 3x \cdot 9 + 9 \cdot 3x + 9 \cdot 9$$
$$9x^2 + 27x + 27x + 81$$
$$9x^2 + 54x + 81$$

Solutions to Practice Problems

1. $13x^2 - 20x + 7y^2 - 5y$. Rewrite the problem using only addition signs. Add like terms.

$$(\underline{7y} + \underline{\underline{-8x}} + 2y^2 + 8x^2) + (5x^2 + \underline{\underline{-12x}}) + (5y^2 + \underline{\underline{-12y}})$$

$$13x^2 - \underline{20x} + 7y^2 - \underline{5y}$$

2. $-2y^4 + 2y^2 - 6$. Rewrite the problem.

$$(7y^4 + -3y^2 + -2) + (4y^2 + -12) + -1(9y^4 + -1y^2 + -8)$$

Use the distributive property to simplify.

$$(7y^4 + -3y^2 + -2) + (4y^2 + -12) + -1 \cdot 9y^4 + -1 \cdot -1y^2 + -1 \cdot -8$$
$$= (7y^4 + -3y^2 + -2) + (4y^2 + -12) + -9y^4 + 1y^2 + 8$$

$$(\underline{7y^4} + \underline{-3y^2} + \underline{\underline{-2}}) + (\underline{4y^2} + \underline{\underline{-12}}) + \underline{-9y^4} + \underline{1y^2} + \underline{\underline{8}}$$

$$\underline{-2y^4} + \underline{2y^2} - 6$$

3. $(10w^8) \cdot (5w^2) = (10 \cdot 5) \cdot (w^8 \cdot w^2) = 50w^{10}$

4. $x^5 w^3 c^{10} \cdot x^{20} w^{10} c = (x^5 \cdot x^{20}) \cdot (w^3 \cdot w^{10}) \cdot (c^{10} \cdot c) = x^{25} w^{13} c^{11}$

5. $\dfrac{x^{20} w^{30} c^{40}}{x^{10} w^{30} c^{10}} = \dfrac{x^{20}}{x^{10}} \cdot \dfrac{w^{30}}{w^{30}} \cdot \dfrac{c^{40}}{c^{10}} = x^{10} c^{30}$. Remember that $\dfrac{w^{30}}{w^{30}}$ is the same as 1.

6. $(w^3 c^4)^5 = (w^3)^5 \cdot (c^4)^5 = w^{15} c^{20}$

7. $\sqrt[10]{w^{50} c^{40} x^{30}} = w^{\frac{50}{10}} c^{\frac{40}{10}} x^{\frac{30}{10}} = w^5 c^4 x^3$

8. $(5x^4)^3 = (5)^3 \cdot (x^4)^3 = 125 x^{12}$

9. $\sqrt[3]{64 c^{36}} = \sqrt[3]{64} \cdot \sqrt[3]{c^{36}} = 4 \cdot c^{\frac{36}{3}} = 4 c^{12}$. To find $\sqrt[3]{64}$, think of what number multiplied by itself three times will give you 64.

10. $c^8 y^5 - c^3 y^7 + 2 c^3 y^5$. Use the distributive property.

$$c^3 y^5 (c^5 - y^2 + 2)$$
$$c^3 y^5 \cdot c^5 + c^3 y^5 \cdot -y^2 + c^3 y^5 \cdot 2$$

Simplify by multiplying.

$$\underline{c^3 y^5 \cdot c^5} + \underline{c^3 y^5 \cdot -y^2} + \underline{c^3 y^5 \cdot 2} = \underline{c^8 y^5} - \underline{c^3 y^7} + \underline{2 c^3 y^5}$$

11. $-12x^2 - x + 35$

$$(5 - 3x)(4x + 7)$$
$$(5 - 3x) \cdot 4x + (5 - 3x) \cdot 7$$
$$4x \cdot 5 + 4x \cdot -3x + 7 \cdot 5 + 7 \cdot -3x$$
$$20x + -12x^2 + 35 + -21x$$
$$12x^2 - x + 35$$

12. $16x^2 + 24x + 9$. $4(x + 3)^2$ means $(4x + 3)(4x + 3)$. You cannot just square the $4x$ and square the 3.

$$(4x + 3)^2$$
$$(4x + 3)(4x + 3)$$
$$(4x + 3) \cdot 4x + (4x + 3) \cdot 3$$
$$4x \cdot 4x + 4x \cdot 3 + 3 \cdot 4x + 3 \cdot 3$$
$$16x^2 + 12x + 12x + 9$$
$$16x^2 + 24x + 9$$

13. $14x^2 + 22xy - 12y^2$

$$(7x - 3y)(2x + 4y)$$
$$(7x - 3y) \cdot 2x + (7x - 3y) \cdot 4y$$
$$2x \cdot 7x + 2x \cdot -3y + 4y \cdot 7x + 4y \cdot -3y$$
$$14x^2 - 6xy + 28xy - 12y^2$$
$$14x^2 + 22xy - 12y^2$$

14. $4x^{27} + 2x^9 + 3x^{18} + 1$

$$\frac{36x^{36} + 18x^{18} + 27x^{27} + 9x^9}{9x^9}$$

$$\frac{36x^{36}}{9x^9} + \frac{18x^{18}}{9x^9} + \frac{27x^{27}}{9x^9} + \frac{9x^9}{9x^9}$$

$$4x^{27} + 2x^9 + 3x^{18} + 1$$

15. $7x(6x^2 + 7x - 5)$. The factors of $42x^3$ are 1, 2, 3, 6, 7, 14, 21, 42, x, x^2, and x^3. The factors of $49x^2$ are 1, 7, 49, x, and x^2. The factors of $35x$ are 1, 5, 7, 35, and x. The greatest common factor is $7x$.

$$7x(\underline{6x^2} + \underline{7x} - \underline{\underline{5}})$$
$$7x \cdot \underline{6x^2} = 42x^3$$
$$7x \cdot \underline{7x} = 49x^2$$
$$7x \cdot \underline{\underline{5}} = 35x$$

16. $35x^2 - 48x - 27$

$$(7x + 3)(5x - 9)$$
$$35x^2 - 63x + 15x - 27$$
$$35x^2 - 48x - 27$$

17. $36x^2 - 84x + 49$

Be Careful! Many students look at this problem, just square the $6x$, and square the 7 for an answer of $36x^2 + 49$. Think of $(6x - 7)^2$ as $(6x - 7)(6x - 7)$.

$$(6x - 7)^2$$
$$(6x - 7)(6x - 7)$$
$$36x^2 - 42x - 42x + 49$$
$$36x^2 - 84x + 49$$

18. $(5x - 3)(x + 1)$

$$5x^2 + 2x - 3 = (5x + \underline{\quad})(1x + \underline{\quad})$$
$$5x^2 + 2x - 3 = (5x + \underline{-3})(1x + \underline{1})$$

19. $(7x - 1)(5x - 1)$

$$35x^2 - 12x + 1 = (\underline{\quad} + -1)(\underline{\quad} + -1)$$
$$35x^2 - 12x + 1 = (7x + -1)(5x + -1)$$

20. $(12x + 5)(12x - 5)$

$$144x^2 - 25 = (12x)^2 - 5^2$$
$$= (12x + 5)(12x - 5)$$

Algebra 1: Algebraic Fractions

REDUCING FRACTIONS AND SIMPLIFYING ALGEBRAIC RATIOS

REDUCING FRACTIONS AND SIMPLIFYING ALGEBRAIC RATIOS

Simplifying algebraic ratios is a more complicated version of reducing fractions. The thinking process used to reduce $\frac{34}{51}$ is similar to the process used to simplify $\frac{6x+18}{x^2-9}$.

To reduce $\frac{34}{51}$, factor both 34 and 51. $34 = 2 \cdot 17$ and $51 = 3 \cdot 17$. Cancel any numbers that are the same in the numerator and the denominator.

$$\frac{34}{51} = \frac{2 \cdot \cancel{17}^1}{3 \cdot \cancel{17}^1} = \frac{2}{3}$$

$\frac{34}{51}$ reduces to $\frac{2}{3}$.

The process used to reduce more difficult fractions is factor the numerator and denominator, then cancel any factors that are the same in the numerator and the denominator. This process is used in simplifying algebraic ratios.

To simplify an algebraic ratio, factor the numerator and denominator, then cancel any factors that are the same in the numerator and the denominator. To simplify $\frac{6x+18}{x^2-9}$, factor $6x + 18$ and $x^2 - 9$. If you need help factoring, review factoring in Chapter 12.

$$6x + 18 = 6 \cdot (x + 3)$$
$$x^2 - 9 = (x + 3) \cdot (x - 3)$$

Usually the times signs are left out from between the factors. If it helps you to remember that they are separate factors, then leave in the times signs. Any factors that are the same in the numerator and the denominator can be canceled. Be sure to cancel entire factors, not part of the factor.

$$\frac{6x+18}{x^2-9} = \frac{6 \cdot \cancel{(x+3)}^1}{\cancel{(x+3)}^1 \cdot (x-3)} = \frac{6}{x-3}$$

Be Careful! Some students try to cancel before factoring. You cannot cancel the 18 in the numerator with the 9 in the denominator.

$$\frac{6x+18}{x^2-9} \neq \frac{6x+\cancel{18}^2}{x^2-\cancel{9}^1}$$

Example

Simplify: $\dfrac{x^2+6x+8}{x^2-2x-8}$

Do not try to cancel the numbers in the fraction until you have completely factored both the numerator and the denominator.

$$x^2 + 6x + 8 = (x + 4) \cdot (x + 2)$$
$$x^2 - 2x - 8 = (x - 4) \cdot (x + 2)$$

Place these factors into the original problem.

$$\frac{x^2+6x+8}{x^2-2x-8} = \frac{(x + 4) \cdot (x + 2)}{(x - 4) \cdot (x + 2)}$$

Look for entire factors to cancel.

$$\frac{x^2+6x+8}{x^2-2x-8} = \frac{(x + 4) \cdot \cancel{(x + 2)}^1}{(x - 4) \cdot \cancel{(x + 2)}^1} = \frac{(x + 4)}{(x - 4)}$$

The answer $\dfrac{(x + 4)}{(x - 4)}$ is the same as $\dfrac{x + 4}{x - 4}$. The 4 in the numerator cannot be canceled with the 4 in the denominator. It may help to keep the parentheses around the factors. That way you will remember that the entire factors are $(x + 4)$ and $(x - 4)$.

Exercises

1. Simplify: $\dfrac{x^2-7x-18}{x^2+4x+4}$

2. Change $\dfrac{45x^3+36x^2}{9x^2+18x}$ into the simplest form.

Solutions to Exercises

1. Factor the numerator and factor the denominator. Do not cancel anything until the numerator and the denominator are factored.

$$x^2 - 7x - 18 = (x - 9) \cdot (x + 2)$$
$$x^2 + 4x + 4 = (x + 2) \cdot (x + 2)$$

Place the factors into the original problem and cancel entire factors.

$$\frac{x^2-7x-18}{x^2+4x+4} = \frac{(x - 9) \cdot \cancel{(x + 2)}^1}{(x + 2) \cdot \cancel{(x + 2)}^1} = \frac{(x - 9)}{(x + 2)}$$

2. Factor the numerator and factor the denominator. When factoring, remember to first look for a common factor.

$$45x^3 + 36x^2 = 9 \cdot x^2 \cdot (5x + 4)$$
$$9x^2 + 18x = 9 \cdot x \cdot (x + 2)$$

Now, place the factors into the original problem and cancel entire factors.

$$\frac{45x^3 + 36x^2}{9x^2 + 18x} = \frac{\cancel{9}^1 \cdot \cancel{x^2}^1 \cdot (5x+4)}{\cancel{9}^1 \cdot \cancel{x}^1 \cdot (x+2)} = \frac{x^1 \cdot (5x+4)}{(x+2)}$$

Remember that $\dfrac{x^2}{x}$ simplifies to $\dfrac{x^2}{x^1} = x^{2-1} = x^1 = x$.

Practice Problems

1. Simplify: $\dfrac{5x^2 + 7x + 2}{5x^2 + 6x + 1}$

2. Which of the following is equivalent to $\dfrac{6x^2 - 18x}{2x^3 - 18x}$?

    A. $\dfrac{3}{(x+3)}$

    B. $\dfrac{3}{x}$

    C. $\dfrac{3}{x} - 1$

    D. $\dfrac{3}{(x-3)}$

(Solutions are found at the end of the chapter.)

You may want to review work with proportions in Chapter 6 before beginning Algebraic Proportions.

ALGEBRAIC PROPORTIONS

An algebraic proportion has algebraic expressions as terms. An example of an algebraic proportion is $\dfrac{4}{x+1} = \dfrac{5}{x+3}$.

Solve: $\dfrac{4}{x+1} = \dfrac{5}{x+3}$

The product of the means equals the product of the extremes.

$$5 \cdot (x+1) = 4 \cdot (x+3)$$

Use the Distributive Property to simplify each side of the equation.

$$5x + 5 = 4x + 12$$

Solve the equation.

$$x + 5 = 12$$
$$x = 7$$

Check:

If you put 7 in place of the x in the original proportion, the proportion is correct.

$$\frac{4}{7+1} \stackrel{?}{=} \frac{5}{7+3}$$

$$\frac{4}{8} = \frac{5}{10}$$

Example

When solving the proportion $\frac{x+6}{x+3} = \frac{x+2}{x+1}$, the process is the same. The product of the means equals the product of the extremes.

$$(x + 3)(x + 2) = (x + 6)(x + 1)$$

Simplify each side by using the FOIL process from Chapter 12.

$$(x + 3)(x + 2) = x^2 + 2x + 3x + 6 = x^2 + 5x + 6$$
$$(x + 6)(x + 1) = x^2 + 1x + 6x + 6 = x^2 + 7x + 6$$

The problem now changes to a simpler problem to solve.

$$x^2 + 5x + 6 = x^2 + 7x + 6$$
$$5x + 6 = 7x + 6$$
$$-2x + 6 = 6$$
$$-2x = 0$$
$$x = 0$$

Some students are bothered when an answer is zero. In this case, zero is an acceptable answer.

Exercises

1. Solve: $\dfrac{2x+11}{3} = \dfrac{2x+3}{-5}$

2. Solve: $\dfrac{x-8}{x+2} = \dfrac{x-7}{x+1}$

Solutions to Exercises

1. The product of the means equals the product of the extremes.

$$3(2x + 3) = -5(2x + 11)$$

Use the Distributive Property.

$$6x + 9 = -10x + -55$$

Solve the equation.

$$16x + 9 = -55$$
$$16x = -64$$
$$x = -4$$

2. The product of the means equals the product of the extremes.

$$(x + 2)(x - 7) = (x - 8)(x + 1)$$

Simplify each side using the FOIL process.

$$(x + 2)(x - 7) = x^2 - 7x + 2x - 14 = x^2 - 5x - 14$$
$$(x - 8)(x + 1) = x^2 + 1x - 8x - 8 = x^2 - 7x - 8$$

The problem now changes.

$$x^2 - 5x - 14 = x^2 - 7x - 8$$
$$-5x - 14 = -7x - 8$$
$$2x - 14 = -8$$
$$2x = 6$$
$$x = 3$$

Practice Problems

3. Solve: $\dfrac{-9}{5x+1} = \dfrac{2}{x+4}$

4. Solve: $\dfrac{x-5}{2x+3} = \dfrac{x-1}{2x+7}$

(Solutions are found at the end of the chapter.)

SAMPLE GQE QUESTIONS FOR CHAPTER 13

1. Simplify: $\dfrac{3x-1}{9x^2-1}$

   A. $\dfrac{1}{3x}$

   B. $\dfrac{1}{3x}+1$

   C. $\dfrac{1}{3x+1}$

   D. $\dfrac{1}{6x+1}$

2. Simplify: $\dfrac{x^2-8x+15}{x^2-6x+5}$

   A. $\dfrac{4}{3}x+3$

   B. $1x^2+\dfrac{4}{3}x+3$

   C. $\dfrac{x-3}{x-1}$

   D. $\dfrac{x+3}{x+1}$

3. Solve: $\dfrac{x+8}{2x+1}=\dfrac{x+4}{2x-3}$

4. A 270-square foot rectangular room has a width of w and an unknown length. If the width were increased by 3 feet and the length stayed the same, the area of the rectangular room would change to 315 square feet. The proportion $\dfrac{270}{w}=\dfrac{315}{w+3}$ could be used to find the width of the room. Solve the proportion to find the width of the original room.

Answers Explained

1. **C** $\dfrac{1}{3x+1}$

   Before you can reduce the fractions, the numerator and denominator must be factored. The numerator will not factor.

   $$9x^2 - 1 = (3x + 1)(3x - 1)$$

   Look for entire factors that will cancel.

   $$\frac{3x-1}{9x^2-1} = \frac{\cancel{(3x-1)}^1}{(3x+1)\cancel{(3x-1)}_1} = \frac{1}{3x+1}$$

2. **C** $\dfrac{x-3}{x-1}$

   Factor the numerator and factor the denominator.

   $$x^2 - 8x + 15 = (x + -3)(x + -5)$$
   $$x^2 - 6x + 5 = (x + -1)(x + -5)$$

   Place both factored forms into the original fraction and cancel.

   $$\frac{x^2-8x+15}{x^2-6x+5} = \frac{(x+-3)\cancel{(x+-5)}^1}{(x+-1)\cancel{(x+-5)}_1} = \frac{x-3}{x-1}$$

3. 7

   The product of the means equals the product of the extremes.

   $$(2x + 1)(x + 4) = (x + 8)(2x - 3)$$

   Simplify each side of the equation.

   $$(2x + 1)(x + 4) = 2x^2 + 8x + 1x + 4 = 2x^2 + 9x + 4$$
   $$(x + 8)(2x - 3) = 2x^2 - 3x + 16x - 24 = 2x^2 + 13x - 24$$

   Put the simplified forms back into the equation and solve.

   $$2x^2 + 9x + 4 = 2x^2 + 13x - 24$$
   $$9x + 4 = 13x - 24$$
   $$-4x + 4 = -24$$
   $$-4x = -28$$
   $$x = 7$$

4. 18 feet wide

   You do not need to understand the story problem to be able to work this problem. The problem asks you to solve the proportion.

   $$315w = 270(w + 3)$$
   $$315w = 270w + 810$$
   $$45w = 810$$
   $$w = 18$$

   The original width was 18 feet.

Solutions to Practice Problems

1. $\dfrac{(5x+2)}{(5x+1)}$

Factor the numerator and factor the denominator.

$$5x^2 + 7x + 2 = (5x + 2) \cdot (x + 1)$$
$$5x^2 + 6x + 1 = (5x + 1) \cdot (x + 1)$$

Place the factored form into the original problem. Cancel entire factors that are the same in the numerator and the denominator.

$$\frac{5x^2 + 7x + 2}{5x^2 + 6x + 1} = \frac{(5x+2) \cdot \cancel{(x+1)}^1}{(5x+1) \cdot \cancel{(x+1)}^1} = \frac{(5x+2)}{(5x+1)}$$

You cannot cancel the 5 in the numerator with the 5 in the denominator.

2. **A** $\dfrac{3}{(x+3)}$

Factor the numerator and factor the denominator. Choices B and C are included as wrong answers for those who canceled before factoring.

$$6x^2 - 18x = 6x \cdot (x - 3)$$
$$2x^3 - 18x = 2x \cdot (x^2 - 9) = 2x \cdot (x + 3) \cdot (x - 3)$$

Place the factored form into the original problem and look for entire factors that can be canceled.

$$\frac{6x^2 - 18x}{2x^3 - 18x} = \frac{\cancel{6}^3 \cdot \cancel{x}^1 \cdot \cancel{(x-3)}^1}{\cancel{2}^1 \cdot \cancel{x}^1 \cdot (x+3) \cdot \cancel{(x-3)}^1} = \frac{3}{(x+3)}$$

Remember the 3's in the answer cannot cancel.

3. $x = -2$

$$2(5x + 1) = -9(x + 4)$$
$$10x + 2 = -9x - 36$$
$$19x + 2 = -36$$
$$19x = -38$$
$$x = -2$$

4. $x = -8$

$$(2x + 3)(x - 1) = (x - 5)(2x + 7)$$

Simplify each side using the FOIL process.

$$(2x + 3)(x - 1) = 2x^2 - 2x + 3x - 3 = 2x^2 + x - 3$$
$$(x - 5)(2x + 7) = 2x^2 + 7x - 10x - 35 = 2x^2 - 3x - 35$$

Put the simplified forms into the equation and solve.

$$2x^2 + x - 3 = 2x^2 - 3x - 35$$
$$x - 3 = -3x - 35$$
$$4x - 3 = -35$$
$$4x = -32$$
$$x = -8$$

Chapter 14 | Algebra 1: Quadratic, Cubic, and Radical Equations

GRAPHING CUBIC FUNCTIONS

A cubic equation is an equation where the highest power of the variable is 3. Examples include $y = x^3 + 4$, $y = x^3 + 3x^2 + 2x - 4$, and $y = x^3$. The graph of a cubic equation is not a straight line.

Example

Graph $y = \dfrac{1}{2}x^3$ using $-2, -1, 0, 1$, and 2 as values for x.

Substitute -2 for x in the equation $y = \dfrac{1}{2}x^3$.

$$y = \frac{1}{2}(-2)^3$$

$$y = \frac{1}{2} \cdot (-8)$$

$$y = -4$$

The ordered pair is $(-2, -4)$. Follow the same process to find the other ordered pairs, $(-1, -\dfrac{1}{2})$, $(0, 0)$, $(1, \dfrac{1}{2})$, and $(2, 4)$. Graph the points to get the curve shown below.

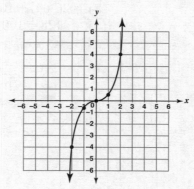

Exercises

1. Graph $y = x^3 + 1$ using $-2, -1, 0, 1$, and 2 as values for x.

2. Using the equation $y = 1 - 2x^3$, what is the value of y when x is 3?

3. Using the equation $y = 1 - x^3$, what is the value of y when x is -2?

4. Using the equation $y = 2x^3 + 100$, what is the value of y when x is -4?

Solutions to Exercises

1.

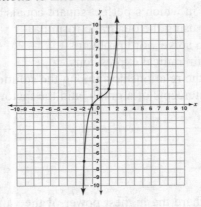

The graph of $y = x^3 + 1$ is shown above. The ordered pairs formed using the given values of x are $(-2, -7)$, $(-1, 0)$, $(0, 1)$, $(1, 2)$, and $(2, 9)$.

2. y is -53 when x is 3

$$y = 1 - 2x^3$$
$$y = 1 - 2 \cdot 3^3$$
$$y = 1 - 2 \cdot 27$$
$$y = 1 - 54$$
$$y = -53$$

3. y is 9 when x is -2

$$y = 1 - x^3$$
$$y = 1 - (-2)^3$$
$$y = 1 - (-8)$$
$$y = 9$$

4. y is -28 when x is -4

$$y = 2x^3 + 100$$
$$y = 2(-4)^3 + 100$$
$$y = 2(-64) + 100$$
$$y = -128 + 100$$
$$y = -28$$

Practice Problems

1. Graph $y = \frac{1}{2}x^3 - 1$ using $-2, -1, 0, 1,$ and 2 as values of x.

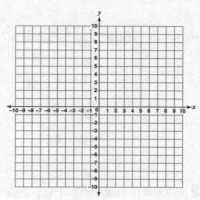

(Solution is found at the end of the chapter.)

GRAPHING RADICAL FUNCTIONS

A radical equation is an equation where the variable is under a radical sign. Review the work with radicals in Chapter 2 and Chapter 4. Examples of radical equations include $y = \sqrt{x}$, $y = \sqrt[3]{x} + 2$, $y = 3\sqrt[4]{x}$, and $y = \sqrt[5]{x+1}$. The graph of a radical equation is not a straight line.

Example

To graph $y = \sqrt{x}$ using $0, 1, 4,$ and 9 as values of x, find the value of y for each given value of x.

$$y = \sqrt{x}$$
$$y = \sqrt{0}$$
$$y = 0$$

Use the same process to find the corresponding y-values for the rest of the given x-values. The ordered pairs include $(0, 0), (1, 1), (4, 2),$ and $(9, 3)$. Notice that you were not given any negative numbers as values of x because the square root of a negative number is not defined in the set of real numbers. The graph is shown below.

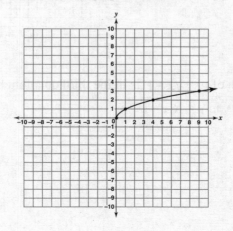

The graph of $y = \sqrt{x}$ is not in the second or third quadrant. You cannot use negative numbers for values of x in $y = \sqrt{x}$, so there are no points to the left of $(0, 0)$.

Example

To graph $y = \sqrt[3]{x-1}$ using 9, 2, 1, 0, and –7 as values of x, find the value of y for each value of x.

$$y = \sqrt[3]{x-1}$$
$$y = \sqrt[3]{9-1}$$
$$y = \sqrt[3]{8}$$
$$y = 2$$

Use the same process to find the corresponding y-values for the other values of x. Be careful when using 0 and –7 for values of x.

$$y = \sqrt[3]{x-1}$$
$$y = \sqrt[3]{0-1}$$
$$y = \sqrt[3]{-1}$$

Remember $\sqrt[3]{-1}$, the cube root of –1, means what number multiplied by itself three times gives you –1.

$$y = -1$$

The ordered pairs include $(9, 2)$, $(2, 1)$, $(1, 0)$, $(0, -1)$, and $(-7, -2)$. The graph is shown below.

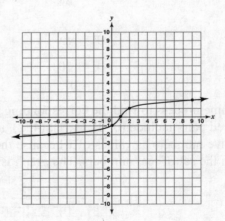

The graph of $y = \sqrt[3]{x-1}$ is in the third quadrant because you can use negative numbers as values of x in $y = \sqrt[3]{x-1}$.

Exercises

1. Graph $y = \sqrt{x-1}$ using 1, 2, 5, and 10 as values for x.

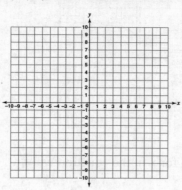

2. Graph $y = \sqrt[3]{x} + 2$ using -8, -1, 0, 1, and 8 as values for x.

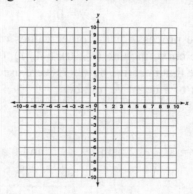

Solutions to Exercises

1.

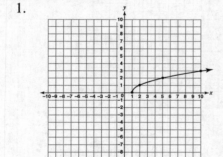

Substitute the values of x into the equation $y = \sqrt{x-1}$ and find the corresponding values for y.

$$y = \sqrt{1-1}$$
$$y = \sqrt{0}$$
$$y = 0$$

Use the same process to find the corresponding values for y using the other given values for x. Points on the curve include $(1, 0)$, $(2, 1)$, $(5, 2)$, and $(10, 3)$. The graph of $y = \sqrt{x-1}$ is shown above. Notice that numbers smaller than 1 were not given as values for x because they would lead to square roots of negative numbers.

2.

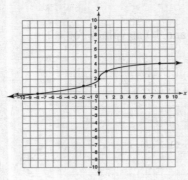

Substitute −8 for x in the equation $y = \sqrt[3]{x} + 2$ and find the corresponding value for y.

$$y = \sqrt[3]{-8} + 2$$
$$y = -2 + 2$$
$$y = 0$$

Use the same process for the other values of x to find their corresponding values of y. Points on the curve include (−8, 0), (−1, 1), (0, 2), (1, 3), and (8, 4). The graph is shown above. Notice that negative numbers were used as values for x because it is possible to take the cube root of a negative number.

Practice Problems

2. Graph $y = \sqrt{x} + 2$ using 9, 4, 1, and 0 as values for x.

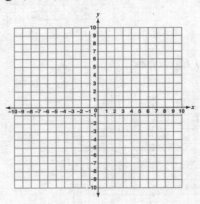

3. Graph $y = 1 - \sqrt[3]{x}$ using −8, −1, 0, 1, and 8 as values for x.

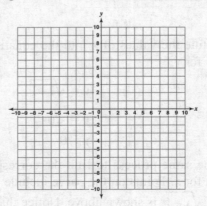

(Solutions are found at the end of the chapter.)

SOLVING QUADRATIC EQUATIONS BY FACTORING

Solving a quadratic means to find values for x so that the equation $x^2 + 7x + 6 = 0$ will be true. Before learning to solve quadratic equations by factoring, review factoring quadratics in Chapter 12.

> When you multiply two numbers together to get zero, one of the two numbers must be zero.
>
> $$\text{If } a \cdot b = 0, \text{ then either } a = 0 \text{ or } b = 0.$$

This property allows us to solve problems like $x^2 + 7x + 6 = 0$.

To solve $x^2 + 7x + 6 = 0$, change $x^2 + 7x + 6$ into two expressions multiplied together. In other words, factor $x^2 + 7x + 6$.

$$x^2 + 7x + 6 = 0$$
$$(x + 6)(x + 1) = 0$$

Now you have two expressions multiplied together to give you zero. Set each factor of $(x + 6)(x + 1)$ equal to 0.

$$x + 6 = 0 \quad \text{or} \quad x + 1 = 0$$
$$x = -6 \quad \text{or} \quad x = -1$$

The solutions to $x^2 + 7x + 6 = 0$ are -6 and -1. That means substituting $x = -6$ or $x = -1$ into the equation $x^2 + 7x + 6 = 0$ will make the equation true.

Check: If you substitute -6 for x, $x^2 + 7x + 6 = 0$ is true.

$$(-6)^2 + 7(-6) + 6 \stackrel{?}{=} 0$$
$$36 + -42 + 6 \stackrel{?}{=} 0$$
$$0 = 0$$

If you substitute -1 for x, $x^2 + 7x + 6 = 0$ is true.

$$(-1)^2 + 7(-1) + 6 \stackrel{?}{=} 0$$
$$1 + -7 + 6 \stackrel{?}{=} 0$$
$$0 = 0$$

Example

Solving $2x^2 = 9x + 5$ takes more work because there is not a zero on one side of the equation. Before factoring a quadratic, the equation must equal zero.

$$2x^2 = 9x + 5$$
$$2x^2 - 9x = 5$$
$$2x^2 - 9x - 5 = 0$$

Now factor $2x^2 - 9x - 5$.

$$(2x + 1)(x - 5) = 0$$

Set each factor equal to 0.

$$2x + 1 = 0 \text{ or } x - 5 = 0$$

Solve each equation.

$$2x = -1 \text{ or } x = 5$$

$$x = -\frac{1}{2} \text{ or } x = 5$$

Exercises

1. Solve for x: $3x^2 - 20x - 7 = 0$

2. Solve for x: $x^2 + 2x = 35$

Solutions to Exercises

1. Since the equation $3x^2 - 20x - 7 = 0$ already has one side equal to zero, factor $3x^2 - 20x - 7$.

$$3x^2 - 20x - 7 = 0$$
$$(3x + 1)(x - 7) = 0$$

Set each factor equal to zero and solve.

$$3x + 1 = 0 \text{ or } x - 7 = 0$$
$$3x = -1 \text{ or } x = 7$$
$$x = -\frac{1}{3} \text{ or } x = 7$$

2. The equation $x^2 + 2x = 35$ must be changed so that one side is zero.

$$x^2 + 2x = 35$$
$$x^2 + 2x - 35 = 0$$

Factor and solve.

$$x^2 + 2x - 35 = 0$$
$$(x + 7)(x - 5) = 0$$
$$x + 7 = 0 \text{ or } x - 5 = 0$$
$$x = -7 \text{ or } x = 5$$

Practice Problems

4. Solve for x: $(7x + 6)(4x - 9) = 0$

5. Solve for x: $3x^2 + 2x - 5 = 0$

6. Solve: $x^2 + 25 = 10x$

(Solutions are found at the end of the chapter.)

SOLVING SPECIAL QUADRATIC EQUATIONS USING SQUARE ROOTS

The solutions to the equation $x^2 = 16$ are $x = 4$ or $x = -4$ because both $(4)^2 = 16$ and $(-4)^2 = 16$. Solve the problem $x^2 = 16$ by taking the square root of both sides.

$$x^2 = 16$$
$$\sqrt{x^2} = \pm\sqrt{16}$$

When taking the square root of both sides, write a plus-or-minus sign, $\pm$, in front of the square root of the number. You will then get two answers for the problem.

$$x = 4 \text{ or } x = -4$$

Taking the square root of both sides can solve special quadratic equations in the form $(a)^2 = b$.

To solve $x^2 = 20$, take the square root of both sides.

$$x^2 = 20$$
$$\sqrt{x^2} = \pm\sqrt{20}$$
$$x = \sqrt{20} \text{ or } x = -\sqrt{20}$$

Review simplifying square roots in Chapter 4.

$$x = 2\sqrt{5} \text{ or } x = -2\sqrt{5}$$

Example

Solve: $(x - 5)^2 = 12$

Since the left side is a perfect square, solve the equation by taking the square root of both sides.

$$(x - 5)^2 = 12$$
$$\sqrt{(x-5)^2} = \pm\sqrt{12}$$
$$x - 5 = \pm\sqrt{12}$$

Simplify $\sqrt{12}$ into $2\sqrt{3}$.

$$x - 5 = \pm2\sqrt{3}$$

Separate the equation into two problems.

$$x - 5 = 2\sqrt{3} \text{ or } x - 5 = -2\sqrt{3}$$

Solve each equation.

$$x = 5 + 2\sqrt{3} \text{ or } x = 5 - 2\sqrt{3}$$

You cannot add 5 and $2\sqrt{3}$.

Exercises

1. Solve: $(2x + 1)^2 = 100$

2. Solve: $(x - 5)^2 = 18$

Solutions to Exercises

1. Take the square root of both sides of $(2x + 1)^2 = 100$.

$$\sqrt{(2x+1)^2} = \pm\sqrt{100}$$
$$2x + 1 = \pm10$$

Separate the equation into two equations.

$$2x + 1 = 10 \text{ or } 2x + 1 = -10$$

Solve each equation.

$$2x = 9 \text{ or } 2x = -11$$
$$x = \frac{9}{2} \text{ or } x = -\frac{11}{2}$$

2. Take the square root of both sides of the equation $(x - 5)^2 = 18$.

$$\sqrt{(x-5)^2} = \pm\sqrt{18}$$
$$x - 5 = \pm\sqrt{18}$$

Simplify $\sqrt{18}$ into $3\sqrt{2}$.

$$x - 5 = \pm 3\sqrt{2}$$

Break into two separate equations and solve.

$$x - 5 = 3\sqrt{2} \text{ or } x - 5 = -3\sqrt{2}$$
$$x = 5 + 3\sqrt{2} \text{ or } x = 5 - 3\sqrt{2}$$

Practice Problems

7. Solve: $(x + 6)^2 = 81$

8. Solve for x: $(x - 7)^2 = 200$

9. Solve: $(5x + 1)^2 = 36$

(Solutions are found at the end of the chapter.)

SOLVING QUADRATIC EQUATIONS USING THE QUADRATIC FORMULA

Sometimes it is not possible to solve a quadratic equation by factoring or by taking the square root of both sides. In that case, the quadratic formula is used. The quadratic formula is given on the reference sheet.

$$x = \frac{-b \pm \sqrt{b^2 - 4ac}}{2a} \text{ where } ax^2 + bx + c = 0, a \neq 0, \text{ and } b^2 - 4ac \geq 0.$$

The equation $5x^2 + 7x + 1 = 0$ cannot be solved by factoring or by taking the square root of both sides. The equation $5x^2 + 7x + 1 = 0$ is in the form $ax^2 + bx + c = 0$ where $a = 5$, $b = 7$, and $c = 1$. Substitute those values for a, b, and c in the quadratic formula.

$$x = \frac{-b \pm \sqrt{b^2 - 4ac}}{2a}$$

$$x = \frac{-7 \pm \sqrt{(7)^2 - 4 \cdot 5 \cdot 1}}{2 \cdot 5}$$

Simplify using the order of operations.

$$x = \frac{-7 \pm \sqrt{49 - 20}}{10}$$

$$x = \frac{-7 \pm \sqrt{29}}{10}$$

There are two answers for this problem, $x = \dfrac{-7 + \sqrt{29}}{10}$ or $x = \dfrac{-7 - \sqrt{29}}{10}$.

Exercises

1. Solve: $-2x^2 + 5x + 4 = 0$

2. Use the quadratic formula to solve $5x^2 + 2 = 9x$.

Solutions to Exercises

1. The equation $-2x^2 + 5x + 4 = 0$ is in the form $ax^2 + bx + c = 0$ where $a = -2$, $b = 5$, and $c = 4$. Substitute those values for a, b, and c into the quadratic formula.

$$x = \frac{-b \pm \sqrt{b^2 - 4ac}}{2a}$$

$$x = \frac{-5 \pm \sqrt{(5)^2 - 4 \cdot -2 \cdot 4}}{2 \cdot -2}$$

Simplify using the order of operations.

$$x = \frac{-5 \pm \sqrt{25 + 32}}{-4}$$

$$x = \frac{-5 \pm \sqrt{57}}{-4}$$

Break the solutions into two separate answers.

$$x = \frac{-5 + \sqrt{57}}{-4} \text{ or } x = \frac{-5 - \sqrt{57}}{-4}$$

2. The equation $5x^2 + 2 = 9x$ is not in the correct form. Get one side equal to zero by subtracting $9x$ from both sides of the equation. Now $5x^2 - 9x + 2 = 0$ is in the form $ax^2 + bx + c = 0$ where $a = 5$, $b = -9$, and $c = 2$. Substitute those values for a, b, and c into the quadratic formula.

$$x = \frac{-b \pm \sqrt{b^2 - 4ac}}{2a}$$

$$x = \frac{-(-9) \pm \sqrt{(-9)^2 - 4 \cdot 5 \cdot 2}}{2 \cdot 5}$$

Simplify.

$$x = \frac{9 \pm \sqrt{81 - 40}}{10}$$

$$x = \frac{9 \pm \sqrt{41}}{10}$$

Break the solutions into two separate answers.

$$x = \frac{9 + \sqrt{41}}{10} \text{ or } x = \frac{9 - \sqrt{41}}{10}$$

Practice Problems

10. Solve: $x^2 + 5x + 1 = 0$

11. Use the quadratic formula to solve $4x^2 - x = 6$.

(Solutions are found at the end of the chapter.)

GRAPHING QUADRATIC FUNCTIONS

Examples of quadratic functions include $y = 2x^2 - 1$, $y = 3x^2 + 5x - 4$, and $y = \dfrac{1}{2}x^2$. Graphs of quadratic functions are not straight lines.

Example

Graph the quadratic function $y = 2x^2 - 1$ for values of x including -2, -1, 0, 1, and 2. Substitute -2 for x in the equation $y = 2x^2 - 1$ and find the corresponding value for y.

$$y = 2(-2)^2 - 1$$
$$y = 2(4) - 1$$
$$y = 8 - 1$$
$$y = 7$$

The ordered pair $(-2, 7)$ will be included on the graph. Use the same process to find the y-value that corresponds to each given x-value. Other points on the graph include $(-1, 1)$, $(0, -1)$, $(1, 1)$, and $(2, 7)$. Graphing all five of the points will give the shape shown below.

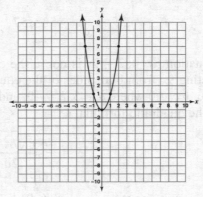

This shape is called a parabola. The graphs of quadratic functions are parabolas.

Exercises

1. Graph the equation $y = -3x^2 + 5$ on the coordinate plane below. Be sure to include the following values of x: -2, -1, 0, 1, and 2.

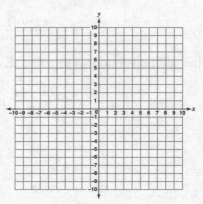

2. Graph the equation $y = x^2 - 2x - 8$ on the coordinate plane below. Be sure to include the following values of x: -3, -1, 0, 1, 2, and 4.

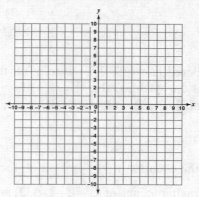

Solutions to Exercises

1.

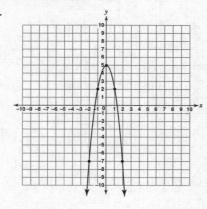

The graph of $y = -3x^2 + 5$ is the parabola shown above. Substitute -2 for x in the equation $y = -3x^2 + 5$.

$$y = -3(-2)^2 + 5$$
$$y = -3(4) + 5$$
$$y = -12 + 5$$
$$y = -7$$

The ordered pair $(-2, -7)$ is a point on the graph. Use the same process to find the y-values for each x-value given. Other points included on the graph are $(-1, 2)$, $(0, 5)$, $(1, 2)$, and $(2, -7)$.

2.

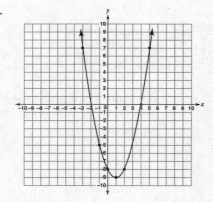

The graph of $y = x^2 - 2x - 8$ is the parabola shown above. Substitute -3 for x in the equation $y = x^2 - 2x - 8$.

$$y = (-3)^2 - 2(-3) - 8$$
$$y = 9 + 6 - 8$$
$$y = 7$$

The ordered pair $(-3, 7)$ is a point on the graph. Use the same process to find the y-values for each x-value given. Other points included on the graph are $(-1, -5)$, $(0, -8)$, $(1, -9)$, $(2, -8)$, and $(4, 0)$.

Practice Problem

12. Graph the equation $y = -\dfrac{1}{2}x^2 + 5$ on the coordinate plane below. Be sure to include the following values of x: -4, -2, 0, 2, and 4.

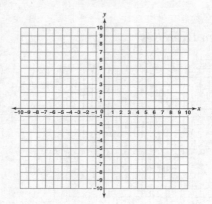

(Solution is found at the end of the chapter.)

SOLUTIONS, ZEROS, x-INTERCEPTS, AND FACTORS

Definition: The zeros of a function are the solutions when the function equals zero. In other words, to find the zero of a function, put 0 in place of the y and solve.

To find the zeros of the function $y = x^2 - 2x - 3$, solve the equation $x^2 - 2x - 3 = 0$.

$$x^2 - 2x - 3 = 0$$
$$(x - 3)(x + 1) = 0$$
$$x - 3 = 0 \text{ or } x + 1 = 0$$
$$x = 3 \text{ or } x = -1$$

The zeros of the function $y = x^2 - 2x - 3$ are $x = 3$ or $x = -1$.
The factors of $x^2 - 2x - 3$ are $(x - 3)$ and $(x + 1)$.
A quadratic equation that is related to the graph of $y = x^2 - 2x - 3$ is the equation $x^2 - 2x - 3 = 0$.
The solutions to the equation $x^2 - 2x - 3 = 0$ are $x = 3$ and $x = -1$.

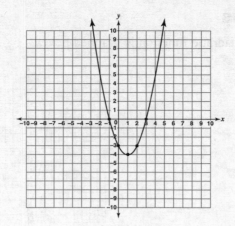

The graph of the quadratic equation $y = x^2 - 2x - 3$ is shown above.

The x-intercepts, the points where the graph crosses the x-axis, are $(-1, 0)$ and $(3, 0)$.

Notice that the zeros of the function are the same as the x-intercepts of the graph. The x-intercepts of the graph are the same as the solutions to the equation when y is equal to zero. The factors of the quadratic are used to find the solution to the quadratic.

Example

The zeros of a function are $x = -12$ and $x = 8$. The factors of the function would be $(x + 12)$ and $(x - 8)$. The x-intercepts of the graph of the function are $(-12, 0)$ and $(8, 0)$. The solutions to the function when y equals zero are $x = -12$ and $x = 8$.

Exercises

1. If the factors of a quadratic expression are $(x - 5)$ and $(x - 3)$, what are the solutions when that quadratic expression is set equal to zero?

2. The solutions to a quadratic equation are $x = -7$ and $x = -4$. What are the x-intercepts of the graph of the quadratic equation?

Solutions to Exercises

1. The quadratic expression set equal to zero would be $(x - 5)(x - 3) = 0$.

$$(x - 5)(x - 3) = 0$$
$$x - 5 = 0 \text{ or } x - 3 = 0$$
$$x = 5 \text{ and } x = 3$$

2. $(-7, 0)$ and $(-4, 0)$

The x-intercepts of the graph of a quadratic function are the same as the solutions to the quadratic equation.

Practice Problems

13. A quadratic function is graphed below.

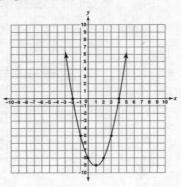

What are the zeros of this quadratic function? What are the solutions to the equation when y is zero?

14. A quadratic function has factors $(x + 21)$ and $(x - 23)$. What are the x-intercepts when the function is graphed?

(Solutions are found at the end of the chapter.)

WORD PROBLEMS WITH QUADRATIC EQUATIONS

Quadratic equations are used to describe many real world problems. To solve word problems using quadratics, you may need to substitute into an equation to find an answer or you may need to solve the quadratic equation.

Example
Fireworks using a rocket are shot into the air at 192 feet per second. The relationship between the time that the rocket is in the air and the height of the rocket is given by the equation $y = -16x^2 + 192x$. The time the rocket is in the air is x seconds. The height of the rocket is y feet.

To find the height of the rocket after 5 seconds, substitute 5 for the x in the equation $y = -16x^2 + 192x$.

$$y = -16(5)^2 + 192(5)$$
$$y = -16(25) + 960$$
$$y = -400 + 960$$
$$y = 560$$

The rocket is 560 feet in the air after 5 seconds.

Example
A patio is 4 feet wide and 8 feet long. A walkway is built around the patio. The area of the walkway is 64 square feet. The width w of the walkway can be found by solving the equation $w^2 + 6w - 16 = 0$.

$$w^2 + 6w - 16 = 0$$
$$(w + 8)(w - 2) = 0$$
$$w + 8 = 0 \text{ or } w - 2 = 0$$
$$w = -8 \text{ or } w = 2$$

The width of the walkway is 2 feet. The answer $w = -8$ does not make sense in this situation because you cannot have a negative width.

Exercises

1. Sam and Bruce began walking away from school. Sam walked north at x miles per hour. Bruce walked east at $x + 2$ miles per hour. In one hour Sam and Bruce are 10 miles apart. The equation $x^2 + (x + 2)^2 = 10^2$ can be solved to find how fast Sam walked. Ms. Math T. Cher simplified the equation into $x^2 + 2x - 48 = 0$. Solve the simplified equation to find how fast Sam walked.

2. At the Mathville Pharmacy the recommended dosage of a certain type of medicine is determined by the patient's weight. The formula used to determine dosage is $d = \dfrac{1}{5}w^2 + 4w$, where d represents the dosage in milligrams and w represents the patient's weight in kilograms. What is the dosage for a patient who weighs 40 kilograms?

Solutions to Exercises

1. Factor and solve $x^2 + 2x - 48 = 0$.

$$x^2 + 2x - 48 = 0$$
$$(x + 8)(x - 6) = 0$$
$$x + 8 = 0 \text{ or } x - 6 = 0$$
$$x = -8 \text{ or } x = 6$$

Sam walked 6 miles per hour. Walking -8 miles per hour does not make any sense.

2. Substitute 40 for the w in the formula $d = \dfrac{1}{5}w^2 + 4w$.

$$d = \frac{1}{5}(40)^2 + 4(40)$$

$$d = \frac{1}{5}(1{,}600) + 160$$

$$d = 320 + 160$$
$$d = 480$$

The dosage is 480 milligrams.

Practice Problems

15. Old Faithful is a geyser in Yellowstone National Park. Old Faithful sends a stream of boiling water into the air. The relationship between the height of the water and the time it is in the air is given by the equation $h = -16x^2 + 150x$. The height of the water is represented using the letter h. The time the water is in the air is represented with an x. To find how long the water is in the air, solve the equation $-16x^2 + 150x = 0$. How long is the water in the air?

16. Pretend a major league pitcher threw a baseball on the planet Mathiter. Scientists use the equation $h = -x^2 + 146.67x + 4$ to show the relationship between the height of the ball and the time the ball is in the air. The letter h represents the number of feet high the ball is above the ground. The letter x represents the number of seconds the ball is in the air. Find the height of the ball on planet Mathiter after 70 seconds.

(Solutions are found at the end of the chapter.)

SOLVING RADICAL EQUATIONS

A radical equation is an equation with a radical sign over a variable. Examples of radical equations include $\sqrt{x+3} = 7$, $\sqrt{x+2} = x$, and $\sqrt[3]{x+7} = 4$. Solving a radical equation means finding the value of x that makes the equation true.

To solve $\sqrt{x+3} = 7$, square both sides of the equation. Squaring both sides sometimes creates extra answers that are wrong. These extra wrong answers are called extraneous solutions. To be sure that you have correct answers, you have to check all of the answers.

$$\sqrt{x+3} = 7$$
$$\left(\sqrt{x+3}\right)^2 = (7)^2$$
$$x + 3 = 49$$
$$x = 46$$

Check the answer in the original problem.

$$\sqrt{x+3} = 7$$
$$\sqrt{46+3} \overset{?}{=} 7$$
$$\sqrt{49} \overset{?}{=} 7$$
$$7 = 7$$

The answer is correct.

Example
To solve $x = \sqrt{4x+5}$, square both sides.

$$x = \sqrt{4x+5}$$
$$(x)^2 = \left(\sqrt{4x+5}\right)^2$$
$$x^2 = 4x + 5$$

Solve $x^2 = 4x + 5$. First change the equation so that one of the sides is zero.

$$x^2 = 4x + 5$$
$$x^2 - 4x - 5 = 0$$

Factor or use the quadratic formula. Use factoring to solve this problem.

$$x^2 - 4x - 5 = 0$$
$$(x - 5)(x + 1) = 0$$
$$x - 5 = 0 \text{ or } x + 1 = 0$$
$$x = 5 \text{ or } x = -1$$

Be sure to check your answers in the original problem.

$5 \overset{?}{=} \sqrt{4(5)+5}$	$-1 \overset{?}{=} \sqrt{4(-1)+5}$
$5 \overset{?}{=} \sqrt{20+5}$	$-1 \overset{?}{=} \sqrt{-4+5}$
$5 \overset{?}{=} \sqrt{25}$	$-1 \overset{?}{=} \sqrt{1}$
$5 = 5$	$-1 \neq 1$
The answer 5 is correct.	The answer -1 is incorrect.

Example

To solve the radical equation $\sqrt[3]{x-3} = 4$, cube both sides of the equation. To eliminate a square root in a radical equation, square both sides. To eliminate the cube root in a radical equation, cube both sides. To eliminate the fourth root in a radical equation, take both sides to the fourth power.

$$\sqrt[3]{x-3} = 4$$

$$\left(\sqrt[3]{x-3}\right)^3 = (4)^3$$

$$x - 3 = 64$$

$$x = 67$$

You do not need to check your answer for extraneous solutions when you cube both sides. You only need to check your answers when you raise both sides to an even power.

Exercises

1. Solve for x: $\sqrt{x-4} = 10$

2. Solve: $x = \sqrt{6x-8}$

3. Solve: $\sqrt[4]{2x+3} = 2$

Solutions to Exercises

1. Square both sides of the equation and solve.

$$\sqrt{x-4} = 10$$

$$\left(\sqrt{x-4}\right)^2 = (10)^2$$

$$x - 4 = 100$$

$$x = 104$$

Since you squared both sides of a radical equation, you must check your answer.

$$\sqrt{x-4} = 10$$

$$\sqrt{104-4} \stackrel{?}{=} 10$$

$$\sqrt{100} \stackrel{?}{=} 10$$

$$10 = 10$$

The answer 104 is correct.

2. Square both sides of the equation.

$$x = \sqrt{6x-8}$$

$$(x)^2 = \left(\sqrt{6x-8}\right)^2$$

$$x^2 = 6x - 8$$

Solve $x^2 = 6x - 8$

$$x^2 - 6x + 8 = 0$$

$$(x-4)(x-2) = 0$$

$$x - 4 = 0 \text{ or } x - 2 = 0$$

$$x = 4 \text{ or } x = 2$$

Since you squared both sides of the equation, check both answers in the original equation.

$$x = \sqrt{6x-8}$$

$$4 \overset{?}{=} \sqrt{6(4)-8}$$

$$4 \overset{?}{=} \sqrt{24-8}$$

$$4 \overset{?}{=} \sqrt{16}$$

$$4 = 4$$

The answer 4 is correct.

$$x = \sqrt{6x-8}$$

$$2 \overset{?}{=} \sqrt{6(2)-8}$$

$$2 \overset{?}{=} \sqrt{12-8}$$

$$2 \overset{?}{=} \sqrt{4}$$

$$2 = 2$$

The answer 2 is correct.

3. Take both sides of the radical equation to the fourth power.

$$\sqrt[4]{2x+3} = 2$$

$$\left(\sqrt[4]{2x+3}\right)^4 = (2)^4$$

$$2x + 3 = 16$$

$$2x = 13$$

$$x = \frac{13}{2}$$

Since both sides of the radical equation were raised to an even power, check your answer in the original problem.

$$\sqrt[4]{2x+3} = 2$$

$$\sqrt[4]{2\left(\frac{13}{2}\right)+3} \overset{?}{=} 2$$

$$\sqrt[4]{13+3} \overset{?}{=} 2$$

$$\sqrt[4]{16} \overset{?}{=} 2$$

$$2 = 2$$

The answer $\frac{13}{2}$ is correct.

Practice Problems

17. Solve: $x = \sqrt{3x+10}$

18. Solve: $\sqrt[5]{3x-1} = 1$

(Solutions are found at the end of the chapter.)

SAMPLE GQE QUESTIONS FOR CHAPTER 14

1. On the coordinate plane below, draw the graphs of $y = x^2 + 1$ and $y = x^3 - 2$ using the following values for x: $-2, -1, 0, 1,$ and 2.

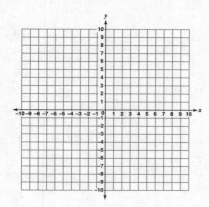

2. On the coordinate plane below, draw the graph of $y = \sqrt{x + 4}$ using the following values of x: $-4, -3, 0,$ and 5.

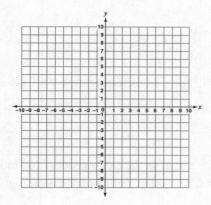

3. Solve: $2x^2 + 5x + 3 = 0$

Show All Work

4. Solve: $(2x - 7)^2 = 25$

Show All Work

5. Solve: $(x + 1)^2 = 7$

Show All Work

6. What is the solution to the equation $5x^2 - 7x - 1 = 0$?

A. $x = \dfrac{7 \pm \sqrt{69}}{10}$

B. $x = \dfrac{-7 \pm \sqrt{69}}{10}$

C. $x = \dfrac{7 \pm \sqrt{29}}{10}$

D. $x = \dfrac{-7 \pm \sqrt{29}}{10}$

Use the graph below to answer questions 7 and 8.

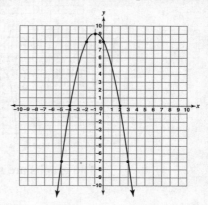

7. What are the zeros of the function graphed above?

A. $x = 4$ and $x = -2$
B. $x = -4$ and $x = 2$
C. $x = -8$
D. $x = 8$

8. What are the factors of the quadratic function graphed above?

A. $(x + 4)$ and $(x - 2)$
B. $(x - 4)$ and $(x + 2)$
C. $(x + 4)$ and $(x + 2)$
D. $(x - 4)$ and $(x - 2)$

9. A baton twirler throws her baton into the air. The equation $h = -16x^2 + 30x + 5$ describes the relationship between the height of the baton and the time the baton is in the air. The letter h represents the height of the baton in feet. The letter x represents the time the baton is in the air in seconds. How high is the baton after 2 seconds?

10. Solve $\sqrt{3 - 2x} = 5$ for x.

Show All Work

Answers Explained

1.

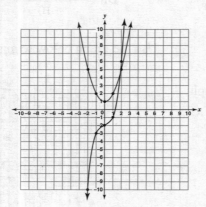

The graphs are shown above. Substitute -2 for x in the equation $y = x^2 + 1$.

$$y = (-2)^2 + 1$$
$$y = 4 + 1$$
$$y = 5$$

The point $(-2, 5)$ is on the graph of $y = x^2 + 1$. Use the same process to find the corresponding y-values for the given x-values. The other points on the graph are $(-1, 2)$, $(0, 1)$, $(1, 2)$, and $(2, 5)$.

Substitute -2 for x in the equation $y = x^3 - 2$.

$$y = (-2)^3 - 2$$
$$y = -8 - 2$$
$$y = -10$$

The point $(-2, -10)$ is on the graph of $y = x^3 - 2$. Use the same process to find the corresponding y-values for the given x-values. The other points on the graph are $(-1, -3)$, $(0, -2)$, $(1, -1)$, and $(2, 6)$.

2.

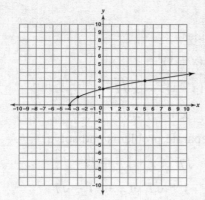

The graph of $y = \sqrt{x+4}$ is shown above. Substitute -4 for x in the equation $y = \sqrt{x+4}$.

$$y = \sqrt{-4+4}$$
$$y = \sqrt{0}$$
$$y = 0$$

The ordered pair $(-4, 0)$ is on the graph of $y = \sqrt{x+4}$. Use the same process to find the y-values for each given x-value. The other ordered pairs are $(-3, 1)$, $(0, 2)$, and $(5, 3)$.

3. $x = -\dfrac{3}{2}$ or $x = -1$

Factoring is the easiest way to do this problem.

$$2x^2 + 5x + 3 = 0$$
$$(2x + 3)(x + 1) = 0$$
$$2x + 3 = 0 \text{ or } x + 1 = 0$$
$$2x = -3 \text{ or } x = -1$$

$$x = -\dfrac{3}{2} \text{ or } x = -1$$

4. $x = 6$ or $x = 1$

Take the square root of both sides.

$$(2x - 7)^2 = 25$$
$$\sqrt{(2x - 7)^2} = \pm\sqrt{25}$$
$$2x - 7 = \pm 5$$

Break the equation into two separate problems and solve.

$$2x - 7 = 5 \quad \text{or } 2x - 7 = -5$$
$$2x = 12 \text{ or } 2x = 2$$
$$x = 6 \quad \text{or } x = 1$$

5. $x = -1 + \sqrt{7}$ or $x = -1 - \sqrt{7}$

Take the square root of both sides.

$$(x + 1)^2 = 7$$
$$\sqrt{(x+1)^2} = \pm\sqrt{7}$$
$$x + 1 = \pm\sqrt{7}$$

Break the equation into two separate problems and solve.

$$x + 1 = \sqrt{7} \text{ or } x + 1 = -\sqrt{7}$$
$$x = -1 + \sqrt{7} \text{ or } x = -1 - \sqrt{7}$$

6. **A** $x = \dfrac{7 \pm \sqrt{69}}{10}$

The answer choices show that you must use the quadratic formula. The equation $5x^2 - 7x - 1 = 0$ is in the form $ax^2 + bx + c = 0$, where $a = 5$, $b = -7$, and $c = -1$.

$$x = \frac{-b \pm \sqrt{b^2 - 4ac}}{2a}$$

$$x = \frac{-(-7) \pm \sqrt{(-7)^2 - 4(5)(-1)}}{2(5)}$$

$$x = \frac{7 \pm \sqrt{49 - (-20)}}{10}$$

$$x = \frac{7 \pm \sqrt{69}}{10}$$

7. **B** $x = -4$ and $x = 2$

The zeros of the function graphed are the places where the graph crosses the x-axis. The graph crosses the x-axis at $x = -4$ and $x = 2$.

8. **A** $(x + 4)$ and $(x - 2)$

The solutions are $x = -4$ and $x = 2$. So, the equations used to get the solutions are $x + 4 = 0$ and $x - 2 = 0$. The factors are $(x + 4)$ and $(x - 2)$.

9. 1 foot high

Substitute 2 for x in the equation $h = -16x^2 + 30x + 5$.

$$h = -16(2)^2 + 30(2) + 5$$
$$h = -16(4) + 60 + 5$$
$$h = -64 + 60 + 5$$
$$h = 1$$

10. $x = -11$

Square both sides of the equation and solve.

$$\sqrt{3-2x} = 5$$

$$\left(\sqrt{3-2x}\right)^2 = (5)^2$$

$$3 - 2x = 25$$
$$-2x = 22$$
$$x = -11$$

Check your answer because you squared both sides.

$$\sqrt{3-2x} = 5$$

$$\sqrt{3-2(-11)} \stackrel{?}{=} 5$$

$$\sqrt{3+22} \stackrel{?}{=} 5$$

$$\sqrt{25} \stackrel{?}{=} 5$$

$$5 = 5$$

The solution $x = -11$ is correct.

Solutions to Practice Problems

1.

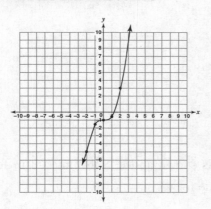

The graph of $y = \dfrac{1}{2}x^3 - 1$ is shown above. Substitute -2 for x in the equation

$y = \dfrac{1}{2}x^3 - 1$ and solve for y.

$$y = \frac{1}{2}(-2)^3 - 1$$

$$y = \frac{1}{2}(-8) - 1$$

$$y = -4 - 1$$
$$y = -5$$

Use the same process to find the y-values for the rest of the given x-values. The ordered pairs that should be graphed are $(-2, -5)$, $(-1, -1\frac{1}{2})$, $(0, -1)$, $(1, -\frac{1}{2})$, and $(2, 3)$.

2.

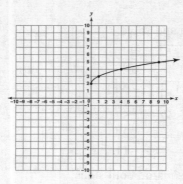

The graph of $y = \sqrt{x} + 2$ is shown above. Substitute 9 for x in the equation $y = \sqrt{x} + 2$ and find the corresponding value of y.

$$y = \sqrt{9} + 2$$
$$y = 3 + 2$$
$$y = 5$$

Use the same process with the other values of x. Points on the curve include (9, 5), (4, 4), (1, 3), and (0, 2). Notice that negative numbers were not given as values for x because you cannot take the square root of a negative number.

3.

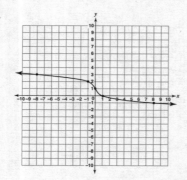

The graph of $y = 1 - \sqrt[3]{x}$ is shown above. Substitute -8 in for x in the equation $y = 1 - \sqrt[3]{x}$ and find the value for y.

$$y = 1 - \sqrt[3]{-8}$$
$$y = 1 - (-2)$$
$$y = 3$$

Use the same process to find the other points on the curve $y = 1 - \sqrt[3]{x}$. Points include $(-8, 3)$, $(-1, 2)$, $(0, 1)$, $(1, 0)$, and $(8, -1)$. Notice that negative numbers are given for values of x because you can take the cube root of a negative number.

4. $x = -\dfrac{6}{7}$ or $x = \dfrac{9}{4}$

Since the equation $(7x + 6)(4x - 9) = 0$ is already factored, set each factor equal to zero and solve.

$$7x + 6 = 0 \text{ or } 4x - 9 = 0$$
$$7x = -6 \text{ or } 4x = 9$$
$$x = -\dfrac{6}{7} \text{ or } x = \dfrac{9}{4}$$

5. $x = -\dfrac{5}{3}$ or $x = 1$

Factor $3x^2 + 2x - 5$.

$$(3x + 5)(x - 1) = 0$$

Set each equation equal to zero and solve.

$$3x + 5 = 0 \text{ or } x - 1 = 0$$
$$3x = -5 \text{ or } x = 1$$

$$x = -\dfrac{5}{3} \text{ or } x = 1$$

6. $x = 5$

The equation $x^2 + 25 = 10x$ must be changed so that one side is zero.

$$x^2 - 10x + 25 = 0$$
$$(x - 5)(x - 5) = 0$$

Set each equation equal to zero and solve.

$$x - 5 = 0 \text{ or } x - 5 = 0$$
$$x = 5$$

7. $x = 3$ or $x = -15$

$$\sqrt{(x+6)^2} = \pm\sqrt{81}$$
$$x + 6 = \pm 9$$
$$x + 6 = 9 \text{ or } x + 6 = -9$$
$$x = 3 \text{ or } x = -15$$

8. $x = 7 + 10\sqrt{2}$ or $x = 7 - 10\sqrt{2}$

$$\sqrt{(x-7)^2} = \pm\sqrt{200}$$

$$x - 7 = \pm\sqrt{200}$$

Simplify $\sqrt{200}$ into $10\sqrt{2}$.

$$x - 7 = \pm 10\sqrt{2}$$

$$x - 7 = 10\sqrt{2} \text{ or } x - 7 = -10\sqrt{2}$$

$$x = 7 + 10\sqrt{2} \text{ or } x = 7 - 10\sqrt{2}$$

9. $x = 1$ or $x = -\dfrac{7}{5}$

$$\sqrt{(5x+1)^2} = \pm\sqrt{36}$$
$$5x + 1 = \pm 6$$
$$5x + 1 = 6 \text{ or } 5x + 1 = -6$$
$$5x = 5 \text{ or } 5x = -7$$

$$x = 1 \text{ or } x = -\dfrac{7}{5}$$

10. $x = \dfrac{-5+\sqrt{21}}{2}$ or $x = \dfrac{-5-\sqrt{21}}{2}$

The equation $x^2 + 5x + 1 = 0$ is in the form $ax^2 + bx + c = 0$ where $a = 1$, $b = 5$, and $c = 1$. Substitute those values for a, b, and c into the quadratic formula.

$$x = \frac{-b \pm \sqrt{b^2 - 4ac}}{2a}$$

$$x = \frac{-5 \pm \sqrt{(5)^2 - 4 \cdot 1 \cdot 1}}{2 \cdot 1}$$

Simplify using the order of operations.

$$x = \frac{-5 \pm \sqrt{25 - 4}}{2}$$

$$x = \frac{-5 \pm \sqrt{21}}{2}$$

Break the solution into two separate answers.

$$x = \frac{-5 + \sqrt{21}}{2} \text{ or } x = \frac{-5 - \sqrt{21}}{2}$$

11. $x = \dfrac{1+\sqrt{97}}{8}$ or $x = \dfrac{1-\sqrt{97}}{8}$

The equation $4x^2 - x = 6$ is not in correct form. Get one side equal to zero by subtracting 6 from both sides.

$$4x^2 - x - 6 = 0$$

The equation $4x^2 - x - 6 = 0$ is in the form $ax^2 + bx + c = 0$ where $a = 4$, $b = -1$, and $c = -6$. Substitute those values for a, b, and c into the quadratic formula.

$$x = \frac{-b \pm \sqrt{b^2 - 4ac}}{2a}$$

$$x = \frac{-(-1) \pm \sqrt{(-1)^2 - 4 \cdot 4 \cdot -6}}{2 \cdot 4}$$

Simplify using the order of operations.

$$x = \frac{1 \pm \sqrt{1 + 96}}{8}$$

$$x = \frac{1 \pm \sqrt{97}}{8}$$

Break the solution into two separate answers.

$$x = \frac{1 + \sqrt{97}}{8} \text{ or } x = \frac{1 - \sqrt{97}}{8}$$

12.

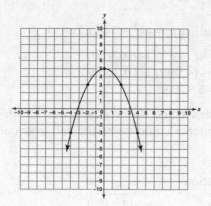

The graph of $y = -\frac{1}{2}x^2 + 5$ is the parabola shown above. Substitute -4 for x in the

equation $y = -\frac{1}{2}x^2 + 5$.

$$y = -\frac{1}{2}(-4)^2 + 5$$

$$y = -\frac{1}{2}(16) + 5$$

$$y = -8 + 5$$

$$y = -3$$

The ordered pair $(-4, -3)$ is on the graph. Other points included on the graph are $(-2, 3)$, $(0, 5)$, $(2, 3)$, and $(4, -3)$.

13. The zeros are $x = 4$ and $x = -2$. The zeros of a function are the same as the x-intercepts of the function.
   The solutions are $x = 4$ and $x = -2$. The solutions are the same as the zeros.

14. $(-21, 0)$ and $(23, 0)$
   Factors of $(x + 21)$ and $(x - 23)$ tell you that the x-intercepts are the solutions to the equations $x + 21 = 0$ and $x - 23 = 0$.

15. $\frac{75}{8}$ or 9.375 seconds

   Factor $-16x^2 + 150x$, set each factor equal to zero, and solve.

$$2x(-8x + 75) = 0$$
$$2x = 0 \text{ or } -8x + 75 = 0$$
$$x = 0 \text{ or } -8x = -75$$
$$x = 0 \text{ or } x = \frac{75}{8}$$

The answer 0 seconds is the time when the water is first shot into the air. The 9.375 seconds is the time when the water returns to the ground.

16. 5,370.9 feet high
Substitute 70 for x in the equation $h = -x^2 + 146.67x + 4$.

$$h = -(70)^2 + 146.67(70) + 4$$
$$h = -(4,900) + 10,266.9 + 4$$
$$h = 5,370.9$$

A major league baseball pitcher could throw a ball over 5,370.9 feet high. In other words, on planet Mathiter a pitcher could throw the ball over one mile high.

17. $x = 5$
Square both sides of the equation.

$$x = \sqrt{3x+10}$$

$$(x)^2 = \left(\sqrt{3x+10}\right)^2$$

$$x^2 = 3x + 10$$
$$x^2 - 3x - 10 = 0$$
$$(x - 5)(x + 2) = 0$$
$$x - 5 = 0 \text{ or } x + 2 = 0$$
$$x = 5 \text{ or } x = -2$$

Since you squared both sides of the radical equation, check both answers.

$$5 \overset{?}{=} \sqrt{3(5)+10} \qquad\qquad -2 \overset{?}{=} \sqrt{3(-2)+10}$$

$$5 \overset{?}{=} \sqrt{15+10} \qquad\qquad -2 \overset{?}{=} \sqrt{-6+10}$$

$$5 \overset{?}{=} \sqrt{25} \qquad\qquad -2 \overset{?}{=} \sqrt{4}$$

$$5 = 5 \qquad\qquad\qquad -2 \neq 2$$

The answer 5 is correct.   The answer –2 is incorrect.

18. $x = \dfrac{2}{3}$

Take both sides of the radical equation to the fifth power.

$$\sqrt[5]{3x-1} = 1$$

$$\left(\sqrt[5]{3x-1}\right)^5 = (1)^5$$

$$3x - 1 = 1$$
$$3x = 2$$

$$x = \dfrac{2}{3}$$

You do not need to check the answer $x = \dfrac{2}{3}$ because you took both sides of the equation to the fifth power, not an even power.

| # Problem Solving

INDIANA'S ACADEMIC MATHEMATICS STANDARDS ADDRESSED:

8.7.1 Analyze problems by identifying relationships, telling relevant from irrelevant information, identifying missing information, sequencing and prioritizing information, and observing patterns.

8.7.3 Decide when and how to divide a problem into simpler parts.

8.7.4 Apply strategies and results from simpler problems to solve more complex problems.

8.7.5 Make and test conjectures using inductive reasoning.

8.7.6 Express solutions clearly and logically using the appropriate mathematical terms and notation. Support solutions with evidence in both verbal and symbolic work.

8.7.10 Make precise calculations and check the validity of the results in the context of the problem.

8.7.11 and A1.9.2 Decide whether a solution is reasonable in the context of the original situation.

A1.9.1 Use a variety of problem solving strategies, such as drawing a diagram, making a chart, guess-and-check, solving a simpler problem, writing an equation, and working backwards.

A1.9.4 Understand that the logic of equation solving begins with the assumption that the variable is a number that satisfies the equation, and that the steps taken when solving equations create new equations that have, in most cases, the same solution as the original. Understand that similar logic applies to solving systems of equations simultaneously.

A1.9.6 Distinguish between inductive and deductive reasoning, identifying and providing examples of each.

A1.9.7 Identify the hypothesis and conclusion in a logical deduction.

A1.9.8 Use counterexamples to show that statements are false, recognizing that a single counterexample is sufficient to prove a general statement false.

PROBLEM SOLVING

Many problem solving strategies have been used in solving story problems in Chapters 1 through 14. Those strategies include solving a simpler problem, making an organized list, working backwards, guessing and checking, looking for a pattern, drawing a diagram, and creating a chart. These strategies are combined with mathematical concepts to create the problem-solving situations on the GQE. No new mathematics will be introduced for problem solving.

All of the story problems on the GQE will be different than ones you have seen before. You need to have a clear plan when working story problems.

Problem Solving Plan

Always read the problem several times. Underline the question that needs to be answered.

Try to rephrase the question in your own words. Would a drawing or a list make the problem simpler to understand?

Think about what steps are needed to solve the problem. Most of the time there will be more than two steps.

Follow your steps to find the solution. Show your work in an organized way. Points are given for your work. If you do not show your work, you earn half credit.

Check to see if the solution answers the question you underlined and if your solution makes sense.

Example

Mr. Keywire Man is in charge of roping off a seating area for the Mathville choir show. A diagram of the seating area is shown below.

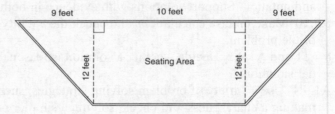

The rope is to be placed along the double lines on the perimeter of the seating area. The rope costs $2.50 per foot. Sales tax is 6%. <u>What is the cost of the rope needed?</u>

Read the question again.

Steps to solve the problem

Find the length of the rope needed. Multiply that answer by the cost per foot. Add the sales tax to get the total cost.

Find the length of the hypotenuse of the triangle on the left.

$$9^2 + 12^2 = c^2$$
$$81 + 144 = c^2$$
$$c^2 = 225$$
$$c = 15$$

Find the perimeter of the seating area.

$$\text{Perimeter} = 9 + 10 + 9 + 15 + 10 + 15$$
$$\text{Perimeter} = 68$$

The perimeter of the seating area is 68 feet.

Find the cost of the rope.

$$\text{Cost} = 68 \cdot 2.50 = 170$$

The cost of the rope is $170.00.

Find the sales tax. Sales tax is 6% of the cost of the rope.

$$\text{Tax} = .06 \cdot 170 = 10.2$$

Sales tax is $10.20.
Find the total price.

$$\text{Total} = 170.00 + 10.20 = 180.20$$

The total cost for the rope is $180.20.

Exercises

1. The distance that a bicycle wheel travels during 1 revolution of the wheel is equal to the circumference of the wheel. Lance's bicycle wheel has a radius of 13.5 inches. How many revolutions will Lance's wheel make in 1 mile? Round your answer to the nearest whole number of revolutions. Read the problem again. Underline the question.

2. The Eisenhower Memorial tunnel in Colorado is 8,941 feet long. Bray Kneeguy determines that the Eisenhower Memorial tunnel is about 3 miles long. Explain why Bray Kneeguy's solution is or is not reasonable.

Solutions to Exercises

1. Steps to solve the problem: Find the circumference of the wheel in inches. Change the distance of 1 mile into inches. Divide the distance by the circumference of the wheel.

   Find the circumference of Lance's wheel.

   $$C = 2\pi r = 2 \cdot 3.14 \cdot 13.5 = 84.78$$

   The circumference is 84.78 inches.

   Change 1 mile into inches.

   $$1 \text{ mile} \cdot \frac{5280 \text{ feet}}{1 \text{ mile}} \cdot \frac{12 \text{ inches}}{1 \text{ foot}} = 63{,}360 \text{ inches}$$

   The distance of 1 mile is 63,360 inches.

   Find the number of revolutions.

   $$63{,}360 \div 84.78 = 747.346... \approx 747$$

   The number of revolutions to travel 1 mile is 747 revolutions. If you use the π button on your calculator, you will get 746.967… which also rounds to 747 revolutions.

2. Bray Kneeguy's solution is not reasonable. There are 5,280 feet in one mile, so there would be more than 15,000 feet in 3 miles. The tunnel is only 8,941 feet long.

   The more story problems that you try; the better you are at solving them. Practice! Practice! Practice!

Practice Problems

1. Mathola Soda Company filled 5 delivery trucks with soda in 12-ounce cans. Each delivery truck holds 5,344 cans. How many gallons of soda are in the 5 delivery trucks? Read the problem again. Underline the question.

2. The fundraising committee of the Math Club at Mathville High School has designed a paperweight in the shape of a rectangular prism, as shown below.

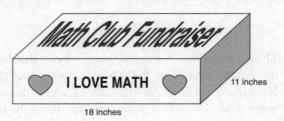

The length of the paperweight is 18 inches, and the width of the paperweight is 11 inches. The volume of the paperweight is 1,386 cubic inches. All of the faces of the paperweight need to be painted. How many square inches of paperweight must be painted? Read the problem again. Underline the question.

3. A salmon can swim up to 23 miles per hour. The Mathville River is 2,340 miles long. Gail determines that a salmon could swim 2,340 miles in a little less than one day. Explain why Gail's solution is or is not reasonable.

4. Sam is packing sweatshirts to sell at an amusement park. He needs to buy cartons in which to pack the sweatshirts.

   Each carton costs $5 plus 6% sales tax. Each carton holds 22 sweatshirts. Sam has 250 sweatshirts to pack. He states it will cost him $58.30. He is wrong.

   Explain why Sam is incorrect. Include the number of boxes Sam could buy using the amount of money he determined. What is the least amount of money Sam actually needs to pay for the needed cartons?

(Solutions are found at the end of the chapter.)

IF-THEN STATEMENTS, HYPOTHESIS, CONCLUSION, AND COUNTEREXAMPLE

An example of an if-then statement is if today is Monday, then tomorrow is Tuesday. The "if" part of the statement, the hypothesis, is the part that is accepted as true. It is accepted that today is Monday. The "then" part of the statement, the conclusion, is a deduction that comes from the hypothesis. It can be concluded that tomorrow is Tuesday.

Not all if-then statements are true. If a number is divisible by 2, then it is divisible by 4. This statement is false.

> To show that a statement is false, give one example that satisfies the hypothesis but not the conclusion. This example is called a counterexample.

A counterexample to the statement could be 6. Six is divisible by 2 but is not divisible by 4.

Example

If a number is an integer, then it is a whole number. To decide if this statement is true or false, think of all of the different kinds of numbers that are integers. Decide whether all of those numbers are also whole numbers.

The statement is false. Any negative integer satisfies the hypothesis that it is an integer but does not satisfy the conclusion that it is a whole number. A counterexample to show the statement is false would be –5. The number –5 is an integer but not a whole number.

Exercises

1. Classify each statement as true or false. If false, give a counterexample.

   A. If you add together two odd numbers, then your answer is odd.
   B. If two numbers are even, then their product is even.

2. Identify the conclusion of the statement "If I study Algebra, then I am a happy person."

3. Identify the hypothesis of the statement "If I learn Algebra, then I am really smart."

Solutions to Exercises

1. A. False. A counterexample could be $5 + 7 = 12$. Two odd numbers are added together, but the answer is even.
   B. True

2. The conclusion is "I am a happy person."

3. The hypothesis is "I learn Algebra."

Practice Problems

5. If $a = x + 3$, then $a^2 = x^2 + 9$.

   A. Is the statement true or false? If false, give a counterexample.
   B. Identify the hypothesis.
   C. Identify the conclusion.

(Solutions are found at the end of the chapter.)

DEDUCTIVE AND INDUCTIVE REASONING

If a person is 16 years old, then that person can apply for a driver's license.
Carol is 17 years old.
You can conclude that Carol can apply for a driver's license.

The reasoning process used in those three statements is called deductive reasoning. You use an if-then statement along with given information to make a conclusion.
We use deductive reasoning in everyday life.

If-then statements: If there are dark clouds in the sky, then it usually rains. If it rains, then I need my umbrella.
Given information: There are dark clouds in the sky.
Conclusion: I will need my umbrella.

Deductive reasoning is used everywhere in mathematics. Proofs in geometry take given statements and list steps with justifications to prove a conclusion is true. Solving equations in algebra begins with an equation to solve, creates other equivalent equations, and, finally, writes an equation that is the solution. Each step in solving the equation can be justified by a mathematical property.

Example

If two angles are vertical angles, then they are equal in measure.
$\angle 1$ and $\angle 3$ are vertical angles.

Using deductive reasoning, you can conclude that $\angle 1$ and $\angle 3$ are equal in measure.

Inductive reasoning does not follow the same pattern as deductive reasoning.
Look at the multiplication problems.

$$11 \cdot 11 = 121$$
$$111 \cdot 111 = 12{,}321$$
$$1{,}111 \cdot 1{,}111 = 1{,}234{,}321$$
$$11{,}111 \cdot 11{,}111 = 123{,}454{,}321$$

You could guess or conjecture that the answer to $111{,}111 \cdot 111{,}111$ is $12{,}345{,}654{,}321$.

The reasoning process used here is inductive reasoning. You observed several specific cases and arrived at a general conclusion. You know that the specific multiplication problems are true. You use the pattern that you found to make an educated guess or conjecture.

Detectives frequently use inductive reasoning to solve cases. They observe a series of similar crimes. By examining the evidence at the crimes, they hope to find a pattern or rule that applies to all of the crimes. Using that pattern, they arrive at a conclusion about who might have committed the crimes. Their conclusion is an educated guess or conjecture. This conjecture can be right or wrong.

Inductive reasoning is not used as frequently in mathematics. Explorations in geometry use inductive reasoning to make conjectures. The actual proofs of the conjectures are usually done with deductive reasoning.

Example

$$1 + 3 = 2^2$$
$$1 + 3 + 5 = 3^2$$
$$1 + 3 + 5 + 7 = 4^2$$
$$1 + 3 + 5 + 7 + 9 = 5^2$$

Notice the pattern.
Conjecture: The sum of the first 6 odd numbers is 6^2. The sum of the first 20 odd numbers should be 20^2. The sum of the first n odd numbers is n^2.

Exercises

1. Carl finds that on his way to the ballpark for the first 5 games of the baseball season, there is a person outside the stadium selling peanuts. Carl concludes that there is a person outside of the stadium selling peanuts before every baseball game. Did Carl use deductive or inductive reasoning?

2. All rectangles have diagonals that are congruent. A square is a rectangle. You conclude that the diagonals of a square are congruent. Did you use deductive or inductive reasoning?

3. Use inductive reasoning to find the pattern.

$$(a + 1)^2 = a^2 + 2a + 1$$
$$(a^2 + a + 1)^2 = a^4 + 2a^3 + 3a^2 + 2a + 1$$
$$(a^3 + a^2 + a + 1)^2 = a^6 + 2a^5 + 3a^4 + 4a^3 + 3a^2 + 2a + 1$$

Show the next equation in the pattern.

Solutions to Exercises

1. Carl used inductive reasoning. He observed several baseball games and made a conclusion about every baseball game.

2. You used deductive reasoning.
If a shape is a rectangle, then its diagonals are congruent.
A square is a rectangle. Therefore, a square's diagonals are congruent.

3. $(a^4 + a^3 + a^2 + a + 1)^2 = a^8 + 2a^7 + 3a^6 + 4a^5 + 5a^4 + 4a^3 + 3a^2 + 2a + 1$

Practice Problems

6. Jean Yus eats shellfish on Saturday and she breaks out in hives Saturday night. Jean Yus eats shellfish a week later and breaks out in hives again. Jean Yus's doctor concludes that Jean Yus is allergic to shellfish. What type of reasoning, deductive or inductive, did the doctor use?

7. Smar Tee knows that to be a member of the orchestra, you must play an instrument. James is in the orchestra. Smar Tee concludes that James must play an instrument. What type of reasoning, deductive or inductive, did Smar Tee use?

(Solutions are found at the end of the chapter.)

SAMPLE GQE QUESTIONS FOR CHAPTER 15

1. At the Mathville Amusement Park, the Logarithmic Ride holds 25 people in each boat. It takes 8 minutes for the ride and the loading/unloading of people. There are 253 people who want to take the ride. Your friend says that it will take about 1 hour and 20 minutes for all 253 people to be finished with the ride. Explain why your friend's observation is not reasonable. Include in your explanation the actual amount of time it will take for 253 people.

2. Mathville Manufacturing is packing facial tissue boxes into cases. The facial tissue box is 4 inches by 4 inches by 5 inches. The case holding the individual boxes is 20 inches by 20 inches by 5 inches. A diagram of the facial tissue box and the case are shown below.

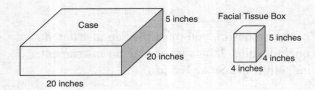

Mathville Manufacturing decides to double the length, double the width, and double the height of the case.

Mathville Manufacturing's competitor, Not-So-Smart Production, says that the new case will only hold twice as many boxes.

Explain why Not-So-Smart Production's observation is not reasonable. Include in your explanation the number of facial tissue boxes that fit in the original case and the number that will fit in the new size case.

3. Whenever Sally sees George at the movies, George is buying popcorn. Sally concludes that George always buys popcorn at the movies. What type of reasoning, deductive or inductive, did Sally use?

4. If two lines are parallel, $a // b$, then alternate interior angles are congruent, $\angle 1 \cong \angle 2$. If $\angle 2$ and $\angle 3$ are vertical angles, then they are congruent, $\angle 2 \cong \angle 3$. If $\angle 1 \cong \angle 2$ and $\angle 2 \cong \angle 3$, then $\angle 1 \cong \angle 3$.

   What type of reasoning, deductive or inductive, is being used in the argument?

5. If a number is elohw, then it is a regetni. If a number is a regetni, then it is lanoitar.

   Using deductive reasoning, what conclusion can be made?

   A. If a number is a regetni, then it is elohw.
   B. If a number is lanoitar, then it is a regetni.
   C. If a number is elohw, then it is lanoitar.
   D. No conclusion can be made.

6. What is the hypothesis in the following if-then statement?

   If a student is in algebra class, then the student is having a great time.

7. The statement below is false. Give a counterexample to show that it is false.

   If a real number is squared, then the answer is always positive.

8. Is the statement below true or false? If false, give a counterexample.

   If you take half of a number, your answer is smaller than the original number.

9. Use inductive reasoning to find the pattern.

   $2^2 + 3^2 + 6^2 = 7^2$
   $3^2 + 4^2 + 12^2 = 13^2$
   $4^2 + 5^2 + 20^2 = 21^2$

   Show the next equation in the pattern.

Answers Explained

1. It will take 11 trips of the Logarithmic Ride for all 253 people to go. 11 trips is 88 minutes or 1 hour and 28 minutes. Your friend thought you would need only 10 trips for 1 hour and 20 minutes.

2. Mathville Manufacturing could put 25 boxes in the original case. There would be 5 rows of 5.
   The new case would be 40 inches by 40 inches by 10 inches. There would be 10 rows of 10 that are stacked 2 high or 200 boxes. If the dimensions of a rectangular prism are doubled, then the volume is 8 times as large.

$$V \text{ of original case} = l \cdot w \cdot h$$
$$V \text{ of new case} = (2l) \cdot (2w) \cdot (2h)$$
$$= 8 \cdot (l \cdot w \cdot h)$$

3. Sally used inductive reasoning. She took several individual events to make a conjecture about all events.

4. Deductive reasoning is used. A conclusion is drawn from several if-then statements.

5. **C** If a number is elohw, then it is lanoitar. The words in this problem did not make sense to you, but the reasoning can still be done. To help you understand, the words are just whole, integer, and rational written backwards. If a number is whole, then it is an integer. If a number is an integer, then it is rational. Now try the reasoning. If a number is a whole number, then it is rational.

6. The hypothesis is a student is in algebra class.

7. A counterexample is 0. $0^2 = 0$, which is not a positive number.

8. False. A counterexample is half of any negative number. Half of –8 is –4, which is larger than –8.

9. $5^2 + 6^2 + 30^2 = 31^2$ is the next one in the pattern. Any equation of the form $a^2 + (a + 1)^2 + [a(a + 1)]^2 = [a(a + 1) + 1]^2$ will work.

Solutions to Practice Problems

1. 2,505 gallons
   Steps to solve the problem: Find the number of ounces in the cans in 1 delivery truck. Change the ounces into gallons. Multiply the number of gallons in 1 delivery truck by 5.
   Find the number of ounces in 1 delivery truck.

$$5{,}344 \cdot 12 = 64{,}128 \text{ ounces}$$

The number of ounces in the cans in 1 delivery truck is 64,128 ounces.
Change 64,128 ounces to gallons.

$$64{,}128 \text{ ounces} \cdot \frac{1 \text{ cup}}{8 \text{ ounces}} \cdot \frac{1 \text{ pint}}{2 \text{ cups}} \cdot \frac{1 \text{ quart}}{2 \text{ pints}} \cdot \frac{1 \text{ gallon}}{4 \text{ quarts}}$$

$$= \frac{64{,}128}{8 \cdot 2 \cdot 2 \cdot 4} \text{ gallons} = \frac{64{,}128}{128} \text{ gallons} = 501 \text{ gallons}$$

There are 501 gallons in 64,128 ounces.
Multiply by 5.

$$501 \cdot 5 = 2,505$$

There are 2,505 gallons in 5 delivery trucks.

2. 802 square inches. Steps to solve the problem: Find the height of the rectangular prism. Use the surface area formula.
Use the formula for the volume of a rectangular prism to find the height.

$$1,386 = 18 \cdot 11 \cdot h$$
$$1,386 = 198h$$
$$h = 7$$

The height will be 7 inches.
Use the formula for the surface area of a rectangular prism.

$$SA = 2 \cdot 18 \cdot 11 + 2 \cdot 7 \cdot 11 + 2 \cdot 18 \cdot 7$$
$$SA = 396 + 154 + 252$$
$$SA = 802$$

3. Gail's solution is not reasonable. It would take about 100 hours for a salmon to swim 2,300 miles. One day is 24 hours. It would take a salmon much longer than one day to swim 2,300 miles.

4. Steps to solve the problem: Find the number of cartons that Sam will need. Find the cost of those cartons. Add the sales tax. To determine the number of cartons Sam needs, divide the number of sweatshirts by the number that fill a carton.

$$\text{Number of cartons} = 250 \div 22 = 11.363636$$

Sam will need to buy 12 cartons, not 11. Each carton costs $5 plus 6% tax.

$$\text{Cost of cartons} = 12 \cdot 5 = 60$$

The cartons alone will cost $60. Then, add on the sales tax.

$$\text{Total Cost} = 60 + .06 \cdot 60 = 63.60$$

The true cost is $63.60. Sam did his calculations with 11 cartons instead of 12.

$$\text{Cost of cartons} = 11 \cdot 5 = 55$$
$$\text{Total Cost} = 55 + .06 \cdot 55 = 58.30$$

5. A. False. A counterexample is $a = 5$ and $x = 2$. You know that $5 = 2 + 3$, but $5^2 \neq 2^2 + 9$.
   B. The hypothesis is $a = x + 3$.
   C. The conclusion is $a^2 = x^2 + 9$.

6. Jean Yus's doctor used inductive reasoning. The doctor took several examples of an event and made a generalization from the examples.

7. Smar Tee used deductive reasoning.
   If a person is in the orchestra, then that person plays an instrument.
   James is in the orchestra. Therefore, James plays an instrument.

PRACTICE TEST 1 ANSWER SHEET

Fill in the bubble completely.
Erase carefully if answer is changed.

1. Ⓐ Ⓑ Ⓒ Ⓓ
2. Ⓐ Ⓑ Ⓒ Ⓓ
3. Ⓐ Ⓑ Ⓒ Ⓓ
4. Ⓐ Ⓑ Ⓒ Ⓓ
5. Ⓐ Ⓑ Ⓒ Ⓓ
6. Ⓐ Ⓑ Ⓒ Ⓓ
7. Ⓐ Ⓑ Ⓒ Ⓓ
8. Ⓐ Ⓑ Ⓒ Ⓓ
9. Ⓐ Ⓑ Ⓒ Ⓓ
10. Ⓐ Ⓑ Ⓒ Ⓓ
11. Ⓐ Ⓑ Ⓒ Ⓓ
12. Ⓐ Ⓑ Ⓒ Ⓓ
13. Ⓐ Ⓑ Ⓒ Ⓓ
14. Ⓐ Ⓑ Ⓒ Ⓓ
15. Ⓐ Ⓑ Ⓒ Ⓓ
16. Ⓐ Ⓑ Ⓒ Ⓓ
17. Ⓐ Ⓑ Ⓒ Ⓓ
18. Ⓐ Ⓑ Ⓒ Ⓓ
19. Ⓐ Ⓑ Ⓒ Ⓓ
20. Ⓐ Ⓑ Ⓒ Ⓓ
21. Ⓐ Ⓑ Ⓒ Ⓓ
22. Ⓐ Ⓑ Ⓒ Ⓓ
23. Ⓐ Ⓑ Ⓒ Ⓓ
24. Ⓐ Ⓑ Ⓒ Ⓓ
25. Ⓐ Ⓑ Ⓒ Ⓓ
26. Ⓐ Ⓑ Ⓒ Ⓓ
27. Ⓐ Ⓑ Ⓒ Ⓓ

28.

29.

30.

51. Ⓐ Ⓑ Ⓒ Ⓓ
52. Ⓐ Ⓑ Ⓒ Ⓓ
53. Ⓐ Ⓑ Ⓒ Ⓓ
54. Ⓐ Ⓑ Ⓒ Ⓓ
55. Ⓐ Ⓑ Ⓒ Ⓓ
56. Ⓐ Ⓑ Ⓒ Ⓓ
57. Ⓐ Ⓑ Ⓒ Ⓓ
58. Ⓐ Ⓑ Ⓒ Ⓓ
59. Ⓐ Ⓑ Ⓒ Ⓓ
60. Ⓐ Ⓑ Ⓒ Ⓓ
61. Ⓐ Ⓑ Ⓒ Ⓓ
62. Ⓐ Ⓑ Ⓒ Ⓓ
63. Ⓐ Ⓑ Ⓒ Ⓓ

64.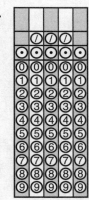

65.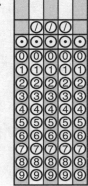

66.

67.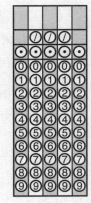

PRACTICE TEST 1

Directions:

For questions 1–27, choose the best answer and fill in the corresponding space on your answer sheet. Be sure to use a #2 pencil. You CANNOT use a calculator on this section of the test.

1. Which one of the following gives an answer that is different from the others?

   A. $-48 \div 4$
   B. $-3 \cdot 4$
   C. $-8 - 4$
   D. $-21 + 9$

2. Simplify: $-27 - (-13)$

   A. -40
   B. 40
   C. 14
   D. -14

3. Simplify: $-20\frac{7}{10} - 4\frac{1}{2}$

   A. $-24\frac{7}{12}$

   B. $-16\frac{5}{8}$

   C. $-25\frac{1}{5}$

   D. $-24\frac{1}{5}$

4. Which is the best choice as an answer for $83.456 \cdot .51$?

   A. 4.256256
   B. 42.56256
   C. 425.6256
   D. $.4256256$

5. Simplify: $\left(\dfrac{8^8}{8^2}\right)^5$

   A. 8^{30}
   B. 8^{20}
   C. 1^{30}
   D. 1^{20}

6. Simplify: 11^{-2}

   A. -22
   B. -121
   C. $\dfrac{1}{22}$
   D. $\dfrac{1}{121}$

7. Multiply: $(9 \times 10^{-3}) \times (6 \times 10^{-7})$

   A. 54×10^{21}
   B. 54×10^{10}
   C. 5.4×10^{-11}
   D. 5.4×10^{-9}

8. Subtract: $5 \times 10^{-3} - 4 \times 10^{-4}$

   A. 1×10^{-4}
   B. 1×10^{-3}
   C. 4.6×10^{-3}
   D. 4.6×10^{-4}

9. Which one of the following is false?

   A. π is irrational.
   B. $19.\overline{813813}$ is irrational.
   C. $\sqrt{78 + 3}$ is rational.
   D. 0 is rational.

10. $\sqrt{68}$ is closest to which whole number?

    A. 8
    B. 9
    C. 34
    D. 35

11. Solve: $5 - 2x \le 11$

    A. $x \ge -3$
    B. $x \le -3$
    C. $x \ge 3$
    D. $x \le 3$

GO ON

12. Yesterday, Lance rode his bike 84.25 miles. Today, Lance is riding at 14.9 miles per hour. He wants to ride a total of 196 miles in two days. Which equation could be used to find the number of hours Lance must ride today?

A. $14.9 + 84.25 + h = 196$
B. $14.9 + 84.25h = 196$
C. $14.9h + 84.25 = 196$
D. $h + 1\,4.9 \cdot 84.25 = 196$

13. Place the expressions $\sqrt{17} + \sqrt{5}$, $\sqrt{22}$, and $\sqrt{36} - \sqrt{4}$ in order from smallest to largest.

A. $\sqrt{17} + \sqrt{5}, \sqrt{22}, \sqrt{36} - \sqrt{4}$
B. $\sqrt{36} - \sqrt{4}, \sqrt{22}, \sqrt{17} + \sqrt{5}$
C. $\sqrt{22}, \sqrt{36} - \sqrt{4}, \sqrt{17} + \sqrt{5}$
D. $\sqrt{36} - \sqrt{4}, \sqrt{17} + \sqrt{5}, \sqrt{22}$

14. Simplify: $\sqrt{150}$

A. $25\sqrt{6}$
B. $5\sqrt{6}$
C. $6\sqrt{25}$
D. $6\sqrt{5}$

15. Which one of the following is the graph of $x + 3y > 6$?

A.

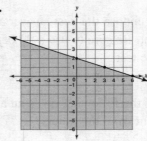

B.

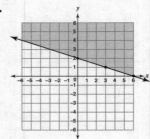

C.

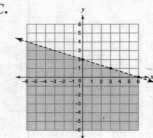

D.

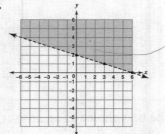

16. Which one of the following is the solution to the system of equations graphed below?

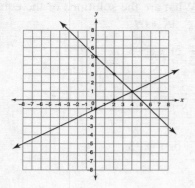

A. $(5, 0)$
B. $(0, 5)$
C. $(4, 1)$
D. $(1, 4)$

17. Which graph is the solution set to the system of inequalities $x + y \geq 2$ and $x - 3y < 0$?

A.

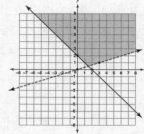

B.

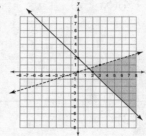

C.

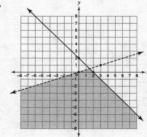

D.

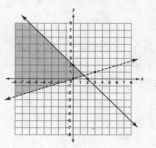

18. Simplify: $(7x^2 - 4x - 2) + (-9x^2 + 3x + 12)$

A. $-63x^4 - 12x^2 - 24$
B. $-2x^2 - x + 10$
C. $-63x^4 + 57x^3 + 84x^2 - 54x - 24$
D. $-2x^4 - x^2 + 10$

GO ON ➡

19. Simplify: $(3x^5w^4) \cdot (7x^{10}w)$

    A. $3x^{15} + 3x^5w + 7x^{10}w^4 + 7x^{10}w^5$
    B. $21x^{50}w^4$
    C. $21x^{15}w^5$
    D. $3x^{50} + 3x^5w + 7x^{50}w^4 + 7x^{50}w^4$

20. Simplify: $(7x^8w)^2$

    A. $14x^{10}w^2$
    B. $14x^{16}w^2$
    C. $49x^{10}w^2$
    D. $49x^{16}w^2$

21. Simplify: $\sqrt[3]{27w^{27}}$

    A. $3w^3$
    B. $3w^9$
    C. $9w^3$
    D. $9w^9$

22. Multiply: $(7x - 4)(5x + 2)$

    A. $12x - 2$
    B. $35x^2 - 8$
    C. $35x^2 + 6x - 8$
    D. $35x^2 - 6x - 8$

23. Simplify: $(4x - 3)^2$

    A. $16x^2 + 9$
    B. $16x^2 - 9$
    C. $16x^2 - 24x + 9$
    D. $16x^2 - 12x + 9$

24. Solve: $6x^2 - 2x - 1 = 0$

    A. $x = \dfrac{1 \pm \sqrt{7}}{6}$

    B. $x = \dfrac{-1 \pm \sqrt{7}}{6}$

    C. $x = \dfrac{1 \pm 2\sqrt{7}}{6}$

    D. $x = \dfrac{-1 \pm 2\sqrt{7}}{6}$

25. Consider the equation $5x^2 - 13x - 6 = 0$. The factors of $5x^2 - 13x - 6$ are $(5x + 2)$ and $(x - 3)$. What are the solutions of the equation $5x^2 - 13x - 6 = 0$?

    A. $x = \dfrac{2}{5}$ or $x = -3$

    B. $x = -\dfrac{2}{5}$ or $x = 3$

    C. $x = \dfrac{5}{2}$ or $x = -3$

    D. $x = \dfrac{5}{2}$ or $x = 3$

26. Solve: $x = \sqrt{11 - 10x}$

    A. $x = -11$ or $x = 1$
    B. $x = 1$ only
    C. $x = -11$ only
    D. There is no solution.

27. Consider the statement.

    When you multiply 8 by a positive fraction, the answer is always smaller.
    Which one of the following is a counter-example that proves the statement is false?

    A. $8 \cdot -\dfrac{1}{2} = -4$

    B. $8 \cdot \dfrac{1}{2} = 4$

    C. $8 \cdot \dfrac{7}{4} = 14$

    D. $8 \cdot -\dfrac{7}{4} = -14$

Practice Test 1

Directions:

For questions 28–30, first solve the problem and then enter your answer on the grid on the answer sheet. Write your answer at the top of the grid. Fill in the ovals below each box. Remember to enter all answers as decimals or improper fractions. Be sure to use a #2 pencil. You CANNOT use a calculator on this section of the test.

---

28. Packages weighing 4.3 pounds, 15 pounds, and .37 pounds are taken to the post office. What is the total weight of the packages?

30. Simplify: $2^{-1} + 8^{\frac{2}{3}}$

29. What is the slope of the line that passes through $(1, -3)$ and $(4, 7)$?

Directions:

For questions 31–50, solve each problem. Be sure to show all work in an organized way. Partial credit is given on these problems. Be sure to use a #2 pencil. You CANNOT use a calculator on this section of the test. **Note:** On the actual exam, you will have space in which to do your work. For this practice test, please use a separate sheet of paper.

---

31. Solve: $5x - 7 = 3x - 23$

Show All Work

Answer $x =$ _____

32. Solve: $-4(2x - 3) - 7 = 14 - 5x$

Show All Work

Answer $x =$ _____

33. Simplify: $12 \div 2 \cdot 3 - 7 + 6 - (2 - 6)$

Show All Work

Answer _____

34. Graph the line $y = -3x + 2$ on the coordinate plane below.

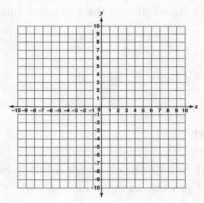

35. What is the equation of a line that passes through the points $(4, -6)$ and $(8, 2)$? Write the equation in slope-intercept form.

Show All Work

Equation _____

GO ON ➡

36. What is the equation of a line that passes through (–5, 3) and has a slope of –4. Write the equation in slope-intercept form.

Show All Work

Equation _____

37. What is the equation of a line that passes through (6, 9) and is parallel to the line $y = \frac{1}{3}x - 40$? Write the equation in slope-intercept form.

Show All Work

Equation _____

38. What is the solution to the system of equations?

$$y = 4 - x$$
$$x + 2y = 3$$

Show All Work

Answer _____

39. What is the solution to the system of equations?

$$2x + 5y = 7$$
$$-3x - 2y = 6$$

Show All Work

Answer _____

40. Gree D. has 40 nickels and quarters hidden away. The value of the coins is $5.40. Write a system of equations that could be used to find the number of nickels n and the number of quarters q that Gree D. has hidden away.

Equations _____

_____

41. Factor: $7x^2 + 20x - 3$

Answer _____

42. Solve: $\dfrac{4}{x+2} = \dfrac{2}{x+6}$

Show All Work

Answer $x =$ _____

43. Solve: $\dfrac{x+2}{-4} = \dfrac{3}{x-5}$

Show All Work

Answer $x =$ _____ or $x =$ _____

44. On the coordinate plane, graph $y = 2 - x^3$ using –2, –1, 0, 1, and 2 as values for x.

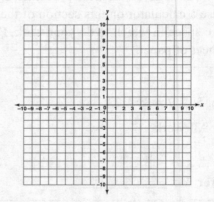

45. On the coordinate plane below, graph $y = \sqrt{x + 3}$ using the following as values for x: –3, –2, 1, and 6.

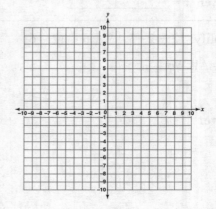

GO ON ➡

46. Solve for x: $3x^2 - 16x + 5 = 0$

Show All Work

Answer $x =$ _____ or $x =$ _____

47. Solve: $(x - 6)^2 = 16$

Show All Work

Answer $x =$ _____ or $x =$ _____

48. On the coordinate plane, draw the graph of $y = 2 - x^2$ using $-3, -2, -1, 0, 1, 2,$ and 3 as values for x.

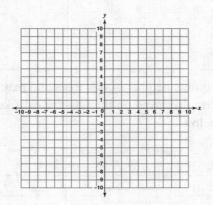

49. Pih Churr, standing 48 feet below ground in a pit, throws a ball up to his brother. Cat Churr is standing at the edge of the pit prepared to grab the ball. The equation $-16x^2 + 64x - 48 = 0$ describes the two times that the ball would be at a height where Cat Churr could catch it. Ms. Supa T. Cher has simplified the equation $-16x^2 + 64x - 48 = 0$ into $x^2 - 4x + 3 = 0$. Solve $x^2 - 4x + 3 = 0$ to find the two times that Cat Churr could catch the ball.

Show All Work

Answer $x =$ _____ or $x =$ _____

50. Ms. X. Tream jumped from the top of a cliff. She will slowly hang glide to the town below. The relationship between her height (H) and the time (t) that she has been in the air is given by the equation $H = 300 - 5t$. Graph $H = 300 - 5t$ on the coordinate plane below.

Ms. X Tream's Hang Gliding

What is the y-intercept? $+300$

On the lines below, explain the meaning of the y-intercept in the hang gliding situation.

The place where he starts from

What is the slope? $\dfrac{-5}{1}$

On the lines below, explain the meaning of the slope in the hang gliding situation.

Every 1 second he is losing 5 feet of height

GO ON ➡

Directions:
For questions 51–63, choose the best answer and fill in the corresponding space on your answer sheet. Be sure to use a #2 pencil. You may use a calculator on this section of the test.

51. Rich Guy deposited $5,000 in the bank. He is earning simple interest at a rate of 4.5 percent. He withdraws his money and the interest at the end of 3 years. How much money will Rich Guy withdraw?

    A. $675
    B. $5,675
    C. $11,750
    D. $6,750

52. Solve for e: $te + n = 4$

    A. $e = \dfrac{4-n}{t}$

    B. $e = 4 - n - t$

    C. $e = \dfrac{4}{t} - n$

    D. $e = \dfrac{4}{t} + n$

53. Using the figure below, determine which statement is correct.

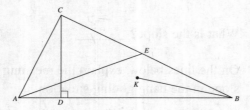

    A. $\overline{CD}$ is a median.
    B. $\overline{AE}$ is a perpendicular bisector.
    C. $\overline{BK}$ is an angle bisector.
    D. None of the above are correct.

54. Tella Scope counted the number of shooting stars, as shown in the table below.

Number of Shooting Stars Sighted

4	6	12	12	13
14	15	15	16	17
18				

What is the upper quartile of the data?

    A. 14
    B. 15
    C. 16
    D. 18

55. The various heights of oak trees and maple trees are displayed on the box-and-whisker plot below.

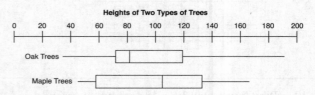

Using the information in the plot, which of the following statements is false?

    A. The median height of the oak trees is less than the median height of the maple trees.
    B. The oak trees measured the tallest and the shortest of all of the trees that were measured.
    C. The lower quartile of the oak trees is more than the lower quartile of the maple trees.
    D. The upper quartile of the oak trees is more than the upper quartile of the maple trees.

GO ON ➡

56. Mr. Dry Ver graphed the speed of his trip over time on the line graph below.

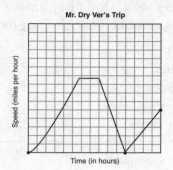

Mr. Dry Ver's Trip

According to the information on the graph, which of the following is true?

A. Mr. Dry Ver did not stop during his entire trip.

B. Mr. Dry Ver stayed at a constant speed for his entire trip.

C. Mr. Dry Ver slowed down during his trip.

D. Mr. Dry Ver did not accelerate during his trip.

57. Which of the following is not a function?

A.

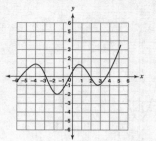

B.

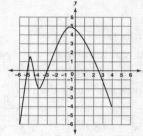

C.

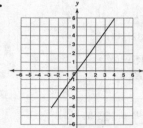

D.

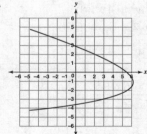

GO ON ➡

58. What is the range of the function graphed below?

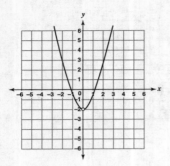

A. all real numbers
B. all positive real numbers
C. all real numbers larger than –2
D. all real numbers greater than 1

---

59. Divide: $\dfrac{27x^6w^{12}+18x^{15}w^{24}}{3x^3w^6}$

A. $24x^3w^6 + 15x^{12}w^{18}$
B. $9x^3w^6 + 6x^{12}w^{18}$
C. $24x^2w^2 + 15x^5w^4$
D. $9x^2w^2 + 6x^5w^4$

---

60. Factor: $15x^8w^{10} + 12x^6w^{12}$

A. $3x^6w^{10}(5x^2 + 4w^2)$
B. $3x^2w^2(5x^4w^5 + 4x^3w^6)$
C. $3x^6w^{10}(12x^2 + 9w^2)$
D. $3x^2w^2(5x^4w^5 + 4x^3w^6)$

---

61. Simplify: $\dfrac{x^2-6x+5}{x^2-4x-5}$

A. $\dfrac{3}{2}x - 1$

B. -1

C. $\dfrac{x-1}{x+1}$

D. $\dfrac{x-5}{x+5}$

---

62. A sleeping person breathes an average of 124 times every 7 minutes. Suh News sleeps 11 hours every night. How many times will Suh News breathe while sleeping for 28 nights?

A. 5,456 times
B. 29,760 times
C. 327,360 times
D. 2,291,520 times

---

63. Consider the statements.

Supa Smart is more intelligent than Jean Yuhs.
Jean Yuhs is more intelligent than Bray Kneeguy.
Nigh Sky is a really great guy.

Using deductive reasoning, which of the following is a valid conclusion?

A. Jean Yuhs is more intelligent than Nigh Sky.
B. Nigh Sky is more intelligent than Supa Smart.
C. Supa Smart is more intelligent than Bray Kneeguy.
D. Bray Kneeguy is more intelligent than everyone.

GO ON ➡

Directions:
For questions 64–67, first solve the problem and then enter your answer on the grid on the answer sheet. Write your answer at the top of the grid. Fill in the ovals below each box. Remember to enter all answers as decimals or improper fractions. Be sure to use a #2 pencil. You may use a calculator on this section of the test.

64. Mr. Clow Enn has 360 slips of paper with integers 1, 2, 3, and 4 written on them. A slip of paper is chosen at random. The probability of choosing each number is shown in the table below.

Paper Probabilities

Number	Probability of Choosing
1	.25
2	.15
3	.2
4	.4

What is the probability of not choosing 2?

65. At work, Mr. Y. Ner sighs 2 times every 5 minutes. How many times does Mr. Y. Ner sigh in an 8-hour workday?

66. Athel Eat runs at an average rate of .75 laps per minute. About how many minutes will it take Athel Eat to run 24 laps?

67. The circle graph below shows the percent of Rich Guy's allowance that he spends on food, movies, and school supplies.

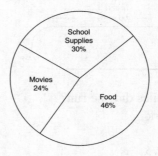

Rich Guy's current weekly allowance is $750. Suppose Rich Guy's weekly allowance increases by 10%. Now how much will Rich Guy get to spend on movies and school supplies?

Directions:

For questions 68–78, solve each problem. Be sure to show all work in an organized way. Partial credit is given on these problems. Be sure to use a #2 pencil. You may use a calculator on this section of the test. *Note*: On the actual exam, you will have space in which to do your work. For this practice test, please use a separate sheet of paper.

68. A soccer ball has a diameter of 22 centimeters. How many cubic centimeters of air does it take to fill the soccer ball?

    Answer _____

69. Kohn Tess Tant ran 20 laps of the track shown below.

    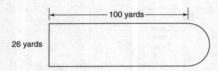

    How many miles did she run?
    Show All Work

    Answer _____

70. Nineteen more than three times a number is the same as four less than twice a number. On the line below, write an equation that could be used to find the number.

    Equation _____

    Now solve the equation you wrote to find the number.
    Show All Work

    Answer_____

71. On the coordinate plane below, draw the rotation of the triangle 90° counterclockwise about the origin.

    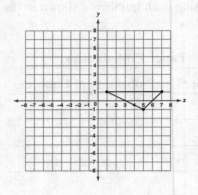

72. Bill Dit wants to put lights along the side of his shed. The length he needs to know is marked with *x* in the drawing below.

    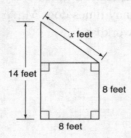

    Find the length of *x*.
    Show All Work

    Answer $x = $ _____

73. What is the area, in square meters, of the shape drawn below?

5 meters

12 meters

Show All Work

Answer _____

74. U R Toys requests a survey determining the favorite toys of 8-year-olds. U R Toys suggests a survey at a local model airplane show. On the lines below, explain how the company's site choice could give a biased sample.

_____

_____

_____

_____

Suppose the survey was given to members of the chess club at a local school. Would the results of that survey be biased? _____
On the lines below, explain your reasoning.

_____

_____

_____

_____

75. Tella Scope kept track of the number of shooting stars he saw for 100 nights. On the 101st night, he counted a record high number of shooting stars. How will the additional night affect the mean number of stars seen? On the lines below, explain your reasoning.

_____

_____

_____

_____

76. Mr. Clow Enn asked his students to tally the number of times they smiled during class that week. The tallied results are given in the table below.

Number of Smiles Per Day

12	18	19	20	24
25	28	41	45	47
47	53	54	57	57
61	63	70	70	81
83	90	100	102	109

Fill in the stem-and-leaf plot below to represent the smile data.

Number of Smiles

Stem	Leaf

 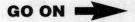

77. At Pizzas R Us there is a special of a medium pizza with two different toppings for $3.99. There are 21 different toppings available. The pizza must be ordered with the two toppings, and the two toppings must be different. How many different pizzas on special are available from Pizzas R Us? _____

What piece of information was not necessary in this problem?

_____

_____

_____

_____

78. The line graph below shows the prices that Spee D Taxi service charges for each mile traveled.

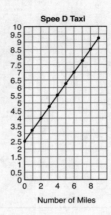

On the lines below, explain what the slope of the graph represents to Spee D Taxi.

_____

_____

_____

_____

On the lines below, explain what the *y*-intercept represents to Spee D Taxi.

_____

_____

_____

_____

STOP

SOLUTIONS: PRACTICE TEST 1

Answer Key

1. **A**	6. **D**	11. **A**	16. **C**	21. **B**	26. **B**	
2. **D**	7. **D**	12. **C**	17. **B**	22. **D**	27. **C**	
3. **C**	8. **C**	13. **B**	18. **B**	23. **C**		
4. **B**	9. **B**	14. **B**	19. **C**	24. **A**		
5. **A**	10. **A**	15. **D**	20. **D**	25. **B**		

28.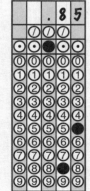

29.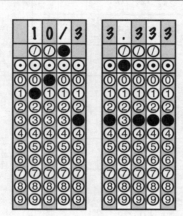

30.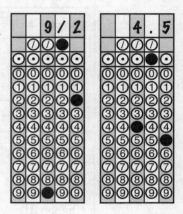

See Answers Explained for 31–50.

51. **B**	54. **C**	57. **D**	60. **A**	63. **C**
52. **A**	55. **D**	58. **C**	61. **C**	
53. **C**	56. **C**	59. **B**	62. **C**	

64.

65.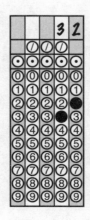

66.

67.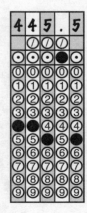

See Answers Explained for 68–78.

Diagnostic Chart

Each question in the Practice Test 1 maps to a chapter in the book. Chapter 15, problem solving, is tested in the story problems throughout this test.

Chapter	Problems on Practice Test 1
Chapter 1	1–4, 28, 51
Chapter 2	5–10
Chapter 3	11, 12, 31, 32, 52, 70
Chapter 4	13, 14, 30, 33, 62, 65
Chapter 5	53, 71, 72
Chapter 6	66, 67, 73
Chapter 7	54, 55, 68, 69, 74–76
Chapter 8	64, 77
Chapter 9	56–58
Chapter 10	15, 29, 34–37, 50, 78
Chapter 11	16, 17, 38–40
Chapter 12	18–23, 41, 59, 60
Chapter 13	42, 43, 61
Chapter 14	24–26, 44–49
Chapter 15	27, 63

Answers Explained

1. **A** Choice A is 12. Choices B, C, and D all are –12.

2. **D** $-27 - (-13) = -27 + (+13) = -14$

3. **C** $-20\frac{7}{10} - 4\frac{1}{2} = -20\frac{7}{10} + -4\frac{1}{2} = -20\frac{7}{10} + -4\frac{5}{10} = -24\frac{12}{10} = -25\frac{2}{10} = -25\frac{1}{5}$

4. **B** 83.456 as a "nice" number is 80 and .51 as a "nice number" is $\frac{1}{2}$. Half of 80 is about 40.

5. **A** $\left(\frac{8^8}{8^2}\right)^5 = (8^{8-2})^5 = (8^6)^5 = 8^{30}$

6. **D** $11^{-2} = \frac{1}{11^2} = \frac{1}{121}$

7. **D** $(9 \times 10^{-3}) \times (6 \times 10^{-7})$
   54×10^{-10}
   $(5.4 \times 10^1) \times 10^{-10}$
   5.4×10^{-9}

8. **C** $5 \times 10^{-3} - 4 \times 10^{-4} = .005 - .0004 = .0046 = 4.6 \times 10^{-3}$

9. **B** Irrational numbers are nonrepeating, nonterminating decimals.

10. **A** $\sqrt{68}$ is between $\sqrt{64} = 8$ and $\sqrt{81} = 9$, but it is closer to $\sqrt{64} = 8$.

11. **A** $5 - 2x \leq 11$
    $-2x \leq 6$

    Reverse the inequality sign because you are multiplying both sides by a negative number.

    $-\frac{1}{2} \cdot -2x \geq -\frac{1}{2} \cdot 6$
    $x \geq -3$

12. **C** Today, Lance rides 14.9 miles every hour, so in h hours he will ride $14.9h$. His total mileage for two days is his miles yesterday, 84.25, and his total miles today, $14.9h$. Add yesterday's mileage and today's mileage to get the total mileage.

13. **B** $\sqrt{36} - \sqrt{4} = 6 - 2 = 4$

    $\sqrt{22}$ is between $\sqrt{16} = 4$ and $\sqrt{25} = 5$, but it is closer to $\sqrt{25} = 5$.

    $\sqrt{17} \approx 4.1$ and $\sqrt{5} \approx 2.1$

    $\sqrt{17} + \sqrt{5} \approx 4.1 + 2.1 = 6.2$

14. **B** $\sqrt{150} = \sqrt{25 \cdot 6} = \sqrt{25} \cdot \sqrt{6} = 5\sqrt{6}$

15. **D** The lines in all the choices are the same. Choice C or choice D are the only choices because they are dotted lines. Pick $(0, 0)$ and substitute into the inequality $x + 3y > 6$. If it makes the inequality true, then shade the side of the dotted line containing $(0, 0)$. If it does not make the inequality true, then shade the side of the dotted line that does not contain $(0, 0)$.

16. **C** The solution to a system of equations is the point where the two lines intersect.

17. **B** The only difference in the choices is the shading. Choose points in each of the four areas. Check each point to see if it makes both inequalities true. The point that makes both inequalities true is in the portion of the graph that should be shaded.

18. **B** Add like terms by adding the coefficients and keeping the variable's power the same.

19. **C** $(3x^5w^4) \cdot (7x^{10}w) = (3 \cdot 7) \cdot (x^5 \cdot x^{10}) \cdot (w^4 \cdot w) = 21x^{15}w^5$

20. **D** $(7x^8w)^2 = (7x^8w^1) \cdot (7x^8w^1) = 49x^{16}w^2$

21. **B** $\sqrt[3]{27w^{27}} = \sqrt[3]{27} \cdot \sqrt[3]{w^{27}} = 3w^{\frac{27}{3}} = 3w^9$

22. **D** $(7x - 4)(5x + 2) = 35x^2 + 14x - 20x - 8 = 35x^2 - 6x - 8$

23. **C** $(4x - 3)^2 = (4x - 3) \cdot (4x - 3) = 16x^2 - 12x - 12x + 9 = 16x^2 - 24x + 9$

24. **A**

$$x = \frac{-b \pm \sqrt{b^2 - 4ac}}{2a}$$

$$x = \frac{-(-2) \pm \sqrt{(-2)^2 - 4 \cdot 6 \cdot -1}}{2(6)}$$

$$x = \frac{2 \pm \sqrt{4 - (-24)}}{12}$$

$$x = \frac{2 \pm \sqrt{28}}{12}$$

$$x = \frac{2}{12} \pm \frac{2\sqrt{7}}{12}$$

$$x = \frac{1}{6} \pm \frac{\sqrt{7}}{6}$$

25. **B**
$$5x^2 - 13x - 6 = 0$$
$$(5x + 2)(x - 3) = 0$$
$$5x + 2 = 0 \text{ or } x - 3 = 0$$
$$5x = -2 \text{ or } x = 3$$
$$x = -\frac{2}{5} \text{ or } x = 3$$

26. **B**
$$x = \sqrt{11 - 10x}$$

$$x^2 = \left(\sqrt{11 - 10x}\right)^2$$

$$x^2 = 11 - 10x$$

$$x^2 + 10x - 11 = 0$$

$$(x + 11)(x - 1) = 0$$

$$x + 11 = 0 \text{ or } x - 1 = 0$$

$$x = -11 \text{ or } x = 1$$

When you square both sides of an equation, you must check your answers to be certain that they are correct.

$-11 \stackrel{?}{=} \sqrt{11 - 10(-11)}$	$1 \stackrel{?}{=} \sqrt{11 - 10(1)}$
$-11 \stackrel{?}{=} \sqrt{11 - (-110)}$	$1 \stackrel{?}{=} \sqrt{11 - 10}$
$-11 \stackrel{?}{=} \sqrt{11 + 110}$	$1 \stackrel{?}{=} \sqrt{1}$
$-11 \stackrel{?}{=} \sqrt{121}$	$1 = 1$
$-11 \neq 11$	$x = 1$ is correct.

$x = -11$ is not correct.

27. **C** Multiply 8 by a positive fraction to get an answer that is larger than 8.

28. **19.67**

$$\begin{array}{r} 4.30 \\ 15.00 \\ +\ .37 \\ \hline 19.67 \end{array}$$

29. $\dfrac{10}{3}$

$$\text{slope} = \frac{\text{change in } y}{\text{change in } x} = \frac{7 - (-3)}{4 - 1} = \frac{10}{3}$$

30. $\dfrac{9}{2}$ or **4.5**

$$2^{-1} = \frac{1}{2^1} = \frac{1}{2}$$

$$8^{\frac{2}{3}} = \left(\sqrt[3]{8}\right)^2 = 2^2 = 4$$

$$2^{-1} + 8^{\frac{2}{3}} = 4 + \frac{1}{2} = 4\frac{1}{2}$$

31. **$x = -8$**

$$5x - 7 = 3x - 23$$
$$2x - 7 = -23$$
$$2x = -16$$
$$x = -8$$

Half credit is given if no work is shown.

32. **$x = -3$**

$$-4(2x - 3) - 7 = 14 - 5x$$
$$-8x + 12 + -7 = 14 + -5x$$
$$-8x + 5 = 14 + -5x$$
$$-3x + 5 = 14$$
$$-3x = 9$$
$$x = -3$$

Half credit is given if no work is shown.

33. **21**

$$12 \div 2 \cdot 3 - 7 + 6 - \underline{(2 - 6)}$$
$$\underline{12 \div 2} \cdot 3 - 7 + 6 - (-4)$$
$$\underline{6 \cdot 3} - 7 + 6 - (-4)$$
$$\underline{18 - 7} + 6 - (-4)$$
$$\underline{11 + 6} - (-4)$$
$$17 - (-4)$$
$$21$$

Half credit is given if no work is shown.

34.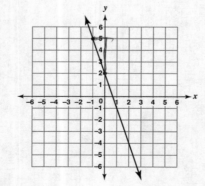

The slope is $-3 = \dfrac{-3}{1} = \dfrac{3}{-1}$ and the y-intercept is 2.

Graph the y-intercept at $(0, 2)$.

From that point, count up 3 and to the left 1 space. Put a point on $(-1, 5)$. Full credit if line graphed correctly. Half credit if correct slope with line drawn *or* if correct intercepts with line drawn *or* if correct slope and correct x- or y-intercept but no line drawn.

35.   slope $= \dfrac{\text{change in } y}{\text{change in } x} = \dfrac{2-(-6)}{8-4} = \dfrac{8}{4} = 2$

$$y - (-6) = 2(x - 4)$$
$$y + 6 = 2x - 8$$
$$y = 2x - 14$$

Full credit is given if both work and correct answer are shown. Partial credit is given if just the answer is given *or* if work is shown and a computation error is made.

36.   Use the point and the slope in point-slope form.

$$y - 3 = -4(x - (-5))$$
$$y - 3 = -4(x + 5)$$
$$y - 3 = -4x - 20$$
$$y = -4x - 17$$

Full credit is given if both work and correct answer are shown. Partial credit is given if just the answer is given *or* if work is shown and a computation error is made.

37.   Parallel lines have the same slope.

$$y - 9 = \frac{1}{3}(x - 6)$$

$$y - 9 = \frac{1}{3}x - 2$$

$$y = \frac{1}{3}x + 7$$

Full credit is given if both work and correct answer are shown. Partial credit is given if just the answer is given *or* if work is shown and a computation error is made.

38.   **(5, −1)**

$$x + 2(4 - x) = 3 \qquad y = 4 - x$$
$$x + 8 - 2x = 3 \qquad y = 4 - 5$$
$$-1x + 8 = 3 \qquad y = -1$$
$$-1x = -5$$
$$x = 5$$

The solution is the ordered pair (5, −1).
Full credit is given if both work and correct answer are shown. Partial credit is given if just the answer is given *or* if work is shown and a computation error is made.

39.   **(−4, 3)**

$$3 \cdot 2x + 3 \cdot 5y = 3 \cdot 7 \qquad 2x + 5(3) = 7$$
$$2 \cdot (-3x) + 2 \cdot (-2y) = 2 \cdot (6) \qquad 2x + 15 = 7$$
$$6x + 15y = 21 \qquad 2x = -8$$
$$-6x - 4y = 12 \qquad x = -4$$
$$11y = 33$$
$$y = 3$$

The solution to the system of equations is (−4, 3).
Full credit is given if both work and correct answer are shown. Partial credit is given if just the answer is given *or* if work is shown and a computation error is made.

40. $n + q = 40$
$.05n + .25q = 5.40$
Half credit is given for one correct equation.

41. $(7x - 1)(x + 3)$
No partial credit is given.

42. **–10**

$$4(x + 6) = 2(x + 2)$$
$$4x + 24 = 2x + 4$$
$$2x + 24 = 4$$
$$2x = -20$$
$$x = -10$$

Full credit is given if both work and correct answer are shown. Partial credit is given if just the answer is given *or* if work is shown and a computation error is made.

43.

$$(x + 2)(x - 5) = -4 \cdot 3$$
$$x^2 - 3x - 10 = -12$$
$$x^2 - 3x + 2 = 0$$
$$(x - 2)(x - 1) = 0$$
$$x - 2 = 0 \text{ or } x - 1 = 0$$
$$x = 2 \text{ or } x = 1$$

Full credit is given if both work and correct answer are shown. Partial credit is given if just the answers are given *or* if one of the answers is correct and work is shown.

44.

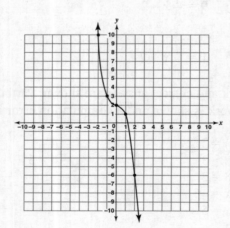

The points on the graph are (–2, 10), (–1, 3), (0, 2), (1, 1), and (2, –6).

Full credit is given only if all 5 points are graphed correctly and the curve is drawn. Half credit if 3 or 4 points are graphed correctly and the curve is drawn.

45.

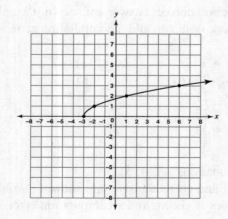

The points on the graph include (–3, 0), (–2, 1), (1, 2) and (6, 3).

Full credit is given only if all 4 points are correctly graphed and the curve is drawn. Partial credit is given if 3 points are correctly graphed and the curve is drawn. No points are given if the curve is drawn to the left of (–3, 0).

46.
$$3x^2 - 16x + 5 = 0$$
$$(3x - 1)(x - 5) = 0$$
$$3x - 1 = 0 \text{ or } x - 5 = 0$$
$$3x = 1 \text{ or } x = 5$$
$$x = \frac{1}{3} \text{ or } x = 5$$

Half credit is given if no work is shown *or* if work is shown and only one answer is correct.

47.
$$(x - 6)^2 = 16$$
$$\sqrt{(x - 6)^2} = \pm\sqrt{16}$$
$$x - 6 = \pm 4$$
$$x - 6 = 4 \text{ or } x - 6 = -4$$
$$x = 10 \text{ or } x = 2$$

Half credit is given if no work is shown *or* if work is shown and only one answer is correct.

48.

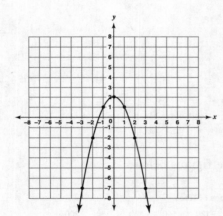

The points on the graph include (–3, –7), (–2, –2), (–1, 1), (0, 2), (1, 1), (2, –2), and (3, –7).

Full credit is given only if all 7 points are graphed correctly and the parabola is drawn. Half credit if 5 or 6 points are graphed correctly and the parabola is drawn.

49.
$$x^2 - 4x + 3 = 0$$
$$(x - 1)(x - 3) = 0$$
$$x - 1 = 0 \text{ or } x - 3 = 0$$
$$x = 1 \text{ or } x = 3$$

Half credit is given if no work is shown *or* if work is shown and only one answer is correct.

50.

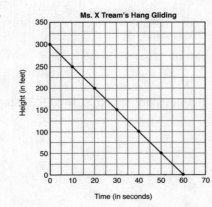

Full credit is given if the line is drawn through the correct points. Half credit is given if the points are correctly graphed but no line is drawn.

The *y*-intercept is 300 and represents the height of the cliff that Ms. X. Tream jumped off.

The slope is –5 and represents the speed at which she is descending, $-5\dfrac{\text{feet}}{\text{minute}}$.

51. **B**

$$i = 5,000 \cdot .045 \cdot 3$$
$$i = 675$$

Rich Guy earns $675 interest. He will withdraw the $675 interest plus the $5,000 principal for a total of $5,675.

52. **A**

$$te + n = 4$$
$$te = 4 - n$$
$$\frac{1}{t}(te) = \frac{1}{t}(4 - n)$$

$$e = \frac{4 - n}{t}$$

53. **C** An angle bisector is a line segment that splits an angle into two congruent angles.

54. **C**

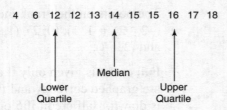

55. **D** Look at all of the labels in the figures below.

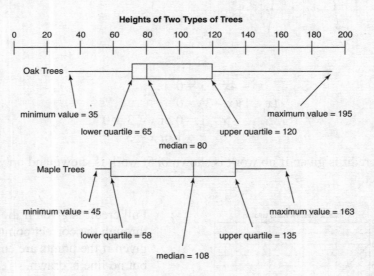

56. **C** Mr. Dry Ver accelerated, stayed at a constant speed, slowed to a stop, and then accelerated again.

57. **D** D is not a function because for one value of x there are two possible y-values.

58. **C** The range is all the values for y.

59. **B**

$$\frac{27x^6w^{12}+18x^{15}w^{24}}{3x^3w^6} = \frac{27x^6w^{12}}{3x^3w^6} + \frac{18x^{15}w^{24}}{3x^3w^6} = 9x^{6-3}w^{12-6} + 6x^{15-3}w^{24-6} = 9x^3w^6 + 6x^{12}w^{18}$$

60. **A** The greatest common factor is $3x^6w^{10}$. Think $3x^6w^{10}$ (_____ + _____).

61. **C**

$$\frac{x^2-6x+5}{x^2-4x-5} = \frac{\cancel{(x-5)}^1(x-1)}{\cancel{(x-5)}^1(x+1)} = \frac{x-1}{x+1}$$

62. **C**

$$\frac{124\text{ times}}{7\text{ \cancel{minutes}}} \cdot \frac{60\text{ \cancel{minutes}}}{1\text{ \cancel{hour}}} \cdot \frac{11\text{ hours}}{1\text{ \cancel{night}}} \cdot \frac{28\text{ \cancel{nights}}}{1} = \frac{2,291,520\text{ times}}{7\text{ \cancel{night}}} = 327,360\text{ times}$$

63. **C** Supa is more intelligent than Jean and Jean is more intelligent than Bray. Information about Nigh is put in to confuse you.

64. **.85** The probability of not choosing 2 is the same as 1 minus the probability of choosing 2, or $1 - .15 = .85$.

65. **192**

$$\frac{2\text{ times}}{5\text{ \cancel{minutes}}} \cdot \frac{60\text{ \cancel{minutes}}}{1\text{ \cancel{hour}}} \cdot 8\text{ \cancel{hours}} = \frac{960\text{ times}}{5} = 192\text{ times}$$

66. **32**

$$24 = .75t$$
$$t = 32$$

67. **445.50** His new allowance will be $825. The total for movies and school supplies is 54%.

$$825 \cdot .54 = 445.5$$

68. **5,572.45**, or **5,575.28** if used π button on calculator

$$V = \frac{4}{3}\pi r^3$$

$$V = \frac{4}{3} \cdot 3.14 \cdot 11^3$$

$$V = \frac{4}{3} \cdot 3.14 \cdot 1,331$$

$$V = 5,572.45$$

69. **5.58** The perimeter of the track is 100 + 26 + 100 + circumference of semicircle.

C of semicircle = $\frac{1}{2} \cdot 3.14 \cdot 13^2$

C of semicircle = $\frac{1}{2} \cdot 3.14 \cdot 169$

C of semicircle = 265.33

The total distance around the track is 100 + 26 + 100 + 265.33 = 491.33 yards.

Number of yards in 20 laps = 20 · 491.33 = 9,826.6

9,826.6 yards · $\dfrac{1 \text{ mile}}{1{,}760 \text{ yards}}$ = 5.58 miles

Full credit is given if work and correct answer are shown. Half credit is given if no work is shown *or* if work is shown and computation error is made.

70. *Nineteen more than three times a number is the same as four less than twice a number* translates to 19 + 3 · x = 2 · x + –4.

No partial credit is given for the equation.

$$19 + 3x = 2x - 4$$
$$19 + x = -4$$
$$x = -23$$

Full credit is given if work and correct answer are shown. Half credit is given if no work is shown *or* if work is shown and computation error is made.

71.

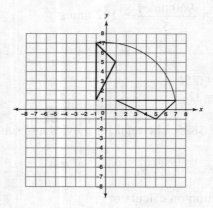

No partial credit is given.

PRACTICE TEST 1 • 353

72. The lengths of the sides are shown below.

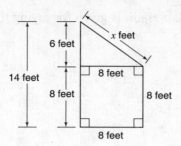

$$a^2 + b^2 = c^2$$
$$6^2 + 8^2 = c^2$$
$$36 + 64 = c^2$$
$$c^2 = 100$$
$$c = 10$$

Full credit is given if work and correct answer are shown. Half credit is given if no work is shown *or* if work is shown and computation error is made.

73. about 116.52 square meters
Find the area of the rectangle and add it to the area of the semicircle to get the area of the entire shape.

$$A = l\,w$$
$$A = 5 \cdot 12$$
$$A = 60 \text{ square meters}$$

$$A \text{ semicircle} = \frac{1}{2}\pi r^2$$

$$A = \frac{1}{2} \cdot 3.14 \cdot 6^2$$

$$A = \frac{1}{2} \cdot 3.14 \cdot 36$$

$$A = 56.52$$

Full credit is given if work and correct answer are shown. Half credit is given if no work is shown *or* if work is shown and computation error is made.

74. Children who attend a model airplane show do not represent all children. Most likely children who enjoy model toys will be over-represented.
No partial credit is given.

The survey at the school would be biased because it does not represent all 8 year olds. Most likely, children who enjoy thinking games or board games would be over-represented.
No partial credit is given.

75. The mean will increase. When you add a number that is higher than all of the other scores, then the mean will increase.
Full credit is given for both answers. Half credit is given for saying the mean increases but having the wrong reason.

76.

Number of Smiles

Stem	Leaf
1	2 8 9
2	0 4 5 8
3	
4	1 5 7 7
5	3 4 7 7
6	1 3
7	0 0
8	1 3
9	0
10	0 2 9

Half credit if the stems are correct and 20 leaves are correct.

77. 420 different pizzas are available. The price of the pizza was not necessary.

$21 \cdot 20 = 420$

78. The slope represents the cost for every mile traveled.
No partial credit is given.

The y-intercept represents how much it costs before you have traveled any miles.
No partial credit is given.

PRACTICE TEST 2 ANSWER SHEET

Fill in the bubble completely.
Erase carefully if answer is changed.

1. Ⓐ Ⓑ Ⓒ Ⓓ
2. Ⓐ Ⓑ Ⓒ Ⓓ
3. Ⓐ Ⓑ Ⓒ Ⓓ
4. Ⓐ Ⓑ Ⓒ Ⓓ
5. Ⓐ Ⓑ Ⓒ Ⓓ
6. Ⓐ Ⓑ Ⓒ Ⓓ
7. Ⓐ Ⓑ Ⓒ Ⓓ
8. Ⓐ Ⓑ Ⓒ Ⓓ
9. Ⓐ Ⓑ Ⓒ Ⓓ
10. Ⓐ Ⓑ Ⓒ Ⓓ
11. Ⓐ Ⓑ Ⓒ Ⓓ
12. Ⓐ Ⓑ Ⓒ Ⓓ
13. Ⓐ Ⓑ Ⓒ Ⓓ
14. Ⓐ Ⓑ Ⓒ Ⓓ
15. Ⓐ Ⓑ Ⓒ Ⓓ
16. Ⓐ Ⓑ Ⓒ Ⓓ
17. Ⓐ Ⓑ Ⓒ Ⓓ
18. Ⓐ Ⓑ Ⓒ Ⓓ
19. Ⓐ Ⓑ Ⓒ Ⓓ
20. Ⓐ Ⓑ Ⓒ Ⓓ
21. Ⓐ Ⓑ Ⓒ Ⓓ

22.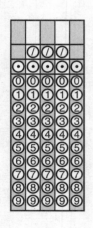

23.

24.

25.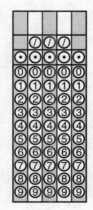

53. Ⓐ Ⓑ Ⓒ Ⓓ
54. Ⓐ Ⓑ Ⓒ Ⓓ
55. Ⓐ Ⓑ Ⓒ Ⓓ
56. Ⓐ Ⓑ Ⓒ Ⓓ
57. Ⓐ Ⓑ Ⓒ Ⓓ
58. Ⓐ Ⓑ Ⓒ Ⓓ
59. Ⓐ Ⓑ Ⓒ Ⓓ

60.

61.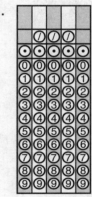

62.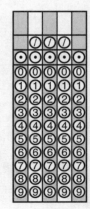

PRACTICE TEST 2

Directions:

For questions 1–21, choose the best answer and fill in the corresponding space on your answer sheet. Be sure to use a #2 pencil. You CANNOT use a calculator on this section of the test.

1. Multiply: $-3\frac{3}{4} \cdot -1\frac{1}{6}$

   A. $-3\frac{1}{8}$

   B. $3\frac{1}{8}$

   C. $4\frac{3}{8}$

   D. $-4\frac{3}{8}$

2. Place the problems in order from the smallest answer to the largest answer.

   $-8 \cdot 2, -8 - (-2), -8 \div 2, -8 + -2$

   A. $-8 \div 2, -8 - 2, -8 - (-2), -8 \cdot 2$
   B. $-8 \div 2, -8 - (-2), -8 + -2, -8 \cdot 2$
   C. $-8 \cdot 2, -8 \div 2, -8 + -2, -8 - (-2)$
   D. $-8 \cdot 2, -8 + -2, -8 - (-2), -8 \div 2$

3. Which is the best estimate for $4\frac{18}{19} \cdot 2\frac{2}{37}$?

   A. close to 6
   B. close to 8
   C. close to 10
   D. close to 12

4. Simplify: $(2x^4)^3$

   A. $6x^7$
   B. $6x^{12}$
   C. $8x^7$
   D. $8x^{12}$

5. Which one of the following is the same as 9^{-2}?

   A. -18
   B. -81

   C. $\frac{1}{18}$

   D. $\frac{1}{81}$

6. Divide: $\dfrac{4.8 \times 10^3}{6 \times 10^6}$

   A. 4.2×10^{-4}
   B. 8×10^{-2}
   C. 8×10^{-4}
   D. 4.2×10^{-2}

7. Simplify: $(5 \times 10^7)^2$

   A. 1×10^{14}
   B. 2.5×10^{15}
   C. 2.5×10^{13}
   D. 2.5×10^{14}

8. Which is the best approximation for $\sqrt{83}$?

   A. 42
   B. 41
   C. 9.1
   D. 8.9

9. Which numbers listed below are rational?

   $\sqrt{29+3}$, $\sqrt{31+5}$, $\pi - 5$, $23.\overline{494949}$

   A. All of them are rational.
   B. None of them are rational.

   C. $\sqrt{29+3}$, $\pi - 5$, and $23.\overline{494949}$ are rational.

   D. $\sqrt{31+5}$ and $23.\overline{494949}$ are rational.

GO ON ➡

Practice Test 2

10. Solve for c: $\dfrac{c}{a} + r = e$

  A. $c = e - r - a$
  B. $c = e - (r \cdot a)$
  C. $c = a(e - r)$
  D. $c = ea - r$

11. Which one of the following is the same as $3 \geq -x + 2$?

  A. $x \leq -1$
  B. $x \leq -5$
  C. $x \geq -1$
  D. $x \geq -5$

12. Simplify: $\sqrt{18} + \sqrt{8}$

  A. $\sqrt{26}$

  B. $5\sqrt{4}$

  C. $6\sqrt{2}$

  D. $5\sqrt{2}$

13. Simplify: $16^{\frac{1}{4}} + 16^{\frac{3}{4}}$

  A. 16
  B. 10

  C. $16^{\frac{3}{16}}$

  D. $16^{\frac{3}{8}}$

14. Subtract: $(5x^2 - 9) - (8x^2 - 8x + 12)$

  A. $-3x^2 - 8x - 21$
  B. $-3x^2 + 8x - 21$
  C. $-3x^2 - 8x - 3$
  D. $-3x^2 + 8x - 3$

15. Divide: $\dfrac{24x^{12}w^{10}}{2x^2w^2}$

  A. $22x^{10}w^8$
  B. $12x^{10}w^8$
  C. $22x^6w^5$
  D. $12x^6w^5$

16. Simplify: $(x^3w^4)^5 \cdot x^{10}$

  A. $x^{18}w^9$
  B. $x^{25}w^{20}$
  C. $x^{25} + w^{20}$
  D. $x^{18} + w^9$

17. Simplify: $\sqrt[4]{16w^{16}}$

  A. $2w^2$
  B. $2w^4$
  C. $4w^2$
  D. $4w^4$

18. Multiply: $(7x + 9)(7x - 9)$

  A. $49x^2 - 81$
  B. $49x^2 + 81$
  C. $49x^2 - 126x - 81$
  D. $49x^2 - 126x + 81$

19. Multiply: $(7x - 4)^2$

  A. $49x^2 - 16$
  B. $49x^2 + 16$
  C. $49x^2 - 56x + 16$
  D. $49x^2 - 28x + 16$

20. Divide: $\dfrac{24x^{24} - 18x^{18}}{2x^6}$

  A. $12x^4 - 9x^3$
  B. $12x^{18} - 9x^{12}$
  C. $22x^4 - 12x^3$
  D. $22x^{18} - 9x^{12}$

21. Factor: $12x^{10}w^3 - 18x^5w^{12} + 6x^5w^3$

  A. $6x^5w^3(2x^5 - 3w^9)$
  B. $6x^5w^3(2x^2 - 3w^4)$
  C. $6x^5w^3(2x^5 - 3w^9 + 1)$
  D. $6x^5w^3(2x^2 - 3w^4 + 1)$

GO ON ➡

Directions:
For questions 22–25, first solve the problem and then enter your answer on the grid on the answer sheet. Write your answer at the top of the grid. Fill in the ovals below each box. Remember to enter all answers as decimals or improper fractions. Be sure to use a #2 pencil. You CANNOT use a calculator on this section of the test.

22. Bill Dit needs to cut a 2-meter piece of balsa wood into smaller pieces. He needs a piece that is .34 meter and a piece that is .5 meter. How much of the 2-meter piece of balsa wood will Bill Dit have left?

23. Shaw Win rode his bike for 9 days. He kept track of his mileage in the table below.

Shaw Win's Bike Rides

Number of Days	Mileage per Day
5	38.6
4	27.1

His goal was to ride 350 miles in 10 days. How many more miles must Shaw Win ride to meet his goal?

24. Evaluate the expression $\dfrac{x^2 - 19}{2x + 1}$ when $x = 7$.

25. Simplify: $8 - 6 + 20 \div (2 + 3) \cdot 5$

GO ON ➡

Directions:

For questions 26–52, solve each problem. Be sure to show all work in an organized way. Partial credit is given on these problems. Be sure to use a #2 pencil. You CANNOT use a calculator on this section of the test. *Note*: On the actual exam, you will have space in which to do your work. For this practice test, please use a separate sheet of paper.

---

26. Solve for *x*: $\sqrt{1-5x} = 6$

    Show All Work

    Answer *x* = _____

---

27. Find the value of *x*: $4 - (x + 3) = 10$

    Show All Work

    Answer *x* = _____

---

28. Solve for *x*: $\dfrac{2x}{3} - 8 = 28$

    Show All Work

    Answer *x* = _____

---

29. Spee D Taxi charges $5 for pick-up at a residence, $1.50 for each piece of luggage, and $.12 per mile traveled. Spee D Taxi picks you up at home. You have 3 pieces of luggage. On the line below, write an equation that could be used to find the number of miles *m* you traveled if your total taxi bill was $12.26.

    Equation _____

    Solve the equation you wrote to find the number of miles you traveled.

    Show All Work

    Answer *m* = _____

30. Look at the line on the coordinate plane below.

    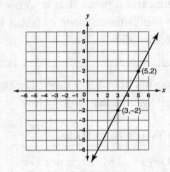

    On the line below, write the equation of the line in slope-intercept form.

    Show All Work

    Equation _____

---

31. Graph the line $2x - y = 4$ on the coordinate plane below.

    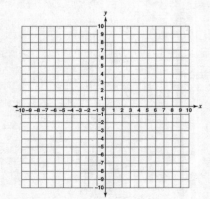

GO ON ➡

32. On the line below, write the equation of a line that passes through the points (2, –4) and (10, 0). Write your answer in slope-intercept form.

Show All Work

Equation _____

33. On the line below, write the equation of the line that passes through (6, 8) and has a slope of –4. Write your answer in slope-intercept form.

Show All Work

Equation _____

34. On the line below, write the equation of the line passing through (4, 6) and perpendicular to the line $y = \dfrac{1}{5}x + 7$. Write your answer in slope-intercept form.

Show All Work

Equation _____

35. The line graph below shows the cost that PrintMorr charges for making campaign posters.

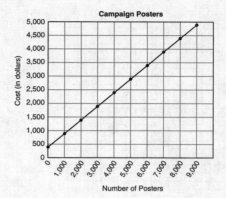

On the lines below, explain the meaning of slope in this printing situation.

_____

_____

_____

_____

On the lines below, explain the meaning of the *y*-intercept in this printing situation.

_____

_____

_____

_____

36. Graph $y < \dfrac{2}{3}x - 2$ on the coordinate plane below.

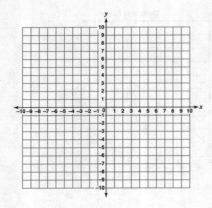

37. Graph the system of equations $x - y = -5$ and $y = -\dfrac{1}{2}x + 2$ on the coordinate plane below.

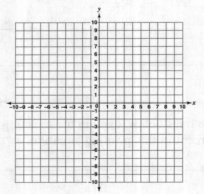

What is the solution of the system of equations? _____

38. On the coordinate plane below, graph the system of inequalities $y \le 2$ and $x > 2y$.

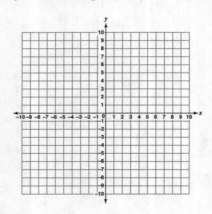

39. Solve the system of equations $x - y = 8$ and $2x + 5y = 2$.

Show All Work

Answer _____

40. What is the solution to the system of equations $x + y = 4$ and $x - y = -12$?

Show All Work

Answer _____

41. Mr. Puz Ilman loves to create story problems for his students. At the county fair there were horses and people in a building. He counted 108 legs and 37 noses. On the lines below, write a system of equations that could be used to find the number of horses h and the number of people p that were in the building.

Equations _____

_____

42. Solve for x: $\dfrac{5}{x-3} = \dfrac{2}{x+3}$

Show All Work

Answer $x =$ _____

43. Solve for x: $\dfrac{x+4}{3} = \dfrac{-6}{x-5}$

Show All Work

Answers $x =$ _____ or $x =$ _____

GO ON ➡

44. On the coordinate plane below, draw the graph of $y = (x - 1)^3$ using $-1, 0, 1, 2,$ and 3 as values of x.

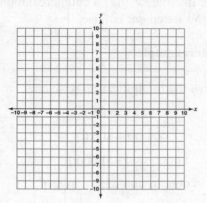

45. On the coordinate plane below, draw the graph of $y = \sqrt{\frac{1}{2}x}$ using the following as values for x: $0, 2,$ and 8.

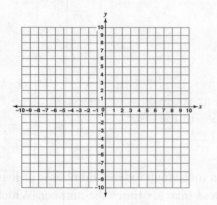

46. Solve $x^2 - 2x - 8 = 0$ for x.

Show All Work

Answers $x = $ _____ or $x = $ _____

47. Solve for x: $x^2 + 18x + 81 = 0$

Show All Work

Answer $x = $ _____

48. On the coordinate plane below, draw the graph of the equation $y = 2x^2 - 3$ using the following as values for x: $-2, -1, 0, 1,$ and 2.

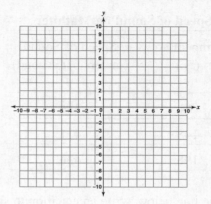

49. Bill Dit placed a garden in the corner of his backyard. The layout of the backyard is shown below.

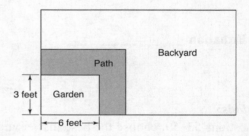

He has 22 square feet of stone to put on the path. He needs to know how wide to make the path. Smar Tee wrote the equation $(w + 3)(w + 6) - 18 = 22$, where w represents the width of the path Bill Dit can make. Solve the equation $(w + 3)(w + 6) - 18 = 22$ to find the width of the path.

Show All Work

Answer $w = $ _____

GO ON ➡

50. The relationship between the speed of sound and the air temperature on the planet Mathiter is given in the table below.

Speed of Sound on Mathiter

Temperature (°F) x	Speed of Sound (meters per second) y
98	386
148	411
198	436

On the line below, write an equation to show the linear relationship between the speed of sound, y, and the temperature, x. Give your answer in slope-intercept form.

Show All Work

Equation _____

51. Consider the statement.

If $(x - 5)^2 = 9$, then the only answer is $x = 8$.

On the line below, give a counterexample to show the statement is false.

Counterexample _____

52. Observe the multiplication problems.

$$9 \cdot 9 = 81$$
$$99 \cdot 99 = 9,801$$
$$999 \cdot 999 = 998,001$$
$$9,999 \cdot 9,999 = 99,980,001$$

Use inductive reasoning to find the next equation in the pattern. Write your answer on the line below.

Answer _____

Directions:

For questions 53–59, choose the best answer and fill in the corresponding space on your answer sheet. Be sure to use a #2 pencil. You may use a calculator on this section of the test.

53. On the day of your birth, an anonymous donor places $20,000 in the bank for you. The account earns 2.5% simple interest. No money is deposited and no money is withdrawn. Twenty years later, you withdraw all of the money including the interest. How much money do you withdraw?

A. $30,000
B. $120,000
C. $15,000
D. $50,000

54. Seven more than twice a number is at least five less than six times the number. Which one of the following inequalities represents this situation?

A. $7 + 2x \le 5 - 6x$
B. $7 + 2x \ge 5 - 6x$
C. $7 + 2x \le 6x - 5$
D. $7 + 2x \ge 6x - 5$

GO ON ➡

55. Use the sketch of circle M to determine which statement is true.

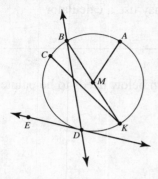

A. $\overline{AM}$ is a chord.
B. $\overline{MB}$ is a diameter.
C. $\overline{BK}$ is a radius.
D. $\angle KMA$ is a central angle.

56. Bowling scores for the Mathville High Strikers are given in the stem-and-leaf plot below.

Bowling Scores

Stem	Leaf
12	3 5 7
13	0
14	3 5 5 7
15	
16	1 3 5 5 5 6
17	8 9

Key
14 | 3 = 143

Which of the following is a false statement?

A. There are 16 scores reported.
B. The mode of the bowling scores is lower than the median of the bowling scores.
C. The lowest bowling score is 123.
D. There are 8 scores higher than 148.

57. Simplify: $\dfrac{3x^2 - 5x - 2}{3x^2 + 7x + 2}$

A. $-\dfrac{5}{7}$

B. $-\dfrac{5}{7}x - 1$

C. $\dfrac{x-2}{x+2}$

D. $\dfrac{3x-1}{3x+1}$

58. Solve for x: $9x^2 - x - 2 = 0$

A. $x = \dfrac{1 \pm \sqrt{73}}{18}$

B. $x = \dfrac{-1 \pm \sqrt{73}}{18}$

C. $x = \dfrac{1 \pm \sqrt{71}}{18}$

D. $x = \dfrac{-1 \pm \sqrt{71}}{18}$

59. The solutions of $x^2 + 5x - 36 = 0$ are $x = -9$ and $x = 4$. What are the zeros of $y = x^2 + 5x - 36$?

A. $(-9, 0)$ and $(4, 0)$
B. $(9, 0)$ and $(-4, 0)$
C. $(0, -9)$ and $(0, 4)$
D. $(0, 9)$ and $(0, -4)$

GO ON ➡

Directions:

For questions 60–68, first solve the problem and then enter your answer on the grid on the answer sheet. Write your answer at the top of the grid. Fill in the ovals below each box. Remember to enter all answers as decimals or improper fractions. Be sure to use a #2 pencil. You may use a calculator on this section of the test.

60. Professor Cy Intist is studying how much sweat a runner produces. At the first trial on the treadmill, Athel Eat produced 3 ounces of sweat for every 20 yards he ran. How many cups of sweat does Athel Eat produce when running one mile?

61. The hypotenuse of a right triangle measures 65 feet. One of the legs is 63 feet long. What is the length of the other leg?

62. The model of a sculpture is made using the scale 1 centimeter represents 1.5 meters. If the surface area of the model is 19 square centimeters, what is the surface area of the sculpture?

63. What is the area of the trapezoid below?

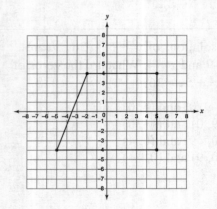

64. The can shown below needs to be painted.

The diameter of the can is 9 inches and the height is 14 inches. How many square inches need to be painted? Round your answer to the nearest square inch.

65. A lean-to shelter, shown in the diagram below, is built in the shape of a triangular prism.

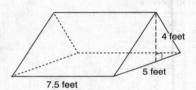

How many cubic feet of air are contained in the lean-to shelter?

66. Smar Tee's homework grades for Algebra class are listed in ascending order.

13 14 15 15 15 16 17 18 18 18 19 20 20

What is the upper quartile of all of Smar Tee's homework grades?

67. A bag of blocks has 53 different colored blocks. There are 25 blue blocks, 10 red blocks, 5 yellow blocks, and the rest are green blocks. A block is drawn at random. What is the probability that the block drawn will be green or red?

68. Ms. T. Cher is arranging 4 books in a straight line on her desk. How many different ways can she line up the books?

GO ON ➡

Directions:

For questions 69–79, solve each problem. Be sure to show all work in an organized way. Partial credit is given on these problems. Be sure to use a #2 pencil. You may use a calculator on this section of the test. *Note*: On the actual exam, you will have space in which to do your work. For this practice test, please use a separate sheet of paper.

69. Look at the quadrilateral on the coordinate plane below.

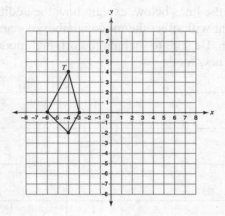

On the same coordinate plane, draw a translation of the quadrilateral 7 units to the right and 4 units down. Then reflect the new quadrilateral across the *x*-axis.

70. Seenyer Prez surveyed 250 couples who attended the prom. The number of pictures taken by the parents of the couples is shown in the circle graph below.

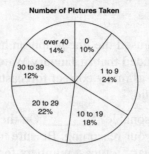

Number of Pictures Taken

Seenyer Prez claims that over 160 couples had more than 15 pictures taken by their parents. On the lines below, explain why this is probably an incorrect statement.

_____

_____

_____

_____

The statement should be changed to claim that over 160 couples had more than how many pictures taken? _____

GO ON ➡

71. Watch Us TV and U Listn Radio provide daily weather forecasts for Mathville. The results are shown in the line graph below.

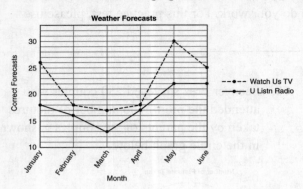

Mr. Ank Erman used the line graph on his show. He concluded that in January Watch Us TV made twice as many correct forecasts as did U Listn Radio.

Is his claim reasonable? _____ On the lines below, explain your reasoning. Be sure to include the January forecast numbers for Watch Us TV and U Listn Radio.

_____

_____

_____

72. Tella Scope kept track of the number of shooting stars he saw for 10 nights, shown in the table below.

Number of Shooting Stars Sighted

9	10	12	12	12
14	15	15	16	20

Tella Scope counted 15 stars on the 11th night. On the lines below, explain how the additional night will affect the mode number of stars seen. Be sure to include the original mode and the new mode.

_____

_____

_____

_____

73. Supa Shopa receives a lot of catalogs in the mail. She kept track of the number of catalogs she received each day. The results are displayed in the table.

Number of Catalogs Per Day

2	3	5	6	8	8	11	12	16

Use the number line below to create a box-and-whiskers plot to display the information.

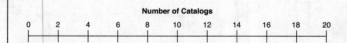

74. Mr. Spee D. Walker graphed the hike that he took yesterday. He used a coordinate plane like the one below.

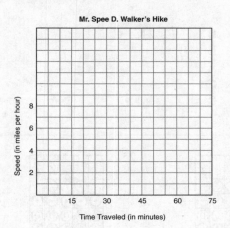

Mr. Spee D. Walker's Hike

He spent 25 minutes accelerating at a steady rate to 4 miles per hour. He spent the next 40 minutes walking at 4 miles per hour. He then spent 5 minutes slowing at a steady rate to a stop. On the coordinate plane above sketch the graph of Mr. Walker's hike.

75. The relation {(3, 5), (7, 2), (4, 5), 7, 9)} is not a function. Which ordered pair could be removed to make the relation a function?

_____

On the lines below, explain your reasoning.

_____

_____

_____

76. Consider the relation

{(4, 9), (6, 3), (8, 2), (1, 7)}.

What is the domain? _____

77. Factor: $3x^2 - 14x - 5$

Answer _____

78. Flowers Warehouse received 15 crates of rose bushes. Each crate held 5 dozen bushes. Flowers Warehouse must unpack all of the rose bushes and repackage them in smaller boxes that hold 8 bushes. How much will Flowers Warehouse pay for the repackaging boxes if each one costs $2.50 and sales tax is 6%?

Show All Work

Answer _____

79. Schoo Buh, Unda Wata, and Lie Kafish compared the use of oxygen tanks while scuba diving. The results are shown on the line graph below.

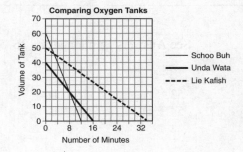

Comparing Oxygen Tanks

A. Who had the largest volume oxygen tank?
B. Who used the oxygen at the slowest rate?
C. Who ran out of oxygen first?

Practice Test 2

SOLUTIONS: PRACTICE TEST 2

Answer Key

1. **C**	6. **C**	11. **C**	16. **B**	21. **C**
2. **D**	7. **B**	12. **D**	17. **B**	
3. **C**	8. **C**	13. **B**	18. **A**	
4. **D**	9. **D**	14. **B**	19. **C**	
5. **D**	10. **C**	15. **B**	20. **B**	

22.

23.

24.

25.

See Answers Explained for 26–52.

53. **A**	55. **D**	57. **C**	59. **A**
54. **D**	56. **B**	58. **A**	

60.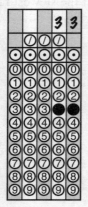

61.

62.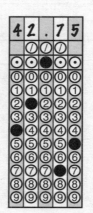

63.

64.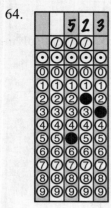

65.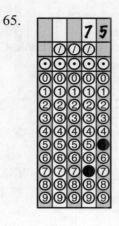

66.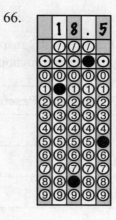

67.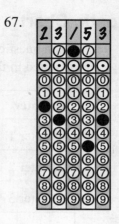

68.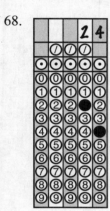

See Answers Explained for 68–79.

Diagnostic Chart

Each question in the Practice Test 2 maps to a chapter in the book. Chapter 15, problem solving, is tested in the story problems throughout this test.

Chapter	Problems on Practice Test 2
Chapter 1	1–3, 22, 23, 53
Chapter 2	4–9
Chapter 3	10, 11, 27–29, 54
Chapter 4	12, 13, 24, 25, 60
Chapter 5	49, 55, 61, 69
Chapter 6	62–65
Chapter 7	56, 66, 70–73, 79
Chapter 8	67, 68
Chapter 9	74–76
Chapter 10	30–36, 50
Chapter 11	37–41
Chapter 12	14–21, 77
Chapter 13	42, 43, 57
Chapter 14	26, 44–48, 58, 59
Chapter 15	51, 52, 78

Answers Explained

1. **C** $-3\frac{3}{4} \cdot -1\frac{1}{6} = -\frac{\overset{5}{\cancel{15}}}{4} \cdot -\frac{7}{\underset{2}{\cancel{6}}} = \frac{35}{8} = 4\frac{3}{8}$

2. **D** $-8 \cdot 2 = -16$
   $-8 + -2 = -10$
   $-8 - (-2) = -8 + 2 = -6$
   $-8 \div 2 = -4$

3. **C** $4\frac{18}{19}$ is a little less than 5 and $2\frac{2}{37}$ is a little more than 2. Multiply 5 and 2 to get 10.

4. **D** $(2x^4)^3 = (2x^4) \cdot (2x^4) \cdot (2x^4) = 8x^{12}$

5. **D** $9^{-2} = \frac{1}{9^2} = \frac{1}{81}$

6. **C** $\frac{4.8 \times 10^3}{6 \times 10^6} = \frac{4.8}{6} \times \frac{10^3}{10^6} = .8 \times 10^{-3} = (8 \times 10^{-1}) \times 10^{-3} = 8 \times 10^{-4}$

7. **B** $(5 \times 10^7)^2 = (5 \times 10^7) \times (5 \times 10^7) = 25 \times 10^{14} = (2.5 \times 10^1) \times 10^{14} = 2.5 \times 10^{15}$

8. **C** $\sqrt{83}$ is between $\sqrt{81} = 9$ and $\sqrt{100} = 10$, but it is closer to $\sqrt{81} = 9$.

9. **D** $\sqrt{29+3} = \sqrt{32}$ is irrational; $\sqrt{31+5} = \sqrt{36} = 6$ is rational;

   $\pi - 5$ is irrational;

   $23.\overline{494949}$ is rational.

10. **C** $\frac{c}{a} + r = e$

    $\frac{c}{a} = e - r$

    $a \cdot \frac{c}{a} = a \cdot (e - r)$

    $c = a(e - r)$

11. **C** $3 \geq -x + 2$
    $-x + 2 \leq 3$
    $-x \leq 1$

    Remember to reverse the negative sign when you multiply both sides by a negative number.

    $-1 \cdot -x \geq -1 \cdot 1$
    $x \geq -1$

12. **D** $\sqrt{18} + \sqrt{8} = 3\sqrt{2} + 2\sqrt{2} = 5\sqrt{2}$

13. **B** $16^{\frac{1}{4}} + 16^{\frac{3}{4}} = \left(\sqrt[4]{16}\right)^1 + \left(\sqrt[4]{16}\right)^3 = 2 + 8 = 10$

14. **B** $(5x^2 - 9) - (8x^2 - 8x + 12)$

$(5x^2 + -9) + -1(8x^2 + -8x + 12)$

$(5x^2 + \underline{-9}) + \underline{-8x^2} + 8x + \underline{-12}$

$-3x^2 + 8x - 21$

15. **B** $\dfrac{24x^{12}w^{10}}{2x^2w^2} = \dfrac{24}{2} \cdot \dfrac{x^{12}}{x^2} \cdot \dfrac{w^{10}}{w^2} = 12x^{12-2}w^{10-2} = 12x^{10}w^8$

16. **B** $(x^3w^4)^5 \cdot x^{10} = x^{15}w^{20} \cdot x^{10} = x^{15+10}w^{20} = x^{25}w^{20}$

17. **B** $\sqrt[4]{16w^{16}} = \sqrt[4]{16} \cdot \sqrt[4]{w^{16}} = 2w^{\frac{16}{4}} = 2w^4$

18. **A** $(7x + 9)(7x - 9) = 49x^2 - 63x + 63x - 81 = 49x^2 - 81$

19. **C** $(7x - 4)^2 = (7x - 4)(7x - 4) = 49x^2 - 28x - 28x + 16 = 49x^2 - 56x + 16$

20. **B** $\dfrac{24x^{24} - 18x^{18}}{2x^6} = \dfrac{24x^{24}}{2x^6} - \dfrac{18x^{18}}{2x^6} = 12x^{24-6} - 9x^{18-6} = 12x^{18} - 9x^{12}$

21. **C** The greatest common factor is $6x^5w^3$. Think $6x^5w^3$ (_____ + _____).

22. **1.16** $2 - (.34 + .5) = 2 - .84 = 1.16$

23. **48.6** $350 - (38.6 \cdot 5 + 4 \cdot 27.1) = 350 - (193 + 108.4) = 350 - 301.4 = 48.6$

24. **2** $\dfrac{(7)^2 - 19}{2(7) + 1} = \dfrac{49 - 19}{14 + 1} = \dfrac{30}{15} = 2$

25. **22**

$8 - 6 + 20 \div \underline{(2 + 3)} \cdot 5$

$8 - 6 + \underline{20 \div 5} \cdot 5$

$8 - 6 + \underline{4 \cdot 5}$

$\underline{8 - 6} + 20$

$2 + 20 = 22$

26. **−7**

$$\sqrt{1-5x} = 6$$

$$\left(\sqrt{1-5x}\right)^2 = 6^2$$

$$1 - 5x = 36$$
$$-5x = 35$$
$$x = -7$$

Since both sides of the equation were squared, you must check your answer.

$$\sqrt{1-5(-7)} \overset{?}{=} 6$$

$$\sqrt{1-(-35)} \overset{?}{=} 6$$

$$\sqrt{36} \overset{?}{=} 6$$

$$6 = 6$$

Full credit is given if work and correct answer are shown. Half credit is given if no work is shown *or* if work is shown and computation error is made.

27. **−9**

$$4 - (x + 3) = 10$$
$$4 + -1(x + 3) = 10$$
$$4 + -1x + -3 = 10$$
$$-1x + 1 = 10$$
$$-1x = 9$$
$$x = -9$$

Full credit is given if work and correct answer are shown. Half credit is given if no work is shown *or* if work is shown and computation error is made.

28. **54**

$$\frac{2x}{3} - 8 = 28$$

$$\frac{2}{3}x = 36$$

$$\frac{3}{2} \cdot \frac{2}{3}x = \frac{3}{2} \cdot 36$$

$$x = 54$$

Full credit is given if work and correct answer are shown. Half credit is given if no work is shown *or* if work is shown and computation error is made.

29. $.12m + 5 + 4.50 = 12.26$ or an equivalent equation. No partial credit is given.

Your total cost includes the $5 for pick-up at your house, $1.50 · 3 for your luggage, and $.12 for every mile you travel. If you travel m miles, it will cost you $5 for pick-up at your house, $1.50 · 3 for your luggage, and $.12m$.

$.12m + 5 + 4.50 = 12.26$

23 miles

$$.12m + 5 + 4.50 = 12.26$$
$$.12m + 9.5 = 12.26$$
$$.12m = 2.76$$
$$m = 23$$

Full credit is given if work and correct answer are shown. Half credit is given if no work is shown *or* if work is shown and computation error is made.

30. $y = 2x - 8$

$$m = \frac{2 - (-2)}{5 - 3} = \frac{4}{2} = 2$$

$$y - 2 = 2(x - 5)$$
$$y - 2 = 2x - 10$$
$$y = 2x - 8$$

Full credit is given if work and correct answer are shown. Half credit is given if no work is shown *or* if work is shown and computation error is made.

31.

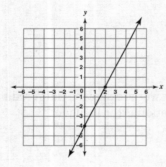

The x-intercept is $(2, 0)$ and the y-intercept is $(0, -4)$.

Full credit if line graphed correctly. Half credit if correct slope with line drawn *or* if correct intercepts with no line drawn.

32. $y = \frac{1}{2}x - 5$

$$\text{slope} = \frac{\text{change in } y}{\text{change in } x} = \frac{0 - (-4)}{10 - 2} = \frac{4}{8} = \frac{1}{2}$$

$$y - (-4) = \frac{1}{2}(x - 2)$$

$$y + 4 = \frac{1}{2}x - 1$$

$$y = \frac{1}{2}x - 5$$

Full credit is given if work and correct answer are shown. Half credit is given if no work is shown *or* if work is shown and computation error is made.

33. $y = -4x + 32$

$$y - 8 = -4(x - 6)$$
$$y - 8 = -4x + 24$$
$$y = -4x + 32$$

Full credit is given if work and correct answer are shown. Half credit is given if no work is shown *or* if work is shown and computation error is made.

34. $y = -5x + 26$. The slope of the given line is $\frac{1}{5}$, so the slope of a perpendicular line is –5.

$$y - 6 = -5(x - 4)$$
$$y - 6 = -5x + 20$$
$$y = -5x + 26$$

Full credit is given if work and correct answer are shown. Half credit is given if no work is shown *or* if work is shown and computation error is made.

35. The slope is the cost per poster. No partial credit is given.
The *y*-intercept is the set-up fee or beginning costs. No partial credit is given.

36.

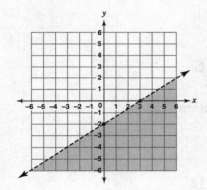

Full credit if dotted line graphed correctly and correct side shaded. Half credit if all is correct but a solid line is used *or* if all is correct but the wrong side is shaded.

37.

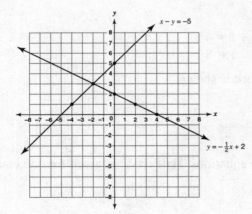

Full credit if both lines are graphed correctly. Half credit if one of the two lines is graphed correctly.

The solution set is (–2, 3).
No partial credit is given.

38.

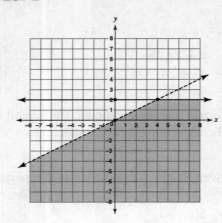

Full credit if all is graphed correctly. Half credit if one of the two lines is graphed correctly and shaded correctly.

39. (6, –2)

$$x = y + 8$$
$$2x + 5y = 2$$
$$2(y + 8) + 5y = 2$$
$$2y + 16 + 5y = 2$$
$$7y + 16 = 2$$
$$7y = -14$$
$$y = -2$$

$$2x + 5(-2) = 2$$
$$2x - 10 = 2$$
$$2x = 12$$
$$x = 6$$

Half credit is given if no work is shown.

40. (–4, 8)

$$x + y = 4$$
$$\underline{x - y = -12}$$
$$2x = -8$$
$$x = -4$$

$$-4 + y = 4$$
$$y = 8$$

Half credit is given if no work is shown.

41. $4h + 2p = 108$
$h + p = 37$

Full credit is given for both equations. Half credit is given for one equation.

42. –7

$$5(x + 3) = 2(x - 3)$$
$$5x + 15 = 2x - 6$$
$$3x + 15 = -6$$
$$3x = -21$$
$$x = -7$$

Half credit is given if no work is shown.

43. $x = 2$ or $x = -1$

$$(x + 4)(x - 5) = 3 \cdot -6$$
$$x^2 - x - 20 = -18$$
$$x^2 - x - 2 = 0$$
$$(x - 2)(x + 1) = 0$$
$$x - 2 = 0 \text{ or } x + 1 = 0$$
$$x = 2 \text{ or } x = -1$$

Full credit is given if both answers and work are correct. Half credit is given if both answers are correct and no work is shown *or* if one answer is correct and work is shown.

44.

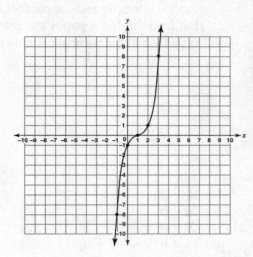

Points on the graph include $(-1, -8)$, $(0, -1)$, $(1, 0)$, $(2, 1)$, and $(3, 8)$.

Full credit is given only if all 5 points are graphed correctly and the curve is drawn. Half credit if 3 or 4 points are graphed correctly and the curve is drawn *or* if all 5 points are graphed correctly and no curve is drawn.

45.

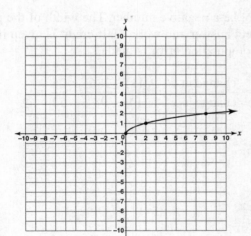

Points on the graph include $(0, 0)$, $(2, 1)$, and $(8, 2)$.

Full credit is given if all 3 points are correctly graphed and the correct curve is drawn. Partial credit is given if 2 points are correctly graphed and the correct curve is drawn. No points are given if the curve is drawn in the second or third quadrant.

46. $x = -2$ or $x = 4$

$$x^2 - 2x - 8 = 0$$
$$(x - 4)(x + 2) = 0$$
$$x - 4 = 0 \text{ or } x + 2 = 0$$
$$x = 4 \text{ or } x = -2$$

Full credit is given if both correct answers and work are shown. Half credit is given if both answers are correct and no work is shown *or* if one answer is correct and work is shown.

47. $x = -9$

$$x^2 + 18x + 81 = 0$$
$$(x + 9)(x + 9) = 0$$
$$x + 9 = 0$$
$$x = -9$$

Full credit is given if correct answer and work are shown. Half credit is given if answer is correct and no work is shown *or* if work is shown and an error in computation is made.

48.

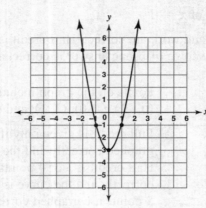

Points on the graph include (–2, 5), (–1, –1), (0, –3), (1, –1), and (2, 5).

Full credit is given only if all 5 points are graphed correctly and the parabola is drawn. Half credit if 3 or 4 points are graphed correctly and the parabola is drawn.

49. 2 feet

$$(w + 3)(w + 6) - 18 = 22$$
$$w^2 + 9w = 22$$
$$w^2 + 9w - 22 = 0$$
$$(w + 11)(w - 2) = 0$$
$$w + 11 = 0 \text{ or } w - 2 = 0$$
$$w = -11 \text{ or } w = 2$$

The width of the path cannot be a negative number. The width of the path is 2 feet. Full credit is given for correct answer and work. Half credit is given if no work is shown *or* if work is shown and a computation error is made.

50. $y = \dfrac{1}{2}x + 337$

$$\text{slope} = \frac{\text{change in } y}{\text{change in } x} = \frac{411 - 386}{148 - 98} = \frac{25}{50} = \frac{1}{2}$$

$$y - 386 = \frac{1}{2}(x - 98)$$

$$y - 386 = \frac{1}{2}x - 49$$

$$y = \frac{1}{2}x + 337$$

Full credit is given for correct answer and work. Half credit is given if no work is shown *or* if work is shown and a computation error is made.

51. $x = 2$. A counterexample is a number that works in the hypothesis but does not work for the conclusion.

$$(2 - 5)^2 = 9$$
$$(-3)^2 = 9$$
$$9 = 9 \qquad \text{No partial credit is given.}$$

52. $99{,}999 \cdot 99{,}999 = 9{,}999{,}800{,}001$. No partial credit is given.

53. **A** $\qquad i = 20{,}000 \cdot .025 \cdot 20$
$\qquad\qquad i = 10{,}000$

You have earned $10,000 simple interest. You withdraw all of the money for a total of $30,000.

54. **D**
Seven more than means $7 +$ *Five less than* means -5
Twice a number means $2 \cdot x$ *Six times the number* means $6 \cdot x$
Is at least means $\geq$

55. **D** $\angle KMA$ is a central angle, an angle that is formed by two points on the circle joined with the center of the circle. In choice A, $\overline{AM}$ is a radius. In choice B, $\overline{MB}$ is a radius. In choice C, $\overline{BK}$ is a diameter.

56. **B**

57. **C** $\dfrac{3x^2 - 5x - 2}{3x^2 + 7x + 2} = \dfrac{\cancel{(3x+1)}^{1}\,(x-2)}{\cancel{(3x+1)}_{1}\,(x+2)} = \dfrac{x-2}{x+2}$

58. **A**

$$x = \frac{-b \pm \sqrt{b^2 - 4ac}}{2a}$$

$$x = \frac{-(-1) \pm \sqrt{(-1)^2 - 4 \cdot 9 \cdot -2}}{2(9)}$$

$$x = \frac{1 \pm \sqrt{1 - (-72)}}{18}$$

$$x = \frac{1 \pm \sqrt{73}}{18}$$

59. **A** The zeros of $y = x^2 + 5x - 36$ are the x-intercepts and are the same as the solutions of $x^2 + 5x - 36 = 0$.

60. **33** $\dfrac{3 \; \cancel{\text{ounces}}}{20 \; \cancel{\text{yards}}} \cdot \dfrac{1{,}760 \; \cancel{\text{yards}}}{1 \text{ mile}} \cdot \dfrac{1 \text{ cup}}{8 \; \cancel{\text{ounces}}} = \dfrac{5{,}280 \text{ cups}}{160 \text{ miles}} = 33$ cups per mile

61. **16**

$$a^2 + b^2 = c^2$$
$$a^2 + 63^2 = 65^2$$
$$a^2 + 3{,}969 = 4{,}225$$
$$a^2 = 256$$
$$a = 16$$

62. **42.75**

$$\frac{\text{Surface Area of model}}{\text{Surface Area of sculpture}} \cdot \left(\frac{\text{side of model}}{\text{corresponding side of sculpture}}\right)^2$$

$$\frac{19}{\text{Surface Area of sculpture}} = \left(\frac{1}{1.5}\right)^2$$

$$\frac{19}{\text{Surface Area of sculpture}} = \frac{1}{2.25}$$

Surface Area of sculpture = 42.75

63. **68**

$$A = \frac{1}{2}(10 + 7) \cdot 8$$

$$A = \frac{1}{2}(17) \cdot 8$$

$$A = 68$$

64. **523**

$$SA = 2\pi r^2 + 2\pi rh$$
$$SA = 2 \cdot 3.14 \cdot (4.5)^2 + 2 \cdot 3.14 \cdot 4.5 \cdot 14$$
$$SA = 127.17 + 395.64$$
$$SA = 522.81$$

65. **75**

$$A = \frac{1}{2} \cdot 5 \cdot 4$$

$$A = 10$$

$$V = 10 \cdot 7.5$$
$$V = 75$$

66. **18.5**

67. $\frac{23}{53}$ The probability of drawing a green or red is the same as the probability of drawing

a red plus the probability of drawing a green. There are $53 - 25 - 10 - 5 = 13$ green blocks.

$$\frac{10}{53} + \frac{13}{53} = \frac{23}{53}$$

68. **24**

69.

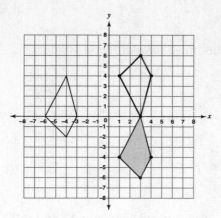

Move each point 7 to the right and down 4 to get the gray quadrilateral. Then reflect the gray quadrilateral across the *x*-axis to get the bold quadrilateral.

Half credit if 3 of 4 vertices are correct.

70. The circle graph does not show how many couples had more than 15 pictures taken. The cut numbers are more than 20 or more than 10.
No partial credit is given.

10

The number of couples who had more than 10 pictures taken is 66 percent of 250 couples: $.66 \cdot 250 = 165$.
No partial credit is given.

71. No.
No partial credit is given.
In January, Watch Us TV made 26 correct forecasts and U Listn Radio made 18 correct forecasts. The line graph is misleading because the scale for correct broadcasts does not show 0.
No partial credit is given.

72. The mode number of stars seen will change from 12 to both 12 and 15.
No partial credit is given.

73.

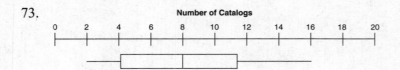

No partial credit is given.

74.

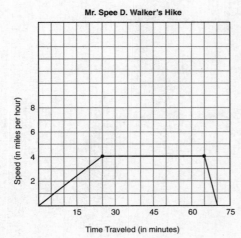

Full credit for all 3 sections graphed correctly. Half credit for 2 of the sections graphed correctly.

75. (7, 9) or (7, 2)
No partial credit is given.
To be a function, all of the *x*-values must have different *y*-values.
No partial credit is given.

76. {4, 6, 8, 1}
No partial credit is given.

77. $(3x + 1)(x - 5)$
No partial credit is given.

78. $299.45
There are 15 crates with 60 rose bushes in each crate or $15 \cdot 60 = 900$ rose bushes.
There will need to be $900 \div 8 = 112.5$ boxes. Flowers Warehouse will purchase
113 boxes. The cost of 113 boxes is $113 \cdot \$2.50 = \$282.50$. Sales tax is
$.06 \cdot 282.50 = 16.95$. The total cost for the boxes is $\$282.50 + 16.95 = \$299.45$.
Full credit is given if all work is shown and the answer is correct. Half credit is given
if no work is shown and the answer is correct *or* if all work is shown but an error in
computation was made.

79. **A.** Schoo Buh
At 0 minutes she had 60 while Lie Kafish started with a little over 50 and Unda Wata
started with 40.
No partial credit is given.

B. Lie Kafish
His line is slanted the least so he was using oxygen slowly.
No partial credit is given.

C. Schoo Buh
Unda Wata hit 0 volume in 16 minutes and Lie Kafish ran out of oxygen in 34 minutes.
No partial credit is given.

Appendix

Eighth Grade Standards

Number Sense

8.1.1   Read, write, compare, and solve problems using decimals in scientific notation.

8.1.2   Know that every rational number is either a terminating or repeating decimal and that every irrational number is a non-repeating decimal.

8.1.3   Understand that computations with an irrational number and a rational number (other than zero) produce an irrational number.

8.1.4   Understand and evaluate negative integer exponents.

8.1.5   Use the laws of exponents for integer exponents.

8.1.6   Use the inverse relationship between squaring and finding the square root of a perfect square integer.

8.1.7   Calculate and find approximations of square roots.

Computation

8.2.1   Add, subtract, multiply, and divide rational numbers (integers, fractions, and terminating decimals) in multi-step problems.

8.2.2   Solve problems by computing simple and compound interest.

8.2.3   Use estimation techniques to decide whether answers to computations on a calculator are reasonable.

8.2.4   Use mental arithmetic to compute with common fractions, decimals, powers, and percents.

Algebra

8.3.1   Write and solve linear equations and inequalities with one variable, interpret the solution or solutions in their context, and verify the reasonableness of the results.

8.3.2   Solve systems of two linear equations using the substitution method and identify approximate solutions graphically.

8.3.3   Interpret positive integer powers as repeated multiplication and negative integer powers as repeated division or multiplication by the multiplicative inverse.

8.3.4   Use the correct order of operations to find the values of algebraic expressions involving powers.

8.3.5   Identify and graph linear functions and identify lines with positive and negative slope.

8.3.6   Find the slope of a linear function given the equation and write the equation of a line given the slope and any point on the line.

8.3.7   Demonstrate an understanding of rate as a measure of one quantity with respect to another quantity.

8.3.8   Demonstrate an understanding of the relationships among tables, equations, verbal expressions, and graphs of linear functions.

8.3.9   Represent simple quadratic functions using verbal descriptions, tables, graphs, and formulas and translate among these representations.

8.3.10   Graph functions of the form $y = nx^2$ and $y = nx^3$ and describe the similarities and differences in the graphs.

Geometry

8.4.1   Identify and describe basic properties of geometric shapes: altitudes, diagonals, angle and perpendicular bisectors, central angles, radii, diameters, and chords.

8.4.2   Perform simple constructions, such as bisectors of segments and angles, copies of segments and angles, and perpendicular segments. Describe and justify the constructions.

8.4.3 Identify properties of three-dimensional geometric objects (e.g., diagonals of rectangular solids) and describe how two or more figures intersect in a plane or in space.

8.4.4 Draw the translation (slide), rotation (turn), reflection (flip), and dilation (stretches and shrinks) of shapes.

8.4.5 Use the Pythagorean Theorem and its converse to solve problems in two and three dimensions.

Measurement

8.5.1 Convert common measurements for length, area, volume, weight, capacity, and time to equivalent measurements within the same system.

8.5.2 Solve simple problems involving rates and derived measurements for attributes such as velocity and density.

8.5.3 Solve problems involving scale factors, area, and volume using ratio and proportion.

8.5.4 Use formulas for finding the perimeter and area of basic two-dimensional shapes and the surface area and volume of basic three-dimensional shapes, including rectangles, parallelograms, trapezoids, triangles, circles, prisms, cylinders, spheres, cones, and pyramids.

8.5.5 Estimate and compute the area and volume of irregular two- and three-dimensional shapes by breaking the shapes down into more basic geometric objects.

Data Analysis and Probability

8.6.1 Identify claims based on statistical data and, in simple cases, evaluate the reasonableness of the claims. Design a study to investigate the claim.

8.6.2 Identify different methods of selecting samples, analyzing the strengths and weaknesses of each method, and the possible bias in a sample or display.

8.6.3 Understand the meaning of, and be able to identify or compute the minimum value, the lower quartile, the median, the upper quartile, the interquartile range, and the maximum value of a data set.

8.6.4 Analyze, interpret, and display single- and two-variable data in appropriate bar, line, and circle graphs, stem-and-leaf plots, and box-and-whisker plots and explain which types of display are appropriate for various data sets.

8.6.5 Represent two-variable data with a scatterplot on the coordinate plane and describe how the data points are distributed. If the pattern appears to be linear, draw a line that appears to best fit the data and write the equation of that line.

8.6.6 Understand and recognize equally likely events.

8.6.7 Find the number of possible arrangements of several objects by using the Basic Counting Principle.

Problem Solving

8.7.1 Analyze problems by identifying relationships, telling relevant from irrelevant information, identifying missing information, sequencing and prioritizing information, and observing patterns.

8.7.2 Make and justify mathematical conjectures based on a general description of a mathematical question or problem.

8.7.3 Decide when and how to divide a problem into simpler parts.

8.7.4 Apply strategies and results from simpler problems to solve more complex problems.

8.7.5 Make and test conjectures by using inductive reasoning.

8.7.6 Express solutions clearly and logically by using the appropriate mathematical terms and notation. Support solutions with evidence in both verbal and symbolic work.

8.7.7 Recognize the relative advantages of exact and approximate solutions to problems and give answers to a specified degree of accuracy.

8.7.8 Select and apply appropriate methods for estimating results of rational-number computations.

8.7.9 Use graphing to estimate solutions and check the estimates with analytic approaches.

8.7.10 Make precise calculations and check the validity of the results in the context of the problem.

8.7.11 Decide whether a solution is reasonable in the context of the original situation.

8.7.12 Note the method of finding the solution and show a conceptual understanding of the method by solving similar problems.

Algebra 1 Standards

A1.1.1 Compare real number expressions.

A1.1.2 Simplify square roots using factors.

A1.1.3 Understand and use the distributive, associative, and commutative properties.

A1.1.4   Use the laws of exponents for rational exponents.

A1.1.5   Use dimensional (unit) analysis to organize conversions and computations.

A1.2.1   Solve linear equations.

A1.2.2   Solve equations and formulas for a specified variable.

A1.2.3   Find solution sets of linear inequalities when possible numbers are given for the variable.

A1.2.4   Solve linear inequalities using properties of order.

A1.2.5   Solve combined linear inequalities.

A1.2.6   Solve word problems that involve linear equations, formulas, and inequalities.

A1.3.1   Sketch a reasonable graph for a given relationship.

A1.3.2   Interpret a graph representing a given situation.

A1.3.3   Understand the concept of a function, decide if a given relation is a function, and link equations to functions.

A1.3.4   Find the domain and range of a relation.

A1.4.1   Graph a linear equation.

A1.4.2   Find the slope, x-intercept and y-intercept of a line given its graph, its equation, or two points on the line.

A1.4.3   Write the equation of a line in slope-intercept form. Understand how the slope and y-intercept of the graph are related to the equation.

A1.4.4   Write the equation of a line given appropriate information.

A1.4.5   Write the equation of a line that models a data set and use the equation (or the graph of the equation) to make predictions. Describe the slope of the line in terms of the data, recognizing that the slope is the rate of change.

A1.4.6   Graph a linear inequality in two variables.

A1.5.1   Use a graph to estimate the solution of a pair of linear equations in two variables.

A1.5.2   Use a graph to find the solution set of a pair of linear inequalities in two variables.

A1.5.3   Understand and use the substitution method to solve a pair of linear equations in two variables.

A1.5.4   Understand and use the addition or subtraction method to solve a pair of linear equations in two variables.

A1.5.5   Understand and use multiplication with the addition or subtraction method to solve a pair of linear equations in two variables.

A1.5.6   Use pairs of linear equations to solve word problems.

A1.6.1   Add and subtract polynomials.

A1.6.2   Multiply and divide monomials.

A1.6.3   Find powers and roots of monomials (only when the answer has an integer exponent).

A1.6.4   Multiply polynomials.

A1.6.5   Divide polynomials by monomials.

A1.6.6   Find a common monomial factor in a polynomial.

A1.6.7   Factor the difference of two squares and other quadratics.

A1.6.8   Understand and describe the relationships among the solutions of an equation, the zeros of a function, the x-intercepts of a graph, and the factors of a polynomial expression.

A1.7.1   Simplify algebraic ratios.

A1.7.2   Solve algebraic proportions.

A1.8.1   Graph quadratic, cubic, and radical equations.

A1.8.2   Solve quadratic equations by factoring.

A1.8.3   Solve quadratic equations in which a perfect square equals a constant.

A1.8.4   Complete the square to solve quadratic equations.

A1.8.5   Derive the quadratic formula by completing the square.

A1.8.6   Solve quadratic equations by using the quadratic formula.

A1.8.7   Use quadratic equations to solve word problems.

A1.8.8   Solve equations that contain radical expressions.

A1.8.9   Use graphing technology to find approximate solutions of quadratic and cubic equations.

A1.9.1   Use a variety of problem solving strategies, such as drawing a diagram, making a chart, guess-and-check, solving a simpler problem, writing an equation, and working backwards.

A1.9.2 Decide whether a solution is reasonable in the context of the original situation.

A1.9.3 Use the properties of the real number system and the order of operations to justify the steps of simplifying functions and solving equations.

A1.9.4 Understand that the logic of equation solving begins with the assumption that the variable is a number that satisfies the equation, and that the steps taken when solving equations create new equations that have, in most cases, the same solution set as the original. Understand that similar logic applies to solving systems of equations simultaneously.

A1.9.5 Decide whether a given algebraic statement is true always, sometimes, or never (statements involving linear or quadratic expressions, equations, or inequalities).

A1.9.6 Distinguish between inductive and deductive reasoning, identifying and providing examples of each.

A1.9.7 Identify the hypothesis and conclusion in a logical deduction.

A1.9.8 Use counterexamples to show that statements are false, recognizing that a single counterexample is sufficient to prove a general statement false.

ISTEP+ Grades 9 and 10 Mathematics Reference Sheet *

Shape	Formulas for Area (A) and Circumference (C)
Triangle	$A = \frac{1}{2}bh = \frac{1}{2} \times$ base $\times$ height
Rectangle	$A = lw =$ length $\times$ width
Trapezoid	$A = \frac{1}{2}(b_1 + b_2) \times h = \frac{1}{2} \times$ sum of bases $\times$ height
Parallelogram	$A = bh =$ base $\times$ height
Square	$A = s^2 =$ side $\times$ side
Circle	$A = \pi r^2 = \pi \times$ square of radius $C = 2\pi r = 2 \times \pi \times$ radius $\pi \approx 3.14$ or $\frac{22}{7}$

Figure	Formulas for Volume (V) and Surface Area (SA)	
Rectangular Prism	$V = lwh =$ length $\times$ width $\times$ height $SA = 2lw + 2hw + 2lh$ $= 2$(length $\times$ width) $+ 2$(height $\times$ width) $+ 2$(length $\times$ height)	
General Prisms	$V = Bh =$ area of base $\times$ height $SA =$ sum of the areas of the faces	
Cylinder	$V = \pi r^2 h = \pi \times$ square of radius $\times$ height $SA = 2\pi r^2 + 2\pi rh$ $= 2 \times \pi \times$ square of radius $+$ $2 \times \pi \times$ radius $\times$ height	$\pi \approx 3.14$ or $\pi \approx \frac{22}{7}$
Sphere	$V = \frac{4}{3}\pi r^3 = \frac{4}{3} \times \pi \times$ cube of radius $SA = 4\pi r^2 = 4 \times \pi \times$ square of radius	
Right Circular Cone	$V = \frac{1}{3}\pi r^2 h = \frac{1}{3} \times \pi \times$ square of radius $\times$ height	
Regular Pyramid	$V = \frac{1}{3}Bh = \frac{1}{3} \times$ area of base $\times$ height	

*Math Reference Sheet reproduced from the ISTEP+ test by permission of the publisher, CTB/McGraw-Hill LLC.

389

Equation of a Line

Slope-Intercept Form:

$y = mx + b$

where m = slope and b = y-intercept

Point-Slope Form:

$y - y_1 = m(x - x_1)$

where m = slope and (x_1, y_1) is a point on a line

Slope of a Line

Let (x_1, y_1) and (x_2, y_2) be two points in the plane.

$\text{slope} = \dfrac{\text{change in } y}{\text{change in } x} = \dfrac{y_2 - y_1}{x_2 - x_1}$ where $x_2 \neq x_1$

Pythagorean Theorem

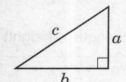

$a^2 + b^2 = c^2$

Distance Formula

$d = rt$

where d = distance, r = rate, and t = time

Temperature Formulas

$°C = \dfrac{5}{9}(F - 32)$

$°\text{Celsius} = \dfrac{5}{9} \times (°\text{Fahrenheit} - 32)$

$°F = \dfrac{9}{5}C + 32$

$°\text{Fahrenheit} = \dfrac{9}{5} \times °\text{Celsius} + 32$

Simple Interest Formula

$i = prt$

where i = interest, p = principal, r = rate, and t = time

Quadratic Formula

$x = \dfrac{-b \pm \sqrt{b^2 - 4ac}}{2a}$

where $ax^2 + bx + c = 0$, $a \neq 0$, and $b^2 - 4ac \geq 0$

Conversions

1 yard = 3 feet = 36 inches
1 mile = 1,760 yards = 5,280 feet
1 acre = 43,560 square feet
1 hour = 60 minutes
1 minute = 60 seconds

1 liter = 1000 milliliters = 1000 cubic centimeters
1 meter = 100 centimeters = 1000 millimeters
1 kilometer = 1000 meters
1 gram = 1000 milligrams
1 kilogram = 1000 grams

1 cup = 8 fluid ounces
1 pint = 2 cups
1 quart = 2 pints
1 gallon = 4 quarts

1 pound = 16 ounces
1 ton = 2,000 pounds

Index